2nd International Conference on Applied Marine Science and Fisheries Technology (MSFT 2017)

IOP Conference Series: Earth and Environmental Science
Volume 89

Langgur City, Indonesia
22 – 25 August 2017

ISBN: 978-1-5108-5027-9
ISSN: 1755-1307

Printed from e-media with permission by:

Curran Associates, Inc.
57 Morehouse Lane
Red Hook, NY 12571

CURRAN ASSOCIATES INC. proceedings.com

Some format issues inherent in the e-media version may also appear in this print version.

Printed by Curran Associates, Inc. (2017)

For permission requests, please contact the Institute of Physics
at the address below.

Institute of Physics
Dirac House, Temple Back
Bristol BS1 6BE UK

Phone: 44 1 17 929 7481
Fax: 44 1 17 920 0979

techtracking@iop.org

Additional copies of this publication are available from:

Curran Associates, Inc.
57 Morehouse Lane
Red Hook, NY 12571 USA
Phone: 845-758-0400
Fax: 845-758-2633
Email: curran@proceedings.com
Web: www.proceedings.com

Table of contents

Volume 89

The 2nd International Conference on Applied Marine Science and Fisheries Technology (MSFT) 2017

22–25 August 2017, Langgur City, Southeast Maluku, Indonesia

Accepted papers received: 5 October 2017
Published online: 22 October 2017

Preface

Papers

IOP Conf. Series: Earth and Environmental Science **89** (2017) 011001 doi:10.1088/1755-1315/89/1/011001

Welcome,

Dear colleagues

On behalf of organizing committee of International Conference and Tual State Fisheries Polytechnic, we are happy and great pleasure to extent a warm welcome to the 2017 MSFT The 2nd International Conference in Langgur, Southeast Maluku, Indonesia.

This international conference was initiated to serve as a platform for the information exchange of research results from various fields of marine science and fisheries technology, including relevant socioeconomic and business disciplines, as well as the sharing of practical experiences in the utilization and management of the marine resources of small islands on an on-going basis. We also hope that the forum will promote the sharing of practical experiences in the utilization and management of the marine resources of small islands on an on-going basis.

With over 50 National and International speakers from, Australia, France, US, Netherland, New Zealand and Indonesia, the conference play its role as bridge for small islands like Southeast Maluku and Maluku in particular and Indonesia in general to the world in order to accelerating sustainable development with applied marine science and fisheries technology.

We look forward for your contribution to a stimulating, lively consideration of the issue arising on the conference.

Sincerely,

Wellem Anselmus Teniwut
MSFT 2017 General Chair

Eugenius Alfred Renjaan
Director of Tual State Fisheries Polytechnic

IOP Conf. Series: Earth and Environmental Science **89** (2017) 011002 doi:10.1088/1755-1315/89/1/011002

Host and Organizer

Tual State Fisheries Polytechnic (POLIKANT)

Advisory Committee

Dr.rer.nat. Ir. Eugenius Alfred Renjaan, MSc,
Director of Tual State Fisheries Polytechnic, INDONESIA

Ismael Marasabessy, S.Pi., MSi,
Chief of Research and Community Service Unit, Tual State Fisheries Polytechnic, INDONESIA

Scientific Committee

Prof. Laszlo Szekely
Karolinska Institute SWEDEN

Prof. Les Christidis
National Marine Science Centre, AUSTRALIA

Prof. Nicklas Adler
Babson University, USA

Assoc.Prof. Fil.Dr.Tomas Cedhagen
Aarhus University, DENMARK

Prof. Zainal Arifin
The Indonesian Institute of Science, INDONESIA

Organizing Committee

General Chair

Wellem Anselmus Teniwut

Secretary

Ida I Dewa Ayu Raka Susanti

Treasury

W.I.Etta Farneubun

Program Chair

Syahibul Kahfi

Technical Program Committee

Pieter Semuel Foudubun
Natalia Lewerisa
Flavi Ruban
Saul A. Serpara
Eka Anto Supeni
Samuel Balubun
Frischila Pentury
Yuliana Kaneu Teniwut
Johny Dobo
Yakobus Watratan
Muamar Kadaafi Aziz
Ever Tehupeory
Petrus J.M Leftungun
Ferdinandus Beruatyaan
Wisye Pattiwael
Flavi Ruban
Merlina Renwarin
Gierly Somnaikubun

Local Arrangement Committee

Abdul Malik Serang
Julianus Notanubun
Simon. M. Picaulima
J.A Rahajaan
Bruri B. Tumiwa
Martha Rettob
Dani S. Royani
Roberto M.K Teniwut
Fien Sudirjo
Jacomina Metungun
Erwin Tanjaya
Benediktus Jeujanan
Cenny Putnarubun
Moses Tjoanda
Anthon D. Kilmanun

IOP Conf. Series: Earth and Environmental Science **89** (2017) 011003 doi:10.1088/1755-1315/89/1/011003

Peer review statement

All papers published in this volume of *IOP Conference Series: Earth and Environmental Science* have been peer reviewed through processes administered by the proceedings Editors. Reviews were conducted by expert referees to the professional and scientific standards expected of a proceedings journal published by IOP Publishing.

Morphometric Relationship, Growth, and Condition Factor of Flyingfish, Hirundicththys oxycephalus during spawning season

Friesland Tuapetel[1,*] Natsir Nessa[2] Syamsu Alam Ali[2] Sudirman[2] B.G. Hutubessy[1] J.W. Mosse[1]

[1]Faculty of Fisheries and Marine Science, PattimuraUniversity, Maluku, Indonesia
[2]Faculty of Fisheries and Marine Science, Hasanuddin University, Makassar

Email: friesland.tuapetel@fpik.unpatti.ac.id

Abstract. Flyingfish, Hirundichthys oxycephalus together with its congener species has become a major artisanal fishery resource from the Seram Sea, an area between Seram and Papua Island. However, biological information about this species has been poorly documented. Our current study focused on the growth and morphometric relationships of spawning cohorts during spawning season (June to October 2013). Flyingfish which were trapped in bale-bale, an egg aggregation device for flying fish, were collected offshore of the Southeastern Seram Sea. Total samples of 1693 with Fork Length (FL) ranged from 156.15mm to 245.52mm and total weight from 53.15gr to 115.45gr. Von Bertalanffy growth functions ($L\infty$) for-pooled sexes were FL = 251mm [1-e (1.82 (t+0.060)], K=1.82 and t0 = 0.060. Morphometrically, significant differences were observed for all individuals between sexes (F=14.20, P=0.0002), sampling locations (F=88.48 P<0.0001) and sampling period (F=138.84, P<0.0001). Condition factor of the fish generally declined in July-August. During spawning season, this fish tended to form single spawning cohort. The results of this study provide a significant understanding of the life history of this valuable fish inhabiting this particular region that rarely receives scientific investigation. For management purposes, harvesting eggs including broodstocks will lead to critical population depletion.

1. Introduction

Flying fish of the family Exocoetidae consist of more than 50 species that are globally distributed over subtropical and tropical regions [5, 7, 21, 18, 50, 53). Ecologically, they play a very significant role in the network food chain in the waters [41, 51, 52]. The most notable external characteristics of this fish are their elongates dorsal and pectoral fins which enable them to jump out of the sea and glide at incredible distances (50-400m) over the sea which gives rise to their common name. They can attain maximum sizes up to 27.0 cm fork length and weighs 115.45 gram. inhabiting mainly in surface waters along the coast and often in the open waters, flying fish feed on zooplankton and some time on small fish. As such, flying fish have been regarded as an intermediary in the bioenergetic transfer. In addition, stomach content analysis of Dolphin and number of tunas have shown that flying fish is the main prey [42].

Like their counter parts of flying fish species, *H. oxycephalus* is one that dominates the catch in some waters [25, 14, 7, 29]. Fishery of this fish includes both harvesting the adults as well as their ovulating eggs commonly found attached to floats some [30, 29, 23]. Given their high price in the market, the eggs has been considered as an export commodity in number of countries including in

Indonesia [3, 23, 29]. Exploitation of this species remain hardly monitored in Indonesia, however, record to date indicates that flying fish of this particular species has been regularly exploited since 2002 along side with other species [47].

Following their wide ranges of distribution including in Indonesian waters, Flying fish, Hirundichthys oxycephalus has been also found in Seram Seacovering an area between North Seram Island of Maluku Province and South Papua Island ranging from Kaimana to Fak-Fak water [47]. Despite their ecological and economic value, not many studies have been conducted on this group of fish and thus information on their basic biological parameters including growth performances remain largely sparse. As a consequence future fisheries management based on certain models such as age-based stock assessment for this particular fish within their distributional range and habitat may rely on studies from else where [6, 37, 36, 13, 11, 26]. The above fisheries model require age and growth data which often becomes problematic in tropical waters including in Indonesia. Otolith has been used to construct growth models of flying fish-from various waters using different pairs suggesting that one appears to be better interpreted than the others [32, 7]. Lenght-based analysis has been also applied to construct growth models in the past for flying fish [19, 35, 34, 4]. However, this appears not to be the case for Indonesia flying fish population. This is in contrast with the fact that flying fish inhabiting this study area has been exploited for decades [48] yet scientific data of their biological properties required for better management remain unavailable. Given the importance of this fish supporting regional fisheries off the coast of Seram Sea, South of Kaimana and Fak Fak waters, this was the first study to understand morphometric parameters, growth, and condition factor of *H.oxycephalus* in this particular region.This study also enables intraspecific growth coefficient comparison between regions derived from different study sites.

2. Methodology

2.1. Collection details

Flying fish of H.oxycephalus samples examined in this study (n=1693) were collected in one season of a commercial fishing trip of five months starting from June until October 2013. Three fishing grounds were designated as Kaimana regional waters, Fak-Fak regional waters and East Seram regional waters, an area that geographically lying between 1260 - 1320East and -30N and -20South (Fig. 1). Flying fish specimens were obtained from the commercial fishing operation that regularly collects flying fish eggs over these three fishing grounds. Following spawning season of flying fish in these areas, Fish Aggregation Devices (FAD) that made of coconut leaves attached to the floating rafts were deployed. The FADs were attached to the fishing boat and drifted away follow the direction of sea current. Total soaking time of the FAD was approximately 20-22 hours starting from 06.00hrs until 04.00hrs in the morning. Following the release of their eggs that attach to the coconut leaves, most fish were then got entangled by their adhesive filament, difficult to release themselves and finally died. During the removal or harvest of these ovulating eggs, dead flying fish were then taken out together and used as specimens for this study.

On board, fish were identified following description of [16] measured to the fork length (FL) using calipers vernier to the nearest 1 mm and weight (Wt) in grams of every individual fish using an electric balance to the nearest 0.01g. Sex of individual fish was determined by visual assessment of the gonads and their respective maturity stages were determined through microscopic examination following [19].

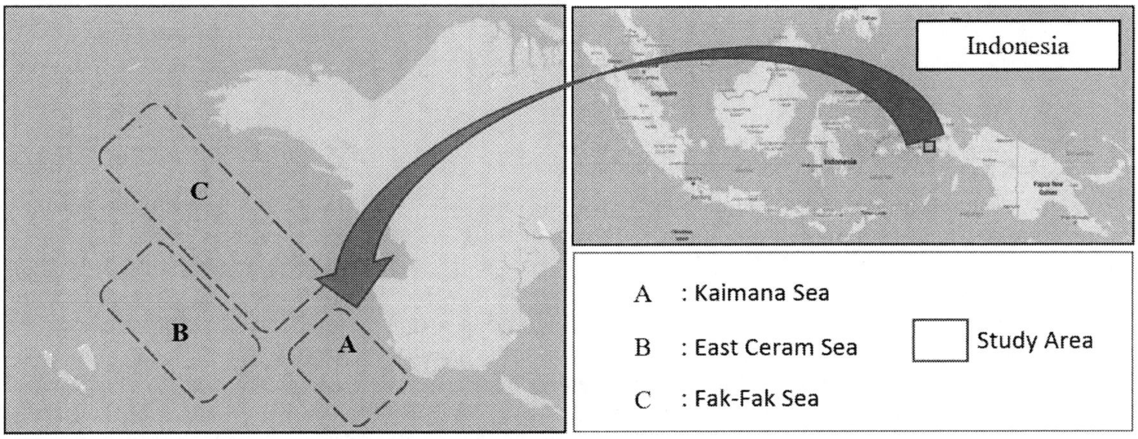

Figure 1. Geographical position of the study sites

2.2. Lenght-weight relationships

Length-weight relationships (LWRs) were estimated using the log form of the allometric growth equation W = aLb, where W = the expected weight; L is Fork Length; a is the intercept or initial growth coefficient and b is the slope of the regression line [27]. The values of constants, a and b were calculated using the least square method.

As has been shown in previous studies, we also assumed that if b was equal to 3, the relationship was considered to be isometric [40]. Correspondingly, if b value differs from 3 indicates allometric growth. Analysis of length-weight relationships was also made independently for both sexes (male and female), locations and on a monthly basis covering the entire sampling periods of five months and compared by using Analysis of Covariance [49]. ANCOVA was then used to examine our notion of whether the regression values resulted differ significantly from the null hypothesis for isometric growth (H0: b = 3) [29]. Lenght frequency distribution of flying fish for combined sexes from the respective sampling location was plotted for the purposes of cohort analysis of the population [20].

2.3. Growth Method

In order to estimate growth parameters of the von Bertalanffy growth function for flying fish of the study area, separate age group data were analyzed applying Bhattacharya methodology [22, 44, 24]. Length frequency data was converted to estimate age composition and enable calculation of growth rate following the function of von Bertalanffy as follows:

$$L_t = L_\infty\left[1 - exp^{k(t-t_0)}\right] \qquad (1)$$

where: Lt=length in millimeter at age(mm); L∞= asymptotic length(mm); K=growth coefficient (year); to = the hypothetical age when theoretical length is equal to zero and t = age. The value of K and L∞ asymptotic were determined as whilst to was calculated using the formula proposed by Pauly (1983) as follows:

$$Log(-t_0) = (-0.3922) - (0{,}2752) * Log(L_\infty) - 1{,}038 * (LogL)) \qquad (2)$$

Further more, an age of the flying fish in this study was estimated based on the size of the group use derivatives formula of von Bertalanffy's follows:

$$t_0 = \left(t + (1/K)\right) * \left(ln(L_\infty - L_t)/(L_\infty)\right) \qquad (3)$$

2.4. Condition Factor

Condition factor is an expression of fish fitness which is stated in number and was analyzed using the formula of the relative condition factor (Kn) [10, 11-12] as follows:

$$K_n = W/\widehat{W} \tag{4}$$

where: Kn = relative condition factor of each individual fish, W = weight of fish (g), $\widehat{W}$ = a L^b (average weight obtained from weight-length relationship), L = body length of fish (mm), a and b = parameter of length-weight relationship.

3. Results

Between June until October 2013, a total of 1693 flying fish were sampled from the three designated fishing sites within the Seram Sea. Since sampling was conducted during spawning season, sex determination was relatively straight forward. Female fish (n = 866) accounted for the majority of samples than male (n = 827), which composed only 48%. In overall, mean Fork Length of the flying fish sampled from East Seram (211.51 ± 16.69 SD) tended to be larger than the mean sizes for those samples from Kaimana (190.58 ± 9.99) and Fak-Fak waters (189.71± 12.19), respectively. Further analysis to examine these mean sizes show significant differences (F-test: 1244.61 P < 0.0001). Interestingly, however, length frequency distribution throughout sampling time (months) indicate the similar consistency of trend. At the early sampling month in June, the small size of Flying fish (< 200mm FL) were present but later disappear in the succeeding months and dominated by larger sizes until in October. This size spectrum might, however, account for only about 55.6% of their maximum size of 457.2mm which is often recorded. Similar size pattern was also recorded under females and males category, as flying fish from East Seram remains larger than the mean size for the fish for two other fishing sites as Kaimana and Fak-Fak (Table 1.).

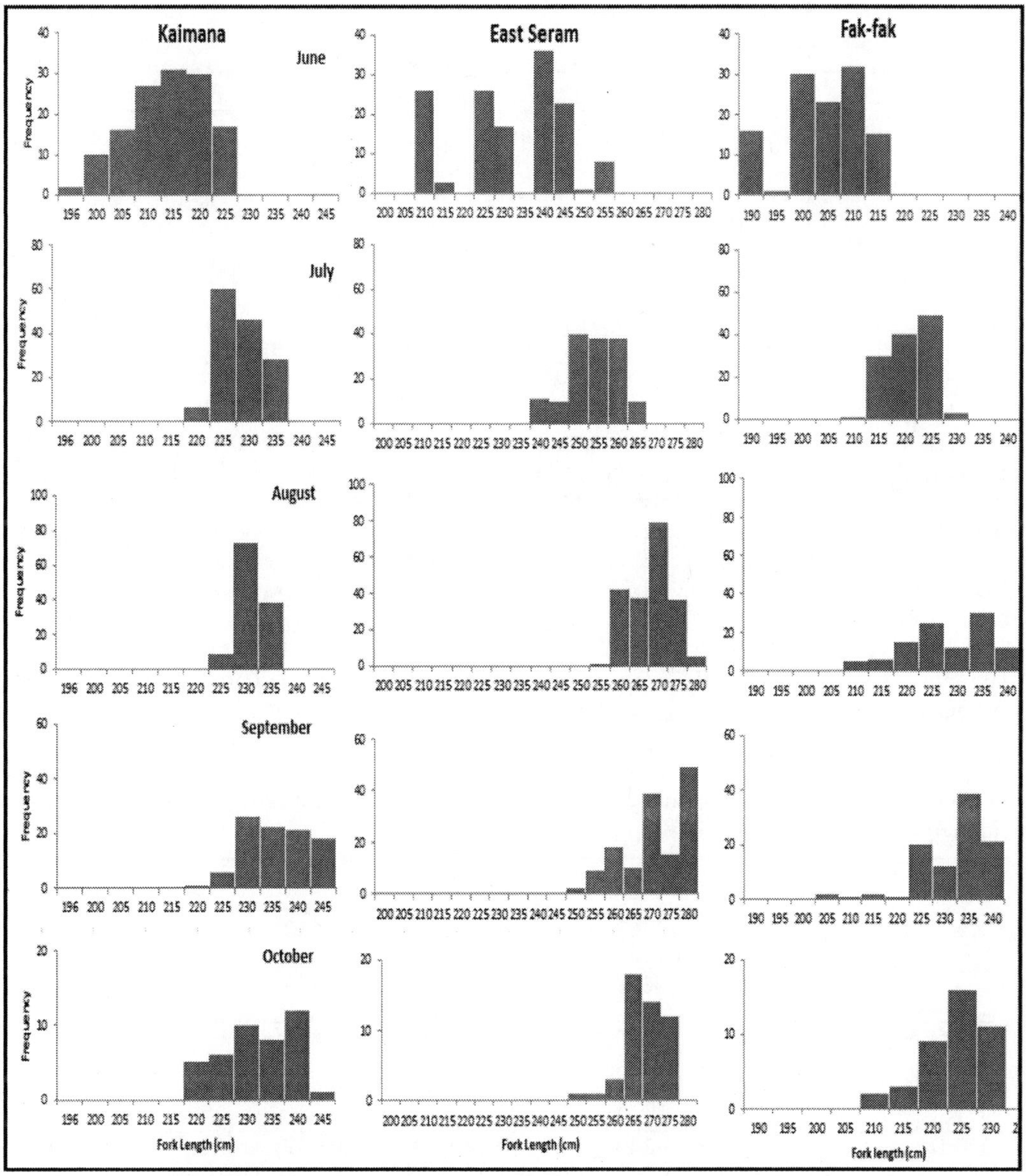

Figure 2. Length frequency distribution of Flying fish, *H.oxycephalus* sampled at the three fishing sites from June to October 2013.

Table 1. Fork Length (mm) of H.oxycephalus by sex category sampled from three different sampling sites.

		Fork Length (mm)		
Site	Sex	Mean ± SD	Range	n
Kaimana	Female	218.5 ± 9.77	208.7-228.3	267
	Male	230.4 ± 5.4	225.0-235.8	262
East Seram	Female	267.3 ± 7.1	260.2-274.4	353
	Male	242.9 ± 16.0	226.9-258.9	325
Fak-Fak	Female	226.5 ± 8.2	218.3-234.7	246
	Male	210.1 ± 10.1	200.0-220.2	240

3.1. *Length-weight relationship*

The length-weight relationship (LWR) calculated for H.oxycephalus comparing females and males, sampling sites and months have shown strong significant differences (Table 2). The overall analysis for length-weight relationship coefficient (b) has shown a negative allometry and significantly different from a value of 3 (Student's t-test: P<0.001), indicating that this species grows towards length rather than in weight. Overall values of b calculated for the three sampling locations regardless of sex also less than 3 (0.529-2.53). Further analysis in corporating sex within location also consistently shown negative allometric growth, b (0.682-1.81) (Table 2). It was also noted that the fish appears to have similar growth pattern at early to mid stages when they attained about 85.1% of their maximum sizes. Following this, the females and males fish tended to perform different growth trajectories in terms of length and weight (Fig. 3).

Table 2. Summary of the length-weight analysis for females and males *H.oxcyphalus* by locations. Two locations showed not significatnt difference (*) and only one (East Seram) showing differences in their morphometeric relationship.

Location		*N*	*a*	*b*	R^2	*F*	*P*
Kaimana	Male	262	-2.396	1.81	0.680	2.72	0.099*
	Female	267	-2.005	1.642	0.704		
East Seram	Male	325	-1.320	1.373	0.833	13.80	<0.0001
	Female	352	-1.941	1.634	0.809		
Fakfak	Male	240	0.241	0.682	0.099	0.10	0.7497*
	Female	246	0.116	0.735	0.171		

3.2. *Growth*

The estimated parameters of the von Bertalanffy growth model of the current study are presented. In overall, the analysis did reveal significant differences in mean size distributions as well as morphometric growth prompting separate analysis of such growth model (Table 4). These were further supported by the fact that although there were no significant differences between females and males of flying fish at Kaimana (F = 2.72; P = 0.009) and Fak-Fak (F = 0.10; P = 0.7497) fishing sites. However, the significant difference of morphometric growth that occurs between females and males at East Seram fishing sites provide strong evidence to perform separate further analysis.

The von Bertalanffy growth function calculated by using Fork Length and estimated age (month) of the flying fish were then fitted for the respective fishing sites as follows:

East Seram: $Lt = 251.30*[1\text{-}e^{(0.4469(t+0.2260)}]$
Kaimana: $Lt = 246.19*[1\text{-}e^{(0.2268(t+0.4587)}]$
Fak-Fak: $Lt = 245.96*[1\text{-}e^{(0.2268(t+0.4587)}]$

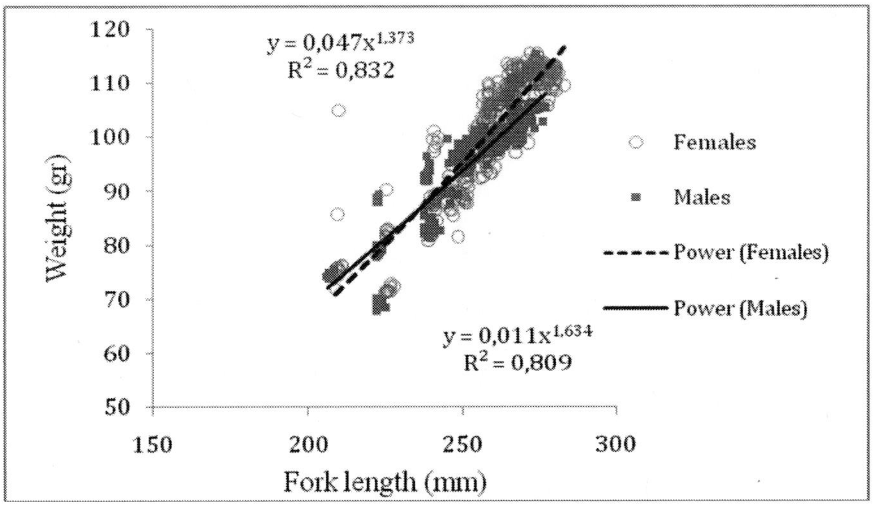

Figure 3. Analysis of Covariance comparing two regression lines of length-weight relationships between females and males of *H.oxycephalus* within locations of East Seram fishing sites.

It appears that growth parameters from the three fishing sites were different. The growth coefficient (K) of the flying fish from East Seram was the largest compared to their counterparts from the two other sites. This could-excel rate maximum growth towards the asymptotic size (L∞) of the flying fish inhabiting East Seram Sea that tends to attain the larger size than their congener species from Kaimana and Fak-fak waters. In order to have a comprehensive comparison, similar parameters from different study sites and flying fish species were also presented. Our overall estimated value for the asymptotic size (L∞) for H.oxycephalusin this study ranges from 246mm-251mm (FL). These sizes are in fact within the range of asymptotic sizes reported from other studies except from the Taiwan (Table 3). Growth coefficient (K) of H.oxycephalusfrom this study ranged from 0.2268-0.4469 per month and in average, however, the value was 1.8/year. This was slightly higher when compared with H.oxycephalus in the Flores Sea and Makassar Strait which had the value of K =1.3/year but remain lower than that of EastTaiwan (K=0.0075 or2.6/year and H.affin is from Barbados with K=0.0085/dayor 2.99/year.

Table 3. Growth coefficient of flying fish (*Exocoetidae*) of this and from other studies.

Location	Growth equation	Gowth Coef/year	Reference
East Seram Sea	Lt = 251[1-e(1.8(t+0.23]	1.8	This study
Flores Sea, Makassar Str.	Lt = 182[1-e(1.30(t+0.074)]	1.3	[3]
Taiwan East	Lt = 254[1-e(2.6(t+0.0042)]	2.6	[7]
Barbados Sea	Lt = 245[1-e(2.9(t+0.0078)]	2.9	[41]

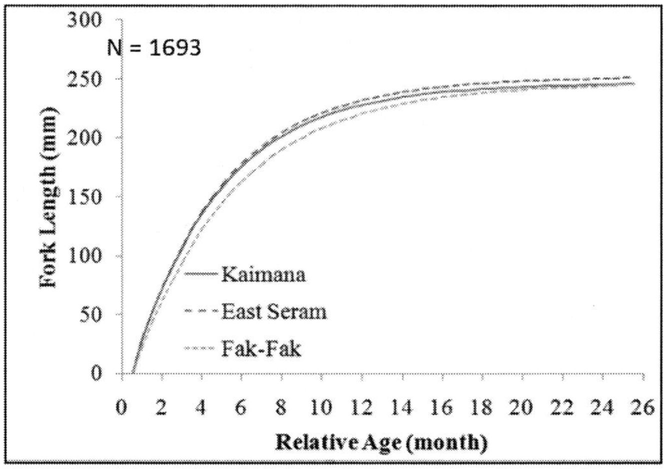

Figure 4. Fork length in mm at relative age and fitted von Bertalanffy growth model for Flying fish (*Hirundichthys oxycephalus*) sampled from Kaimana, East Seram and Fak-Fak fishing site from June – October 2013.

3.3. Condition factor

In general, mean condition factor of flying fish H. Oxycephalus from our study varies slightly starting from the highest value of 1.0015 in June to the lowest value of 1.0004 in July before rebounded again in the following months reaching the value of 1.0007 in October (Figure 5). In terms of the respective study sites, however, interesting variations were reasonably obvious. Condition factor of the flying fish from East Seram waters dropped sharply from June (1.0028) to July (1.0009) and gradually decreased towards the following months reaching the lowest value of (Kn) about 1.0049 in October. This is slightly different for the two other fishing sites (Kaimana and Fak-Fak) whereas despite their values of Kn also drop from June (1.0011 and 1.0004, respectively) to July (1.0003 and 1.0035, respectively), but not in a sharp fashion compared to those flying fish from East Seram waters. If on the other hand, condition factor of Flying fish from the East Seram waters continue to drop and level off in September to October, this was not the case for the Kaimana and Fak-Fak waters as the Kn values showed a trend of slightly increased after July towards the following months. Eventually, condition factor of the flying fish from Fak-Fak waters attain the highest value of about 1.0009but then decline towards October though the value remains higher than previous months.

Discu

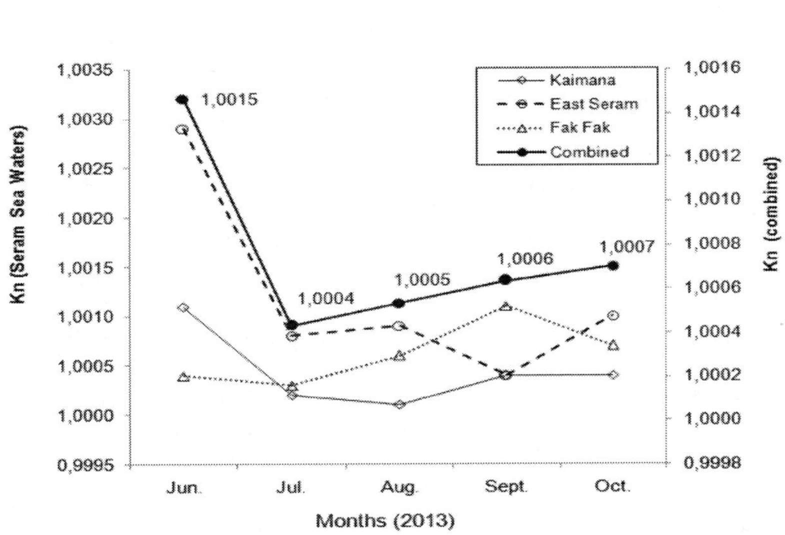

Figure 5. Variations of the values of condition factor of *H. Oxycephalus* sampled from East Seram, Kaimana, and Fak-Fak waters.

3.4. Discussion

Length frequency distribution provides very useful insight pictures of fish population structure which can be applied also in fisheries management. There are numbers of benefits in getting the right and appropriate knowledge of the size structure of fish population as it provides fundamental knowledge of fish population parameters including growth, reproduction, changes in size and recruitment processes. Understanding of these parameters would further provide an initial indicator of potential disruption occurring within the population. Presence of the relatively larger size of Flying fish in this study (208.7mm-274.4mm, FL) and lack of small size range could be due to the gear used throughout sampling process. If in the other studies, drifted gill net or dip net were often used to collect samples [45, 7, 14] however, fish specimens in this study were taken by "Bale-bale" (local name), a Fish Aggregation Device provided to attract Flying fish to attach their ovulated eggs. All samples presence in Bale-bale were infact trapped by the adhesive filament of the ovulated eggs and consequently only a mature and relatively larger size of the fish dominate the catch, mean while the small sizes were absence. Entering the spawning season in June until October is also another reason of catching more larger size of fish. This study appears to be the first providing such record of size spectrum in the region while waiting for further study covering the entire size range and reproductive season.One interesting finding of this study is that the existing size trajectories appear to confirm one cohort of Flying fish population within the respective study sites that arrive to spawn (Figure 2). Although the argument of whether flying fish considered as the batch or synchronous spawners remains [17], the fact is that the reproductive activities vary among and perhaps within regions and quite possible between genera as well as species. [33] Reported two spawning periods of *H.affinisfound* in the eastern Caribbean Sea. Prexocoetus brachypterous, a relatively small (112-117mm SL) Flying fish species from the northeastern Gulf of Mexico has been also reported perform two rhythms [45]. Meanwhile a prolong reproductive activity has been recorded also by several members of Flying fish in Japanese waters including H. Oxycephalus of the current study [15]. The present study strongly indicated that spawning periods of H. Oxycephalus in Seram Sea occurs between June to October as shown by their maturity stages data. In Taiwan however, *H.Oxycephalus* spawns one month earlier in May and last in August [7]. This slight disagreement might be due to different sampling periods during this study which took place for only five months.Until a further and comprehensive study is made, current data suggest that the Flying fish eggs fisheries in Seram Sea must be undertaken with caution. Over harvesting the eggs in the current spawning cohort will very serious implication of the next population.

In overall, our estimation of b values was not within the normal and expected range for fish, i.e b = 2.5-3.5 [11], suggesting that growth of H. oxycephalus in our study is negative allometric. An attempt to analyze length-weight relationship by sex within the respective location also depict a similar pattern of deviations from the hypothetical value of 3 indicating that they put more energy to grow in length than in biomass [10, 1].This was presumably due to more than half of the total fish caught in drift rum pong (bale-bale) were in their spawning phase. This result is in fact in accordance with previous studies reporting on the negative growth of this fish including in Flores Sea and Strait of Makassar [3], in Binuangeun Banten waters. [9] Also found a negative allometric growth of flying fish collected from the waters of West Africa, as well as *Hirundichthys coromandel Ensis* from the waters of India [42], and H. *affinis* from Brazil [28, 31]. Living on the surface of the open waters, certain strategies have to be developed including body shape, mass and parts like wings that enable them to avoid predation. Having body shape like torpedo enable the Flying fish to accelerate the speed and break the surface waters. Supported by the relatively large wing like pectoral fins, less body mass help the fish to jump out the water and glide over a quite distances (about 200m) [9, 15).There was no significant difference in b values between females and males samples taken from Kaimana and Fak-Fak waters suggesting that both populations might experience similar performance in growth. This is not the case however for females and males Flying fish from East Seram waters. Given that number of sample size was reasonably high, the statistical procedure in this analysis could be ignored [6]. The argument of [6] that the exponent b value of fish could be associated with size range appears to be the case for the Flying fish from east Seram Sea as they tend to be larger in size. Whether these facts were associated also with less fishing pressure in the area, food availability (quality, quantity), spawning dynamics (season and time), fatness of the species, water temperature, salinity, gonadal maturity, length range that proposed by [11] remain unclear and require further investigation to cover such a full ranges of all possible factors.

Values of asymptotic length ($L\infty$) estimated from the von Bertalanffy growth model for combined sexes were much higher (326mm, FL) than the maximum length actually observed in the current study (260-274mm FL). This value was also higher than most of the previous studies ever reported [3, 34] as well as its counterpart from Taiwan east waters [7]. The asymptotic value of the two sampling sites (Kaimana and Fak-Fak) were lower, however higher for east Seram sea. Comparing with other previous studies, however, it was slightly higher than in Barbados, Flores and Makassar Strait [3, 34] and in eastern Taiwan [7]. One possible explanation for this may lay on on the lower estimation of the von Bertalanffy parameters K value which is often reported in other species [5]. This is in contrast with estimation of the von Bertalanffy parameters for Flying fish from Barbados and Flores sea and Makasar Strait that tends to grow faster and attain slightly smaller asymptotic length [33, 3]. This slight discrepancy in their growth parameters likely occurred may be because of the estimated parameters that were based on the final stages of growth as samples in this study lacking smaller sizes. Age estimation for Flying fish quite varies and remain an issue including the use of length frequency analysis [8] as well as otolith [34, 7]. It has been often argued that fish length is an unreliable measure for estimating fish age. On the other hand, hard structures like otoliths may also provide unclear pictures of the fish age, especially in many tropical environments. The argument is however, all possible methods could be exercised provided that proper sampling procedures have been followed. For many pelagic species especially in the tropical environment, hards parts like otolith are often difficult to descernagecompared with their counter parts in subtropical or temperate waters. Despite such, the use of fish length included in the current study to estimate flying fish age showed a reasonable estimation of one - two years of age (± 25months) that helps to calculate growth parameters presented here. This was apparently also in accordance with the similar species from Taiwan [7]. Age estimation for its congener species H.affinis [32, 8] also showed consistent age pattern (two years) with the current study. Certainly, ability to grow faster and attain maturity stages at the early age has been considered as remarkable startegyreflecting a phenotypic compensatory response to maintain population [46]. Following spawning, many flying fish *H. Oxycephalus* died on their adhesive filament leaving unsolved questions of whether the event was due to unfit condition after spawning or because of reaching their maximum age limit.

A sharp decrease of condition factor of Flying fish in the present study has been observed coincided with spawning activity. Although the respective values vary among sampling sites, the mean estimated values of condition factor showed a consistent pattern of declining towards during spawning season (Figure 6). Shrink condition of the stomach wall of fish undergoing spawning was observed, begun shrinking and become small. A similar condition has been reported also for fish from the Malaka Strait and the Flores Sea. [3] Reported that condition factor of female fish was also decreased in April, which was predicted to be the peak spawn season of this fish. Similar pattern of declining in Knvalue have been also observed in various fish species and become an indicator of spawning processes like species of the family Triglidae [30, 41] in Trachurusmediterraneusat Eastern Adriatic Sea, and also in Serranus script from North Atlantic sea [54] Given that condition factor is influenced by the reproduction, feeding of individuals and the habitat quality, the value can be useful tool providing information on the general condition of fishes presence in particular habitat.

4. Conclusion

- During spawning season, this fish tended to form single spawing cohort; Morphometric relationship significant differences was observed for all individuals between sexes, sampling locations and sampling periode; Condition factor of the fishgeneraly declined in July-August: indication spawn peak season
- For management purposes, harvesting eggs including broodstocks will lead to critical population depletion.

Acknowledgement
This manuscript was part of the Ph.D. Dissertation submitted by the Author. The Authors are grateful to the anonymous reviewers for critical comments.

References

[1] Abbasi T, and Abbasi SA. 2010. Biomass energy and the environmental impacts associated with its production and utilization. *Renewable and Sustain. Energy Reviews*, *14*(3), 919-937.
[2] Ali SA, Nessa MN, Djawad I, Omar SBA, and Syahailatua A. 2005. The Gonad Maturity Development and Spawning Season of Flying fish (*Hirundichthys oxycephalus* Bleeker, 1852) at the Flores Sea, South Sulawesi. Torani Vol 15 (6): 403-410.
[3] Ali SA. 2005. Biodiversity and Population of Flying Fish (*Hirundichthys oxycephalus*, Bleeker 1852) From Flores Sea and Makassar Strait. [Dissertation]. Agricultural Science Postgraduate Program Hasanuddin University, Makassar. 280p.
[4] Ayoade AA, Sowunmi AA, and Nwachukwu HI. 2004. Gill asymmetry in Labeo ogunensis from Ogun river, Southwest Nigeria. *Revista de biología tropical*, *52*(1), 171-175.
[5] Beven K. 2006. A manifesto for the equifinality thesis. *Journal of hydrology*, *320*(1), 18-36.
[6] Bolger T, and Connolly PL. 1989. The selection of suitable indices for the measurement and analysis of fish condition. *Journal of Fish Biology*, *34*(2), 171-182.
[7] Chang CW, Lin CH, Chen YS, Chen MH, and Chang SK. 2012. Age validation, growth estimation and cohort dynamics of the bony flying fish Hirundichthys oxycephalus off eastern Taiwan. *Aquatic Biology*, *15*(3), 251-260.
[8] Dalzell P. 1993. The fisheries biology of flying fishes (Families: Exocoetidae and Hemiramphidae) from the Camotes Sea, Central Philippines. *Journal of fish biology*, *43*(1), 19-32.
[9] Davenport J. 2003. Allometric constraints on stability and maximum size in flying fishes: implications for their evolution. *Journal of Fish Biology*, *62*(2), 455-463.
[10] Effendie MI. 2002. Fishery biology. *Yayasan Pustaka Nusantara. Yogyakarta.*
[11] Froese R. 2006. Cube law, condition factor and weight–length relationships: history, meta-analysis and recommendations. *Journal of applied ichthyology*, *22*(4), 241-253.
[12] Gerami M H, Abdollahi D, and Patimar R. 2013. Length-weight, length-length relationship and condition factor of Garra rufa in Cholvar River of Iran. *World Journal of Fish and Marine Sciences*, *5*(4), 358-361.

[13] Hossain MY, Ahmed ZF, Leunda PM, Jasmine S, Oscoz J, Miranda R, and Ohtomi J. 2006. Condition, length–weight and length–length relationships of the Asian striped catfish Mystus vittatus (Bloch, 1794)(Siluriformes: Bagridae) in the Mathabhanga River, southwestern Bangladesh. *Journal of Applied Ichthyology*, *22*(4), 304-307.

[14] Hutubessy G. 2011. Encircling gillnet selectivity for oxeye scad (selar boops cuvier, 1833) in the coast of waai, ambon island. Journal of Coastal Development, *14*(2), 125-130.

[15] Ichimaru T. 2007. The life cycle of three species of flyingfish in the north waters of Kyusyu and the recruitment of young flyingfish to the fishing ground. *Bull Nagasaki Prefectural Inst Fisheries*, *33*, 7-110.

[16] Kharin VE, and Saveliev PA. 2011. First occurrence of the bony flying fish Hirundichthys oxycephalus (Exocoetidae) in waters of Russia. *Journal of Ichthyology*, *51*(7), 551.

[17] Khokiattiwong, S., Mahon, R., & Hunte, W. (2000). Seasonal abundance and reproduction of the fourwing flyingfish, Hirundichthys affinis, off Barbados. *Environmental biology of fishes*, *59*(1), 43-60.

[18] Ko HL, Wang YT, Chiu TS, Lee MA, Leu MY, Chang KZ, ... and Shao KT. 2013. Evaluating the accuracy of morphological identification of larval fishes by applying DNA barcoding. *PLoS One*, *8*(1), e53451.

[19] Lewis JB, Brundritt JK, and Fish AG. 1962. The biology of the flyingfish Hirundichthys affinis (Günther). *Bulletin of Marine Science*, *12* (1), 73-94.

[20] Lhatoo, S. D., Johnson, A. L., Goodridge, D. M., MacDonald, B. K., Sander, J. W., & Shorvon, S. D. (2001). Mortality in epilepsy in the first 11 to 14 years after diagnosis: Multivariate analysis of a long-term, prospective, population-based cohort. *Annals of neurology*, *49*(3), 336-344.

[21] Lim SS, Vos T, Flaxman AD, Danaei G, Shibuya K, Adair-Rohani H, ... and Aryee M. 2012. A comparative risk assessment of burden of disease and injury attributable to 67 risk factors and risk factor clusters in 21 regions, 1990–2010: a systematic analysis for the Global Burden of Disease Study 2010. *The lancet*, *380*(9859), 2224-2260.

[22] Mahajan V, Muller E, and Bass FM. 1991. New product diffusion models in marketing: A review and directions for research. In *Diffusion of technologies and social behavior* (pp. 125-177). Springer, Berlin, Heidelberg.

[23] Mahon, Robin, Parker, Christopher, Sinckler, Travis, Willoughby, Stephen, and Johnson, Julieann. 2007. The value of Barbados' fisheries: a preliminary assessment.

[24] Malhotra R, Bhattacharya A, Kumar A, Reddy NV, and Cao J. 2011. A new methodology for multi-pass single point incremental forming with mixed toolpaths. *CIRP Annals-Manufacturing Technology*, *60*(1), 323-326.

[25] Masuda H, Amaoka K, Araga C, Uyeno, Teruya, and Yoshino, Tetsuo. 1985. The fishes of the Japanese Archipelago.(v. 1): Text.(v. 2): Plate.

[26] Mir JI, Sarkar UK, Dwivedi AK, Gusain OP, Pal A, and Jena JK. 2012. Pattern of intrabasin variation in condition factor, relative condition factor and form factor of an Indian Major Carp, Labeo rohita (Hamilton-Buchanan, 1822) in the Ganges Basin, India. *European Journal of Biological Sciences*, *4*, 126-135.

[27] Moutopoulos DK, and Stergiou KI. 2002. Length–weight and length–length relationships of fish species from the Aegean Sea (Greece). *Journal of Applied Ichthyology*, *18*(3), 200-203.

[28] Nóbrega MF, Kinas PG, Lessa R, and Ferrandis E. 2015. Spatial and temporal variation in artisanal catches of dolphinfish Coryphaena hippurus off north-eastern Brazil. *Journal of fish biology*, *86*(2), 785-804.

[29] Nogueira, L. B., Godinho, A. L., & Godinho, H. P. (2014). Early development and allometric growth in hatchery-reared characin Brycon orbignyanus. *Aquaculture research*, *45*(6), 1004-1011.

[30] Olim S, and Borges TC. 2006. Weight–length relationships for eight species of the family Triglidae discarded on the south coast of Portugal. *Journal of Applied Ichthyology*, *22*(4), 257-259.

[31] Oliveira MR, Carvalho MM, Silva NB, Yamamoto ME, and Chellappa S. 2015. Reproductive aspects of the flyingfish, Hirundichthys affinis from the Northeastern coastal waters of Brazil. *Brazilian Journal of Biology*, *75*(1), 198-207.

[32] Oxenford HA, and Hunte W. 1999. Feeding habits of the dolphinfish (Coryphaena hippurus) in the eastern Caribbean. *Scientia Marina*, *63*(3-4), 303-315.

[33] Oxenford HA, Hunte W, Deane R, and Campana SE. 1994. Otolith age validation and growth-rate variation in flyingfish (Hirundichthys affinis) from the eastern Caribbean. *Marine Biology*, *118*(4), 585-592.

[34] Oxenford HA. 1994. Movements of flyingfish (Hirundichthys affinis) in the eastern Caribbean. *Bulletin of Marine Science*, *54*(1), 49-62.

[35] Pauly D. 1983. *Some simple methods for the assessment of tropical fish stocks* (No. 234). Food & Agriculture Org.

[36] Pauly D. 1993. Fishbyte section. *Editorial. Naga. ICLARM Quart*, *16*(26), 26.

[37] Petrakis G, and Stergiou KI. 1995. Weight-length relationships for 33 fish species in Greek waters. *Fisheries research*, *21*(3-4), 465-469.

[38] Potts AC, Thomas AD, and Nichols E. 2003. An economic and social assessment of the flying fish (pelagic) fishery of Tobago, Trinidad and Tobago.

[39] Rennie J. 2001, January. Review of the social and economic status of the flyingfish fishery in Grenada. In *FAO/WECAFC. Report on the Second Meeting of the WECAFC Ad Hoc Flyingfish Working Group of the Eastern Caribbean. Bridgetown, Barbados* (pp. 8-12).

[40] Sangun L, Akamca E, and Akar M. 2007. Weight-length relationships for 39 fish species from the north-eastern Mediterranean coast of Turkey. *Turkish Journal of Fisheries and Aquatic Sciences*, *7*(1).

[41] Šantić M, Pallaoro A, and Jardas I. 2006. Co-variation of gonadosomatic index and parameters of length–weight relationships of Mediterranean horse mackerel, Trachurus mediterraneus (Steindachner, 1868), in the eastern Adriatic Sea. *Journal of Applied Ichthyology*, *22*(3), 214-217.

[42] Satapathy L, Singh D, Ranjan P, Kumar D, Kumar M, Prabhu KV, and Mukhopadhyay K. 2014. Transcriptome-wide analysis of WRKY transcription factors in wheat and their leaf rust responsive expression profiling. *Molecular genetics and genomics*, *289*(6), 1289-1306.

[43] Secor DH, and Dean JM. 1989. Somatic growth effects on the otolith–fish size relationship in young pond-reared striped bass, Morone saxatilis. *Canadian Journal of Fisheries and Aquatic Sciences*, *46*(1), 113-121.

[44] Sparre P, and Venema SC. 2001. Introduction to tropical fish stock assessment. Pt. 1: Manual.- Pt. 2: Exercises. *FAO Fisheries Technical Paper (FAO)*.

[45] Stevens, J. D., Bradford, R. W., & West, G. J. (2010). Satellite tagging of blue sharks (Prionace glauca) and other pelagic sharks off eastern Australia: depth behaviour, temperature experience and movements. *Marine biology*, *157*(3), 575-591.

[46] Trippel EA. 1995. Age at maturity as a stress indicator in fisheries. *Bioscience*, *45*(11), 759-771.

[47] Tuapetel F, Nessa MN, Ali SA, and Sudirman. 2015a. Distribution, Species Composition And Size Of Flying Fish (Exocoetidae) In The Seram Sea. IJSTR.Vol. 4, issue 03: 75-76.

[48] Tuapetel F, Nessa MN, Ali SA, and Sudirman. 2015b. The Utility Level Flying fish (*Exocoetidae*) in Seram Sea. Pros.Marine Sciene and Fishery II, Makassar Okt. 2015: 232-239.

[49] Van Breukelen, GJ. 2006. ANCOVA versus change from baseline had more power in randomized studies and more bias in nonrandomized studies. *Journal of clinical epidemiology*, *59*(9), 920-925.

[50] Van Noord JE, Lewallen EA, and Pitman RL. 2013. Flyingfish feeding ecology in the eastern Pacific: prey partitioning within a speciose epipelagic community. *Journal of fish biology*, *83*(2), 326-342.

[51] Wu CC, Su WC, and Kawasaki T. 2001. Reproductive biology of the dolphin fish Corphaena hippurus on the east coast of Taiwan. *Fisheries science*, *67*(5), 784-793.

[52] Wu GCC, Chiang HC, Chou YW, Wong ZR, Hsu CC, Chen CY, and Yang HY. 2010. Phylogeography of yellowfin tuna (Thunnus albacares) in the Western Pacific and the Western Indian Oceans inferred from mitochondrial DNA. *Fisheries Research*, *105*(3), 248-253.

[53] Xu GH, Zhao LJ, Gao KQ, and Wu FX. 2013. A new stem-neopterygian fish from the Middle Triassic of China shows the earliest over-water gliding strategy of the vertebrates. Proceedings of the Royal Society of London B: Biological Sciences, 280(1750), 20122261.

[54] Zorica B, Sinovčić G, Pallaoro A, and Čikeš Keč V. 2006. Reproductive biology and length–weight relationship of painted comber, Serranus scriba (Linnaeus, 1758), in the Trogir Bay area (middle-eastern Adriatic). *Journal of Applied Ichthyology*, *22*(4), 260-263.

IOP Conf. Series: Earth and Environmental Science **89** (2017) 012002 doi:10.1088/1755-1315/89/1/012002

Present status of grouper fisheries at waters of Kotania Bay, Western Seram District Maluku Province

N V Huliselan[1], M Wawo[1], M A Tuapattinaja[1], D Sahetapy[1]

Department of Aquatic Resources Management
Faculty of Fisheries and Marine Science of Pattimura University

Email: dicky_sahetapy@yahoo.com

Abstract. Study on present status of groupers fishes at waters of Kotania Bay was conducted from 2016 to 2017. Survey and Participatory Rural Appraisal (PRA) method were used to collect and examine data and information concerning species potential and utilization of these species. The result shows that there are 35 species of grouper fishes inhabit Kotania Bay waters. From six genera recorded, *Epinephelus* found to have more varieties species richness compared to other five genera. In general, main habitat of adult grouper is coral reef, whilst mangrove and seagrass are habitat for nursery and grow out. The potency of the *Epinephelus, Cephalopholis* and *Plectropomus* genus tend to decrease started in 2000 up to 2017. At the same period, the production of these genera was also declined. Species potency and production declined was attributable to habitat (coral reef) degradation and high fishing intensity as a result of high market demand. Hand line, bottom long line and trap net are general fishing gear used in harvesting of theses fishes. Fishing activities took place all year round except for bottom long line which only lasted from June to October (East monsoon). Spatial fishing ground distribution is predominantly at coral reef ecosystem of Kotania Bay.

1. Introduction

Kotania bay is one of the unique bay in Western Seram District which has five small islands and three ecosystems of tropical region i.e. coral reef, seagrass bed and mangrove. This bay becomes spawning and feeding ground as well as shelters for many species of economically important finfish and shellfish [1]. Thus, Kotania bay has a great potential to become habitat for high diversity and abundance of marine biota. Some previous studies showed that Kotania bay had variety of marine resources such as fishes [2], crustaceas, echinoderms [3] and molluscs [4]. However, due to population growth and continuous exploitation of marine resources, it seems that these resources have declined and their habitat decreased.

Fish in coral reef can be categorized into three groups namely target species, major species and indicator species [5]. One of the popular and important target reef fish in Kotania Bay is groupers. Exploitation of groupers in this area has been done for long time by artisanal fishers using simple gears and only based on their experience. In the recent year, exploitation of groupers has increased due to the market demand and high value of grouper. As the results, their population becomes decreased and their habitat becomes damaged due to utilization of destructive fishing gear. In addition, lack of government awareness of socialized fishery regulation make traditional fishers in this area do not know how to exploit and utilize their resources environmentally and sustainably.

IOP Conf. Series: Earth and Environmental Science 89 (2017) 012002 doi:10.1088/1755-1315/89/1/012002

Study on groupers fishery in Kotania Bay is lacking. Therefore, this research was carried out to study present status grouper fishery which consists of species richness, spatial distribution of fishing ground as well as fishing gear, utilization pattern and fishing season. Based on the results, proposed strategy management for grouper fishery will be formulated.

2. Methodology

This research was conducted in Kotania Bay, Western Seram District, Maluku Province, Indonesia (Figure 1) from September 2016 to July 2017. Data of fishing ground, fishing activities (types and number of fishing gear as well as fishing frequency), species richness and total individual of grouper caught were obtained through survey on 34 fishing sites. Furthermore, Participatory Rural Appraisal (PRA) method [6] through Focus Group Discussion (FGD) with fishers in villages of Osi island, Kotania and Wael were conducted to get data and information on opinion and condition of grouper fishery, participative mapping of grouper fishing, distribution of fishing gear and fishing season. Data and information on utilization of grouper were also collected through a questionnaire distributed to 10% of fishers in those villages.

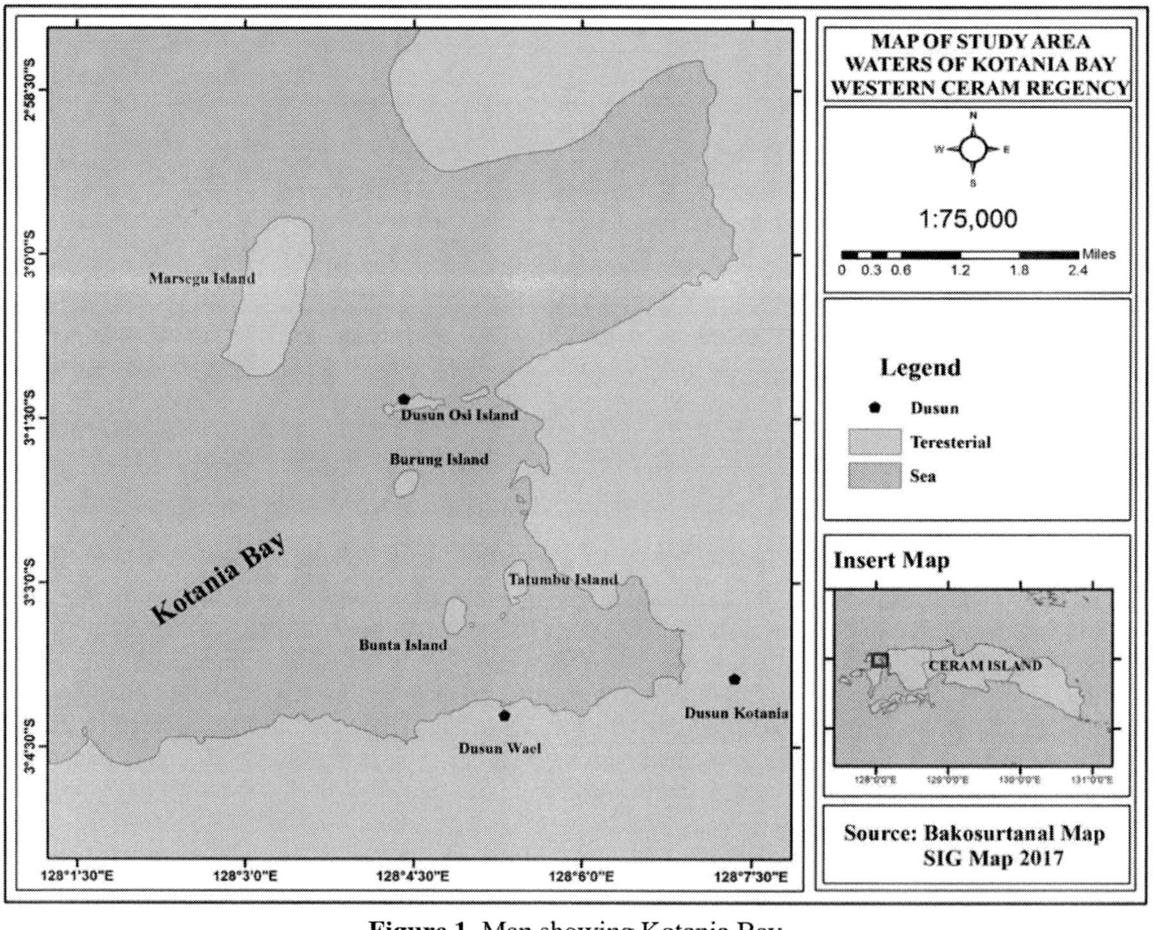

Figure 1. Map showing Kotania Bay

Species of grouper caught by local fishers were identified according to [7], [8], [9], [10] and [11]. Catch data of each species and its production tendency was tabulated, presented in graph and explain descriptively. Data and information on this fish concerning its distribution was also displayed according to its fishing ground. The pattern of grouper utilization, fishery progress and fishing season was described through participatory mapping according to [12]. Grouper fishery strategy and

IOP Conf. Series: Earth and Environmental Science **89** (2017) 012002 doi:10.1088/1755-1315/89/1/012002

management plan was established using alternative method with consideration of problems associated with exploitation of this fishery

3. Results and Discussion

3.1. Species Richness of Groupers

There were 35 species of grouper found during the study in Kotania Bay (Table 1). The high diversity of grouper found in Kotania Bay is not surprising because this bay is was well known as migration area for reef fish including groupers. It can be seen in Table 1 that grouper belong to Serranidae family consists of six genera i.e. *Epinephelus, Cephalopholis, Cromileptes, Plectropomus, Anyperodon* and *Variola.* Genus of *Epinephelus* has more species, followed by *Cephalopholis* and *Plectropomus* while *Variola, Anyperodon* and *Cromileptes* genera have fewer species.

Table 1. Checklist of grouper species at Kotania Bay waters

No	Scientific Name	Common Name	Local Name
1	*Anyperodon leucogrammicus*	White-lined grouper	Kerapu
2	*Cephalopholis aitha*	Rusty rock cod	Kerapu
3	*Cephalopholis argus*	Peacock rock cod	Kerapu Loreng
4	*Cephalopholis boenak*	Dusky-banded cod	Kerapu
5	*Cephalopholis cyanostigma*	Blue-spotted rock cod	Kerapu
6	*Cephalopholis leopardus*	Leopard rock cod	Kerapu
7	*Cephalopholis miniata*	Coral rock cod	Kerapu Tungsing
8	*Cephalopholis nigripinnis*	Black-finange rock cod	Kerapu
9	*Cephalopholis sexmaculata*	Saddled rock cod	Kerapu Sunu
10	*Cephalopholis spiloparaea*	Orange rock cod	Kerapu
11	*Cephalopholis sonnerati*	Tomato rock cod	Kerapu Sunu
12	*Cephalopholis urodeta*	Flag-tail rock cod	Kerapu
13	*Cromileptes altivelis*	Barramundi cod	Kerapu Tikus
14	*Epinephelus bontoides*	Dusky rock cod	Kerapu
15	*Epinephelus coioides*	Brown-spotted grouper	Kerapu Muara
16	*Epinephelus corallicola*	Coral rock cod	Kerapu Loreng
17	*Epinephelus cyanopodus*	Purple rock cod	Kerapu
18	*Epinephelus erythrourus*	Cloudy rock cod	Kerapu
19	*Epinephelus fuscoguttatus*	Flower cod	Kerapu Macan
20	*Epinephelus lanceolatus*	Giant grouper	Kerapu Kertang
21	*Ephinephelus macrospilos*	Snubnose rock cod	Keapu loreng
22	*Ephinephelus maculatus*	Marbled rock cod	Kerapu Loreng
23	*Epinephelus melanostigma*	Black-spotted grouper	Kerapu
24	*Epinephelus merra*	Honeycomb rock cod	Kerapu Loreng
25	*Epinephelus ongus*	White-speckled rock cod	Kerapu Hitam
26	*Epinephelus quoyanus*	Long-finned rock cod	Kerapu Loresng
27	*Epinephelus tukula*	Potato cod	Kerapu
28	*Plectropomus areolatus*	Square-tail coral trout	Kerapu Sunu
29	*Plectropomus laevis*	Footballer cod	Kerapu Sunu
30	*Plectropomus leopardus*	Leopard coral trout	Kerapu Tungsing
31	*Plectropomus maculatus*	Spotted coral trout	Kerapu Lodi
32	*Plectropomus oligacanthus*	Vermicular trout	Kerapu Sunu
33	*Plectropomus pessuliferus*	Violet coral trout	Kerapu Sunu
34	*Variola albimarginata*	Lyre-tail cod	Kerapu
35	*Variola louti*	Common Lyre-tail cod	Kerapu Lodi

Serranidae family is one of the reef fishes group which has high number and important species in coral reef. According to Allen and Adrim (2003) [13], there were 102 species of Serranidae in Indonesia waters. Meanwhile, [9] found 55 species of grouper inhabit coral reef of Indonesia. Based on that statement, it can be concluded that 30.4 - 59.6% of grouper in Indonesia waters can be found in coral reef of Kotania Bay.

Based on fishing sites in Kotania Bay, Marsegu island had the highest species richness of groupers (20 species) compared to Buntal island to Burung island (17 species) and Osi island (16 species). The higher species richness of grouper in Marsegu island is supported by wider area and better condition of coral reef which can be categorized as good to excellent that make this island as ideal habitat for reef fishes. According to Nybakken (2001) [14], high diversity of reef fishes in coral reef is caused by its variety of habitat which consists of coral with many hollow and crevices, sand and different zones.

3.2. Spatial Distribution of Groupers Fishing Ground

There were 36 sites for fishing ground of grouper identified during the research in Kotania Bay (Figure 2). The bottom substrate of those sites are dominated by coral reef, so distribution of grouper depend on distribution of coral reef as their habitat.

Spatial distribution map of fishing ground for grouper gives two important information about the utilization of grouper resource in Kotania Bay, i.e. distribution of fishing gear and distribution of species. It was identified through observation in fishing operation and participative mapping that hand line was used in most of the fishing ground. On the contrary, bottom long line, spear gun and harpoon were only used in the fishing ground close to the villages and those small islands in Kotania Bay.

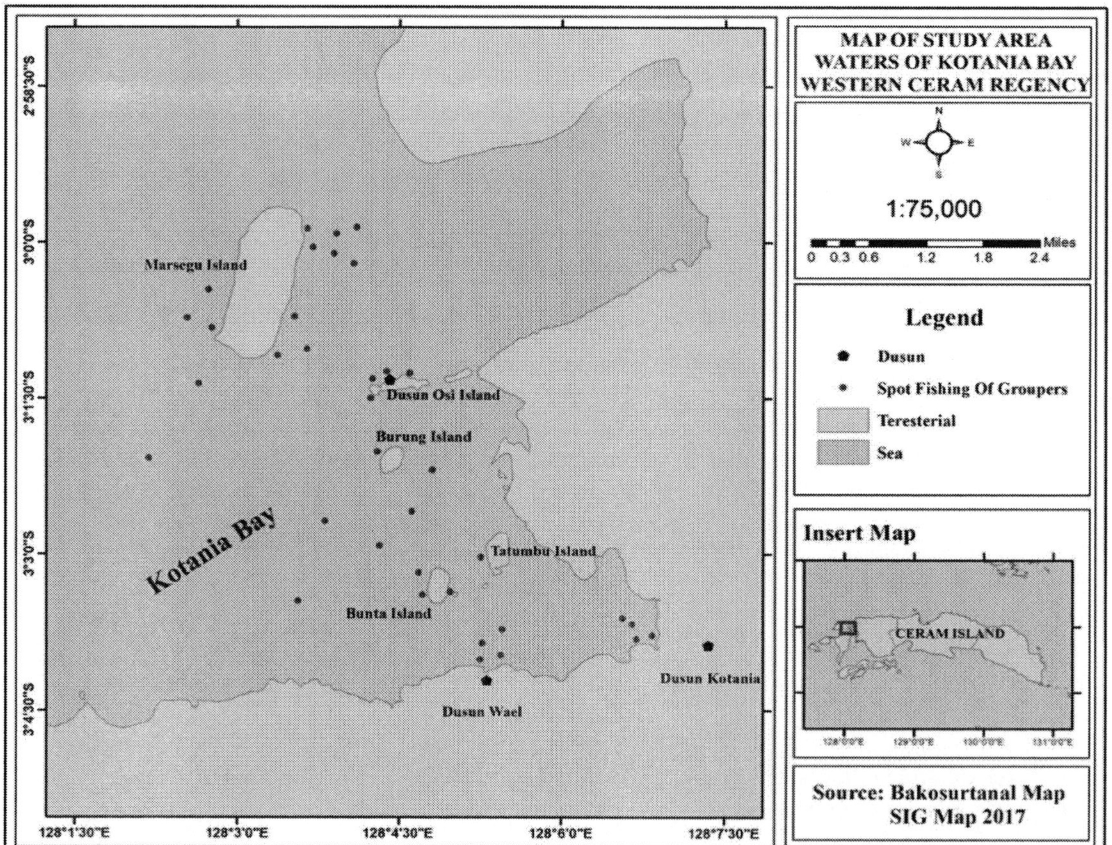

Figure 2. Map showing spatial distribution of gouper fishes fishing gound at Kotania Bay

3.3. Fishing Gears of Grouper Fishes

There were seven fishing gears used by fishers in Kotania Bay to catch grouper namely bottom long line (BLL), hand line (HL), bottom trolling (BT), trap net (TN), bottom gill net (BGN), spear gun (SG) and harpoon (H). Grouper caught by those fishing gears varied in term of number of species (Figure 3) and a number of individual (Figure 4).

It can be seen in Figure 3 that bottom long line (BLL) caught the highest species of groupers, followed by hand line (HL), whilst spear gun (SG) and harpoon (H) caught the least species of groupers. On average, bottom long line also caught more individual of grouper, followed by hand line, while trap net and harpoon-caught less individual (Figure 4). The number of species caught by pots is moderately high, but the average number of individual caught in each fishing operation is low.

The bottom long line has great capacity to catch more groupers because each unit of this fishing gear has about 100 hooks and it can be operated in wider area of coral reef. Hand line is a traditional fishing gear and can be considered as friendly environmentally fishing gear. However, its capacity to catch grouper is quite high. Trap net is also a traditional fishing gear but it is considered as unfriendly environmentally fishing gear because its operation often damages coral reef.

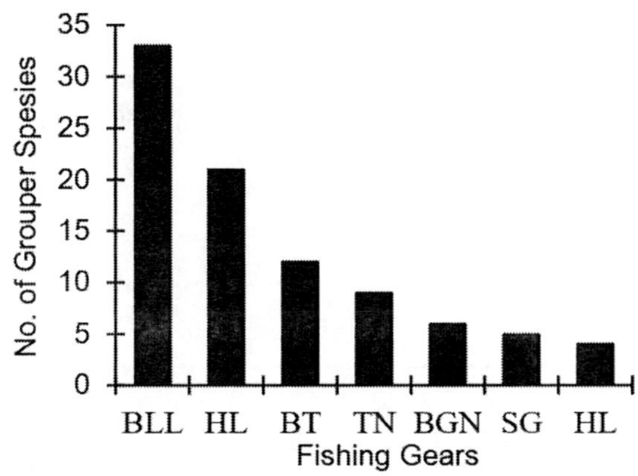

Figure 3. Number of grouper species for each fishing gear

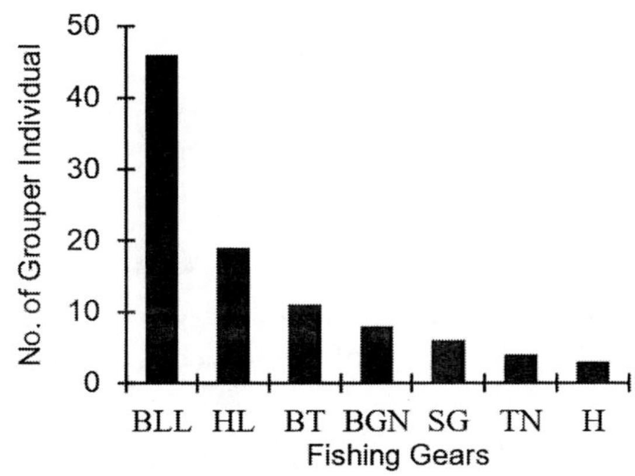

Figure 4. Number of grouper individual for each fishing gear

3.4. Tendency of Decreasing Number of Species and individual of Grouper

Results of the survey and FGD with fishers in Kotania Bay showed that the species number of groupers have decreased (Figure 5). In 2000's, fishers could catch as many as 47 species of groupers in this area which consisted of 25 species of *Ephinephelus*, 12 species of *Cephalopholis*, six species of *Plectropomus*, two species of Variola and one species for both *Cromileptes* and *Anyperodon*. Nowadays, the number of species for *Ephinephelus* and *Cephalopholis* genera in Kotania Bay are 14 and 11 species, respectively. So, during the last 17 years 12 species of groupers are extinct in this area i.e. 11 species of *Ephinephelus* and one species of *Cephalopholis*. Based on FGD with fishers, it was revealed that disappearance of some groupers especially genus of *Ephinephelus* is caused by degradation of coral reef due to fishing pressure and the use of destructive fishing method such as explosive and cyanide during the year 2000 to 2015.

Based on checklist of Indonesian reef fishes and their distribution in Indo-Australian archipelago [12], it is predicted that 11 species of groupers namely *Cephalopholis polleni, Epinephelus areolatus, E. bleekeri, E. bilobatus, E. caeruleopuctatus, E. fasciatus, E. hexagonatus, E. malabaricus, E. polyphekadion, E. tauvina, E. undolosus, E. sexfaciatus* and *Plectropomus pessuliferus* have disappeared from Kotania Bay. Meanwhile, groupers species such as *Cromileptes altivelis*,

Epinephelus fuscoguttatus, E. lanceolatus, Plectropomus areolatus and *P. laevis* found in Kotania Bay are also considered as vulnerable grouper species by IUCN [15].

It was also revealed in the survey and FGD with fishers that declining of species number is followed by decreasing number of individual for six genera of groupers caught from the year 2000 to 2017 in Kotania Bay (Figure 6). Growing number of fishers as well fishing gears, increasing demand and higher price of groupers caused increasing fishing frequency or fishing intensity. For example, fishing intensity for grouper especially for Barramundi Cod (*Cromileptes altivelis*) in Kotania Bay is high because of its high price which is up to IDR 650,000/kg.

According to [16] and [17], decreasing number of species, as well as a number of fish caught by fishers, might be due to overfishing and degradation of habitat. It seems that decreasing number of species and individual of groupers in Kotania Bay is caused by overfishing and degradation of coral reef. Declining number of species and number of individual caused by overfishing could be overcome through management responses directed towards restricting catch [16] or reducing fishing effort (e.g. imposition of catch quotas, minimum size regulations, limited entry, temporary area closures and closed fishing seasons). Meanwhile, degradation of coral reef due to fishing pressure can be done through restoration or rehabilitation and protection of coral reef habitat [16] and [15]. Both approaches are important effort related to grouper fisheries management in Kotania Bay.

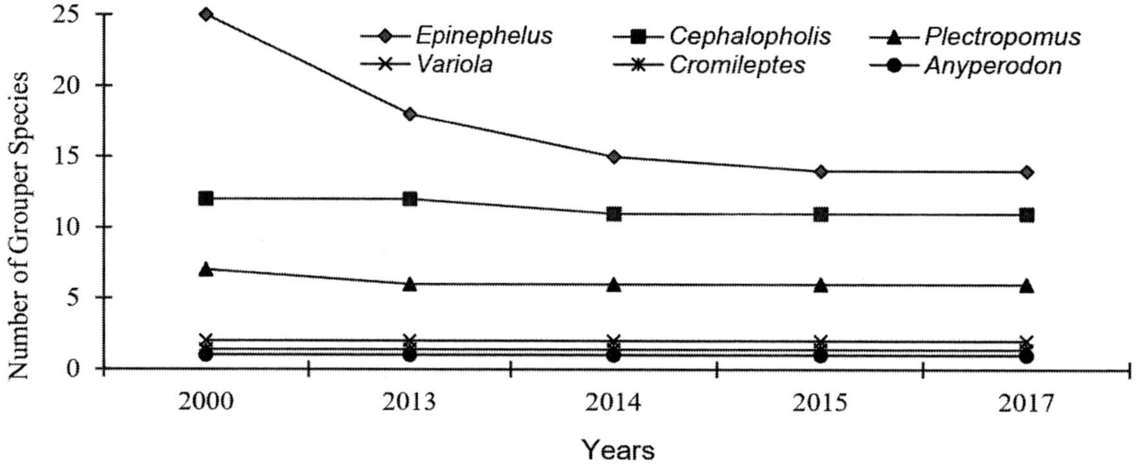

Figure 5. The species number of six grouper genera caught from year 2000 to 2017

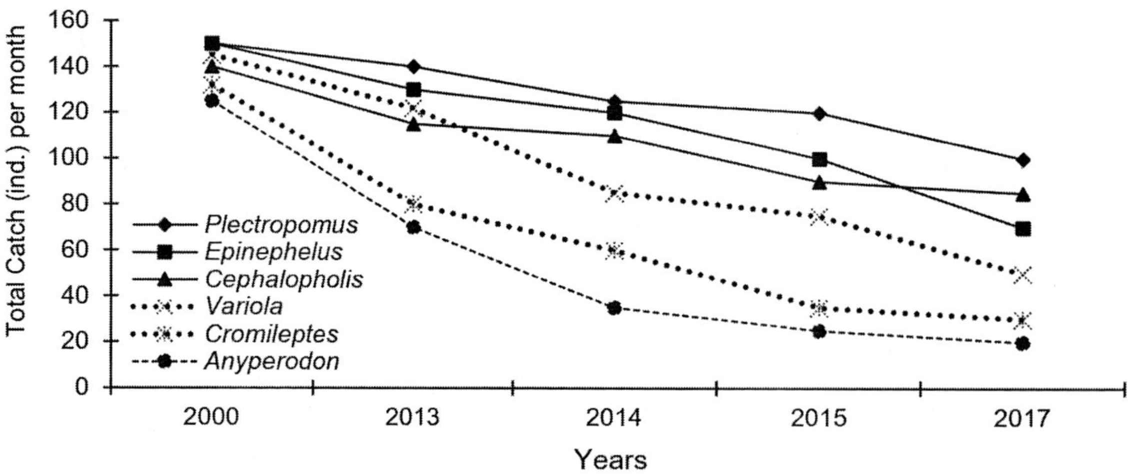

Figure 6. Number of individuals caught per month of six grouper genera from year 2000 to 2017

IOP Conf. Series: Earth and Environmental Science **89** (2017) 012002 doi:10.1088/1755-1315/89/1/012002

3.5. Progress in grouper fishery

Local fisher of Kotania Bay has exploit grouper as a source for self-consumption and economy purposes for a long time. Fishing practices of local fisher in Kotania Bay up to present time is considered as to be traditional fisheries, using simple fishing gears and technique.

Progress of grouper fisheries in term of gear type used and its number as well as price over period can be seen in Table 2. There is progress in this fisheries shown by steady increase of gear number, fisher and the price over time. In specific, trap net, spear gun and harpoon tend to decrease whilst bottom long line and hand line tend to increase and at the same time some fishers have use bottom trolling and bottom gill net as well.

Decrease in trap net may result from fisher understanding in the negative effect of this gear towards sustainability of coral reefs and its maintenance cost, whilst increase in hand line may cause by simple in operationalization, relatively easy in obtaining the material, cheap and easy to repair, ecological friendly as well as its productivity.

Destructive fishing (bomb and cyanide) have been used by some local fisher of Kotania Bay for quite some time. This type of fishing activity will certainly damage the coral reefs, kill finfish, shellfish and other organism associate with coral ecosystem [18]. Most of destructive fishing attributable to limited knowledge on sustainable fishery and less control from related fisheries management agency.

Table 2. Progress of grouper fisheries at Kotania Bay

Year Period	Type of Fishing Gears	Number of Fishing Gears	Number of Fishers	Price (IDR/kg)	Remarks
1990's	Hand Line	35 units	20 persons	5000	Increasing market demand Prices depend on species and size of groupers
	Bottom Long Line	3 units	3 persons		
	Trap Net	30 units	10 persons		
	Harpoon	10 units	10 persons		
	Spear Gun	8 units	8 persons		
	Bomb and cyanide	★	★		
2000's	Hand Line	40 units	30 persons	50000 – 350000	
	Bottom Long Line	5 units	5 persons		
	Trap Net	30 units	10 persons		
	Harpoon	7 units	7 persons		
	Spear Gun	6 units	6 persons		
	Bomb and cyanide	★	★		
2017	Hand Line	45 units	35 persons	100000 – 650000	
	Bottom Long Line	15 units	10 persons		
	Bottom Trolling	11 units	11 persons		
	Bottom Gill Net	12 units	12 persons		
	Trap Net	20 units	8 persons		
	Harpoon	4 units	4 persons		
	Spear Gun	2 units	2 persons		
	Bomb and cyanide	-	-		

Note: ★ unknown quantity - unused (Source: Respondents 2017)

3.6. Grouper fishing Calendar

Information on grouper fishing calendar obtained through focus group discussion and participatory mapping is shown in Table 3. Six fishing gears *i.e.* bottom bill net, bottom trolling, trap net, harpoon and spear gun can be operated through all seasons (West and East monsoon) even on rough weather. Meanwhile bottom long line can be operated between June to October (East monsoon) when weather is calm.

Table 3. Fishing season calendar of grouper at Kotania Bay waters

Monsoon and Fishing Gears	J	F	M	A	M	J	J	A	S	O	N	D	Remarks
Season													
West Monsoon	*	*									*	*	Wavy waters
Transition season I			*	*									Transition season
East Monsoon					*	*	*	*					Waters quite calm
Transition season II									*	*			Transition season
Fishing Gears													
Hand Line	*	*	*	*	*	*	*	*	*	*	*	*	Every month Operation
Bottom Long Line						*	*	*	*	*			Operated when sea calm
Bottom Gill Net	*	*	*	*	*	*	*	*	*	*	*	*	Every month Operation
Bottom Trolling	*	*	*	*	*	*	*	*	*	*	*	*	Every month Operation
Tram Net	*	*	*	*	*	*	*	*	*	*	*	*	Every month Operation
Spear Gun	*	*	*	*	*	*	*	*	*	*	*	*	Every month Operation
Harpoon	*	*	*	*	*	*	*	*	*	*	*	*	Every month Operation

3.7. Grouper fisheries management strategy
Utilization and management of coastal waters by local fisher of Kotania Bay are not following sustainable fisheries management. This in turn will affect sustainability of grouper fisheries and coral reefs as habitat for this fish. Therefore there is a need to establish management plan and strategy in order to sustain both the fish and its habitat. Some recommendations are conservation and rehabilitation of groupers habitats, groupers resource utilization, groupers resource governance, restocking and culture of grouper fishes, as well as awareness and empowerment of fishermen.

4. Conclusions
There are 35 species of groupers which belonging to six genera namely *Epinephelus, Cephalopholis, Cromileptes, Plectropomus, Anyperodon* and *Variola* found during the study in Kotania Bay. Those species are exploited by artisanal fishers using traditional fishing gears such as hand line, bottom trolling, trap net, spear gun, harpoon, bottom long line and bottom gill net at 36 fishing spots. Almost all fishing gears can be operated year round, while bottom long line can only be operated during east monsoon. The number of species as well as the number of individual of groupers caught in Kotania Bay have declined due to overfishing and degradation of coral reef. Approaches that could be proposed to solve those problems are restriction catch as well as fishing gear and rehabilitation as well as protection of coral reef as an important habitat for groupers. There is progress in grouper fisheries shown by steady increase in fishing gear number, fisher and the grouper price over time. Some alternatives management strategies could be recommended to maintain species potency performance and grouper fisheries sustainability are conservation and rehabilitation of groupers habitats, groupers resource utilization control, groupers resource governance, restocking and culture of grouper fishes, as well as awareness and empowerment of the fishermen in Kotania Bay.

References
[1] Marimutu N, Wilson J J and Kumaraguru A K 2010. Reef status in the Mandapam group of Islands, Gulf of Mannar. *Galaxea J Coral Reef Studies* 12: 65-75.
[2] Peristiwady T 1994 Fishes at seagrass of Osi Island and Marsegu Island Western Seram : I. Community structure. *Maluku and Surrounding Waters* (Indonesian Institute of Sciences Press) 7 35-52.
[3] Setyastuti A 2011 Ehchinoderm of Western Seram Coastal Waters Maluku Province. *Maluku and Surrounding Waters* (The Indonesian Institute of Sciences Press) pp 21-31.

[4] Islami, M M 2011 Community structure of molluscs on seagrass habitat of Osi Island Western Seram District Maluku Province *Maluku and Surrounding Waters* (Indonesian Institute of Sciences Press) pp 33-46.
[5] English S, Wilkinson C and Barker V 1997 *Survey manual for tropical marine resources.* 2nd Edition, Australian Institute of Marine Science, Townsville.
[6] COREMAP II 2006 Data collection guide and Rapid Rural Apprasial (PRA) method http// www. coremap2.com RRAPRA.pdf.
[7] Munro I SR 1967 *The Fishes of New Guinea* (Port Moresby New Guinea: Department of Agriculture, Stock and Fisheries).
[8] Kuiter R H 1992. *Tropical Reef-Fishes of The Western Pacifik Indonesia and Adjacent Waters.* (Jakarta : PT. Gramedia Pustaka Utama) p 314.
[9] Kuiter R H and Tanozuka T 2001 *Pictorial Gude to Indonesian Reef fishes* Part 1 Muraenidae to Lutjanidae (Australia: Zoonetics) www.zoonetics.com p 302.
[10] Allen G, Steen R, Humann P and Deloach N 2003 *Reef Fish Identification Tropical Pacific* (Florida USA: New World Publications, Inc) p 457.
[11] White W T, Last P R, Dharmadi, Faisah R, Chodrijah U, Prisantoso B I, Fogonoski J J, Puckridge M and Blaber S J M 2013 *Market Fishes of Indonesia* (Australia : Australian Centre for International Agricultural Research)
[12] Kimani P dan Obura D 2004 Patricipatory Mapping of Terresterial Fishery Resources in Kwale District, Kenya. *Western Indian Ocean J. Mar. Sci.* 3 (2) 209-220.
[13] Allen G R. and Adrim M 2003 Coral Reef Fishes of Indonesia *Zoological Studies* 42 (1) 1-72
[14] Nybakken J W 2001 *Marine Biology An Ecological Approach* 5th ed (USA : Benjamin Cummings, an imprint of Addision Wesley Longman Inc) pp 370-415.
[15] Sadovy de Mitcheson Y, Craig M T, Bertoncini A A, Carpenter K E, Cheung W W L, Choat J H, Cornish A S, Fennessy S T, Ferreira B P, Heemstra P C, Liu M, Myers R F, Pollard D A, Rhodes K L, Rocha L A, Russell B C, Samoilys M A and Sanciangco J 2012 Fishing Groupers Towards Extinction : A Global Assessment of Threats and Extinction Risks in a Billion Dollar Fishery. *Fish and Fisheries* (Blackwell Publishing Ltd) p 18.
[16] Turner S.J, Thrush S F, Hewitt J E, Cummings V J and Funnell G, 1999 Fishing impacts and the degradation or loss of habitat structure. *Fisheries Management and Ecology* (Blackwell Science Ltd) 6 401- 420 .
[17] Dulvy N K., Sadovy de Mitcheson Y, and Reynolds J D 2003. Extinction vulnerability in marine populations. *Fish and Fisheries* 4 25–64.
[18] Burke L, Seling E and Spalding M 2002 *Reef At Risk in Southeast Asia* (USA: World Resources Institute) p 40.

IOP Conf. Series: Earth and Environmental Science **89** (2017) 012003 doi:10.1088/1755-1315/89/1/012003

Small is beautiful: Marine small-scale fisheries catches from the South-West Maluku Regency

BG Hutubessy[1], **JW Mosse**[1], **P Hayward**[2]

[1]Fisheries and Marine Science Faculty of Pattimura University
[2]Southern Cross University Australia

Email: grace.hutubessy@fpik.unpatti.ac.id

Abstract. The fisheries data supplied by fisheries agency have served as the primary tool for regional fisheries statistics. However, it is recognized these data are incomplete and often underestimate actual catches, particularly for small-scale fisheries. There is no widely accepted definition of small-scale fisheries or global data on number of small-scale fishers and their catches. This study reconstructed total marine catches from 1980 to 2015 for South-west Maluku (MBD) regency, by applying an established catch construction approach utilizing all available quantitative and qualitative data, combined with assumption-based estimations and interpolations. As newly established regency since 2009, there is lack of fisheries data available which is needed for fisheries management. Fishers' knowledge is important information taken from to construct long-term fisheries data. Estimated total fish withdrawal from MBD waters was 86,849.66 tonnes during 1980 – 2015, dominated by pelagic fishes. Consistency of estimated total removal and total landings at MBD regency play important role in small-scale fisheries management and this method of visualizing the history of fishery from poor-data condition might be an optimistic effort.

1. Introduction

Nearshore fisheries in the tropical developing countries provide the main source of animal protein which food security remains a pressing issue [2]. With general definition of small-scale fisheries (SSF): those that work from shore or small boats in coastal and inland water [1], these fisheries make important but poorly quantified contribution to national and regional economies and development of millions people [21]. [8] Predicted that 200 million people might depend on SSF and related activities. As a consequence, illegal practices in small-scale fishing methods can be very destructive, such as dynamite and cyanide fishing cannot be avoided (e.g. in South East Asia [19] and Africa [22]). Thus, overfishing can occur with small-scale fisheries [24], and assessment of these fisheries is urgently needed.

Many studies of small-scale fisheries have been conducted, but many problems appeared at first sight. [17] Showed that the disadvantages of SSF by their physical, socio-economic, political and cultural remoteness from urban centers positioned them further to the marginalization. Communities of small-scale fishing in the developing countries mostly located further from political power which then usually receives far less support e.g. from government [7]. Their lacking landing sites and infrastructure create difficulties to market access. Further, small-scale fisheries are distinctively different between countries or even regional, and each fishery is unique [7]. The SSF conditions result less amount of research and data collection efforts compared than large-scale fisheries. Many catches for the SSF are not estimated by fisheries agencies [27]. Hence, an attempt to counter the marginalization of SSF by retroactively estimate catches is needed such conducted by [26].

It is widely recognized that catch statistics are crucial to fisheries management. Incomplete catch statistics are not an issue only for developing countries, but also for developed countries with sufficient human and financial resources to maintain recent statistics [27]. Although the catch statistics routinely collected and published in most countries, deficiency in numerous ways is still occurred [28]. The focus of reported data has been primarily on commercial harvest and has not covered shore-based fisheries [26]. Therefore, SSF has large contribution to the deficiency in fisheries statistic due to its marginalization. Here, we report on work undertaken for the South-West Maluku regency to account for unreported catches.

South-west Maluku (MBD) regency is archipelago regency consisted of 17 subdistricts located at 3 groups of islands. Marine area covers 63,775 km2 or 88.06% of the total area (Fig.1). The regency was recently established in 2009 after regional division from a subdistrict belonged to Southeast-west Maluku (MTB) Regency. The regency of MTB was also as regional division from South-east Maluku (Malra) regency in the year of 2004. Therefore, statistical fisheries data from new regencies is available in short time-series. Domestic fisheries divided into two sectors, small boat-based fisheries and shore-based fisheries; Catch data for both fisheries have been estimated by fisheries department, but the data were likely incomplete because the commercial and noncommercial catches have never been thought to be reported.

Fisheries data collection in Indonesia has always been constrained by numerous issues leading to incomplete surveys. Because the marine fisheries area covers 5.1 million km^2 archipelago waters, including Exclusive Economic Zone, illegal fishing cannot be avoided. The standard statistical system has been introduced since April 1976 for the marine fishery [23], but excluding standard of data collection. Collection of basic data is the responsibility of the Central Bureau of Statistics while the Directorate General of fisheries of Indonesia collects and release fisheries statistics in the Fisheries Statistical reports [18]. The conduct of production survey is supported by insufficient numbers of enumerators which some of them has non-fisheries background [11]. Reduction numbers of auction levy in some regions, while some have none, making it difficult for data collectors to monitor and report the fish landed [11]. As Indonesia has continuously improved capacity and system of its data and fishery statistic, data verification and validation is needed.

The purpose of our study was to assemble available information and data on catches of the small-scale fisheries. The aim was to derive estimates of likely total removal of marine resources for the 1980-2016 period, as an initiative to construct a new 'road map' for small-scale fisheries, where they are put at the front and center of research and management interest, instead of remaining at the margin.

2. Methodology

2.1. Approaches

Successful reconstruction for marine fisheries catches has been conducted in other regions [26-27]. Here, we follow the conceptual framework and approach outlined by these studies to reconstruct marine fisheries catches for South-west Maluku regency. The catch reconstruction approach utilized here consists of six general steps based on [27]:

1. Identification of existing reported catch time series provided by Fisheries and Marine Affair Department and Central Bureau of Statistics from Maluku province
2. Identification of time periods, gears, trips, population, and fish consumption per-capita.
3. Search for available information sources through literature searches and consultation aiming to explain the changes in time series and to supply the missing catch data
4. Development of data anchor points in times for missing data items and their expansion to regencies catch estimates
5. Interpolation for time periods between data anchor points for per-capita consumption rates
6. Estimation on final total catch time series estimates for total catch, combining reported catches and interpolated.

We collected other information from fishers by conducting semi-structured interviews and focus discussion groups. Semi-structured interviews with closed and open questions put to fishers in face-to-face were conducted from April to June 2017. This assessment allows familiarisation with fishing methods, fishing grounds, concerns as well as local culture. We also randomly sampled the fishers from coastal villages based on the gear used and age of fisher. In total, 40 fishers were interviewed. Fishers were guided with books of fish to identify the local names of fish they caught. Personal interviews with fishers can gather large amounts of information regarding fishing practices [16]. The questionnaire was grouped in four sections: personal information including experience and property; knowledge of activities changes at the surrounding coastal areas, efforts and catch currently and in the past.

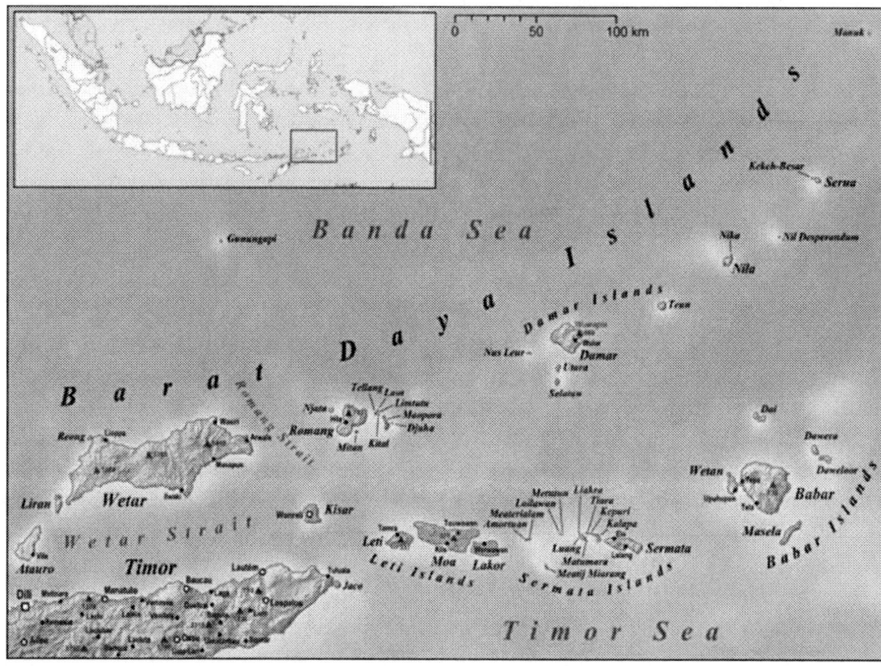

Figure 1. South-west Maluku regency islands located in the eastern Indonesia

2.2. Population Resources

1980-2015: We used human population statistic reported in the book of *Maluku in Numbers*.

1980, 1990, 1992-1996, 2011-2015: From *South-west Maluku in Numbers*, we used human population in the regency as the anchor points. To expand the population data, extrapolation was made by rationing between province and regency for this period. The ratio was fluctuated without showing any trend (level off). Thus, we assume that the ratio is constant and the average of these ratios was used to interpolate some human population between those periods.

2.3. Artisanal Fisheries

1980-2015: We used total marine catches reported for Maluku province and South-east Maluku regency in fisheries statistical annual reports. For South-east-west Maluku (MTB) regency, the available time-series data is 2004-2015, while for South-west Maluku (MBD) regency is 1996 and 2011-2015. Fisheries growth pattern for all regions is used to grow a model for reconstruction long-term time-series catches data for South-west Maluku regency.

2011-2015: Effort time-series data reported for MBD regency including boats, gears and trips, are used to descript the fisheries as small-scale fisheries and to explore possible approach to be used for interpolation.

2.4. Fish consumption per-capita

1993, 1996, 2000, 2003, 2006-2015: The available time-series in the book of *Maluku in Numbers*. Based on the fisher interviews, catches are commercially sold to fulfill local demands. No discard reported. Thus, we assumed that total catch for small-scale fisheries is the reflection fish consumption per-capita (gr day^{-1}) by using human population data.

3. Result and Discussion

3.1. Result

Fisheries statistical reports for MBD regency are available from 2011 (Table 1). Fishing boats in MBD regency is dominated by canoes. Inboard engines are occupied for under 10GTfishing boats. Fishing gears are mostly operated in traditional ways. Gillnets included drift gillnets, encircling gillnets, and bottom gillnets are mostly self-mended by fishers, as well as traditional traps made from bamboos. Line fishing included handlines, longlines and trolled lines is the most gear used. Tuna trolled line activities in the region provided tuna loin to support tuna industry outside the region. In 2013 and 2015, number of trolled lines increased about 4%. Lift nets activities in 2009 and purse seines in 2012 occupied more young people involved in the fisheries. Other gears used in the area include spear fishing, *bameti* or collecting marine organism during low tide, and other types of traditional small traps used occasionally. These general figures of MBD's fisheries convince us to assume that this fishery is categorized as small-scale fisheries.

Table 1. General figures of MBD's Fisheries in numbers. Boats (Can: canoes; W*s*: small wooden boat; W*m*: medium wooden boat; W*l*: large wooden boat; Out: outboard engine; In: inboard engine). Gears (Gn: gillnets; Line: line fishing; Ps: purse seines; Ln: lift nets; Oth: other gears)

Year	Catch	Fishers	Boats						Gears					
			Can	W*s*	W*m*	W*l*	Out	In	Gn	Line	Trap	Ps	Ln	Oth
1996	9,877.9	-	-	-	-	-	-	-	-	-	-	-	-	-
2011	3,286.0	9,366	697	476	84	28	148	146	860	529	84	-	89	194
2012	8,638.0	9,564	698	476	84	28	162	146	860	548	84	14	89	194
2013	4,330.5	9,901	698	476	84	28	176	176	857	565	84	16	89	197
2014	8,653.0	10,004	698	476	84	28	189	150	857	579	84	18	89	194
2015	19,130.4	10,004	589	67	26	4	189	150	857	759	84	18	89	194

The result of interviews and focus group discussions brought some issues to reconstruct long-term time-series catch data. There was no change on fishers' catches or relatively stable fish production. They considered on marine resource demands which increased recently. Previously, these fisheries normally contribute food for household consumption, with small amount of catches for barter or trade. Fish are easy to be caught due to abundantly existed. Nowadays, fishing effort should be expanded following the increase of demands. They are considered commercial fisheries, as catches are landed and sold either by fishers or their family members at the market or through middlemen. Lift nets and purse seine are recently operated in MBD's water. The using of fish aggregation device (*rumpon*) concentrates pelagic fishes and enlarges the catch of purse seine and lift nets. Fish price varies depending on number of catch; it might be lower than Rp. 2000, - (equal to 20 cents USD) per-kg. There is no cold storage available, fisher's self-storage their unsold fish with ice crunch. It is also reported that large ships (>30GT) captured and bought demersal fishes from Mdona Hyera sub-district.

Fish production in Maluku tends to increase annually except for the regencies (Fig. 2). Due to regional division occurred for Malra and MTB regencies, marine fish production dramatically drops in

IOP Conf. Series: Earth and Environmental Science **89** (2017) 012003 doi:10.1088/1755-1315/89/1/012003

the year of 2008. After all, the production increases gradually except for MBD which elevated significantly. There were some missing data for Maluku province and South-east Maluku regency catches, but for other two regencies little catch data series are available.

The re-estimation catch for small-scale fisheries for South-west Maluku regency indicated a substantially different pattern between commercial and subsistence catches (Fig. 3). Subsistence catches remained level off with slight fluctuation for the period 1980-2015. The estimated commercial catches for period 1980-1988 increased steadily and reached a peak in 1998 with 3-fold difference. In 1999, the catch decline to 6172.5t. In the period 2001-2005, the estimation tremendously elevated and reached more than 16 thousand tonnes but then followed by dramatic decline to 3286t in 2011. Another peak of 19 thousand tonnes was reached in 2015.

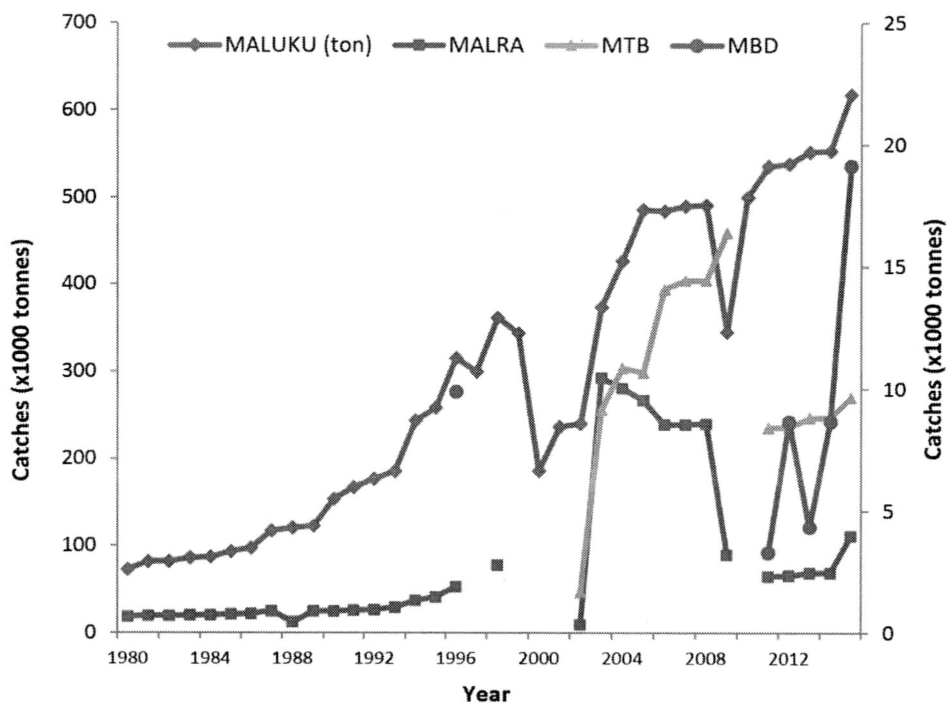

Figure 2. Time-series catches from Fisheries Statistical reports from year 1980 – 2015

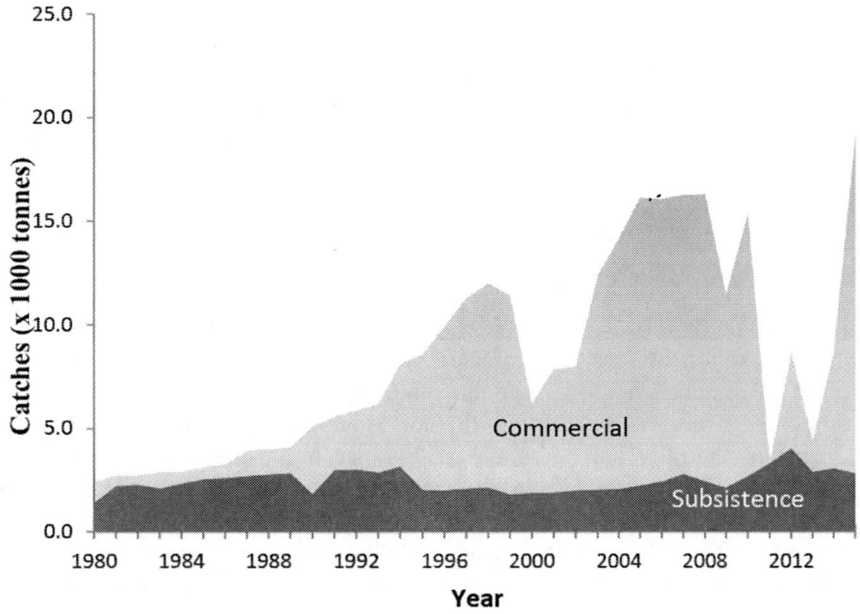

Figure 3. Long-term catch data of small-scale MBD's marine fisheries from 1980-2015 disaggregated into commercial catches (lightly shaded) and subsistence catches (dark shaded) levels.

3.2. Discussion

Without better attempts at accounting for historic extractions of living marine resources for all fisheries sectors, it is not possible to fully understand the formal and informal economic and cultural value of these resources or of the risks of overfishing or ecosystem degradation may represent to South-west Maluku regency [25, 27]. As new regency, exposing the existing resources including fisheries resources which aim to attract investors is a common political and economic intention. However, inaccurate projection or prediction of the available resources may lead to resource depletion after all. In the statistic report of MBD, marine catches dramatically increased in 2015 following extending of gears used. The re-estimation of catches of MBD's indicates two declines in catches of 50% in 1999. This decline might due social conflict occurred in Maluku causing investors fled out the region. During the regency establishment in 2009, when almost all sectors organized their adjustment, catches was recorded about 80% declined. Despite reasonable argues of fish production declined, catch fluctuation might indicate resource recovery after high exploitation. Continues decline and continues increase of fisheries production is an alarm of resource depletion or overfishing.

To support better data collection for SSF, sampling methods should be improved by applying some alternative approaches. The scattered character of SSF, particularly archipelago MBD regency, indicates that sample survey of selected major landing centers is insufficient to provide valid results. Frame survey might structure the information of number, characteristics and spatial distribution of vessels, gears, fishers, landing sites and fishing communities [9]. As a fisheries census [3], frame survey should be updated regularly, and its high cost becomes an obstacle to be conducted. Another alternative approach which is cheaper than frame survey is data collection through filling a simple logbook by fishers and boat owners on a voluntary basis or within co-management scheme [9]. Gathering fisher's knowledge through interviewing some fishers in more general activities [16] also performs beneficial results. The two later alternative approaches were applied in this study. We gathered complement information for the statistical fisheries reports. Since the reports focusing on the fisheries production and its values for export, it indicates that subsistence fisheries are perpetuating. We identified subsistence and commercial household. So, through estimation of total catches based on fish consumption data [20], we determined the subsistence catches.

IOP Conf. Series: Earth and Environmental Science **89** (2017) 012003 doi:10.1088/1755-1315/89/1/012003

The overall finding of our study is likely to be different than other reconstruction small-scale fisheries data have done before. At the country level, small-scale fisheries are exclusively classified as unreported data e.g. [5, 13-14, 25] and their catch estimation is added up to FAO fisheries reports. At regional level, we found small-scale fisheries production can be for commercial and subsistence. The subsistence fisheries comprised only 30% of the total production in the period 1990-2015. While before 1989, fish caught was mostly for consumption. We did not combine the fisheries reports and the subsistence; instead, we deducted those two to count the commercial fish production. Therefore, for further re-estimation catches at country level such as Indonesian archipelago, more evidence are needed to decide which estimation should be added as unreported small-scale fisheries.

Why small-scale fisheries is beautiful? We are confidence that small is beautiful: no take more than need and avoid discards. This is a simple economic understanding which impacts to ecosystem sustainability. Marginalization of SSF is associated with destructive fishing practices [19, 22], non-profit and dispersed to remote areas [6]. On the other hand, fishing practices of SSFs have been reported towards balance fisheries [30]: species or individuals are exploited by their productivity [10, 12]. Since small species naturally more abundance, balance harvesting would substantially increase yield than selectively captured large species [12, 15]. There would be a need to change human consumption patterns towards small fish instead of large. Small fish is more nutritious diet; most essential micronutrients such as calcium, iron, zinc, and vitamin A are concentrated in the part of small fish [4]. Since SSF comprised 85% of Indonesian fisheries focus on data collection, verification and validation should be prioritized.

Acknowledgement

We thank Directorate General of Research and Public service for funding this present study. The books Maluku in Numbers provided by Bureau of Central Statistic Maluku Province and South-east Maluku regency, Fisheries and marine Affair department of Maluku province for providing us with and fisheries statistic reports, are greatly appreciated. Also, Juan Tentua, thank you for your passionate entering data need for this study.

References

[1] Allison, E.H. and Ellis, F. 2001. The livelihoods approach and management of small-scale fisheries. *Marine Policy* 25, 377–388.

[2] Andrew, N. L., Bene, C., Hall, S. J., Allison, E. H., Heck, S., and Ratner, B. D. 2007. Diagnosis and management of small-scale fisheries in developing countries. *Fish and Fisheries*, 8: 227–240.

[3] Bazigos, G. P. 1974. The design of fisheries statistical surveys—inland waters. FAO Fisheries Technical Paper, 133. 122 pp.

[4] Beveridge, M. C. M., Thilsted, S. H., Phillips, M. J., Metian, M., Troell, M., and Hall, S. J. 2013. Meeting the food and nutrition needs of the poor: the role of fish and the opportunities and challenges emerging from the rise of aquaculturea. *Journal of Fish Biology* 83: 1067–1084.

[5] Booth, S. and Watts, P. 2007. Canada's Arctic Marine Catches. *In* Reconstruction of marine fisheries catches for key countries and regions (1950-2005). *Ed.* Zeller and Pauly.

[6] Bene, C., Macfadyen, G., and Allison, E. H. 2007. Increasing the contribution of small-scale fisheries to poverty alleviation and food security. FAO Fisheries Technical Paper, 481. 125 pp.

[7] Chuenpagdee, R. and Pauly, D. 2008. Small is beautiful? A data base approach for global assessment of small-scale fisheries: preliminary reports and hypotheses. *Am.Fish.Soc.Symp.* 49:575-583.

[8] Delgado, C.L., Wada, N. and Rosegrant, M.W. 2003. Fish to 2020. Supply and Demand in Changing Global Markets. International Food Policy Research Institute, Washington, DC, 226 pp.

[9] de Graaf , Grainger G J., Westlund, L, Willmann, R, Mills. Kelleher, K and Koranteng, K. 2011. The status of routin fisheries data collection in South Asia, central America, the South Pacific, and west Africa, with special reference to small-scale fisheries. ICES *Journal of Marine Science*: 68(8): 1743-1750.
[10] Garcia, S.M., Rice, J., and Charles, T. 2016. Balanced harvesting in fisheries: a preliminary analysis of management implications. *ICES Journal of Marine Science*, 73: 1668–1678.
[11] FAO. 2006. Indonesia. www.fao.org/docrep/htm.
[12] Jacobsen, N. S., Gislason, H., and Andersen, K. H. 2014. The consequences of balanced harvesting of fish communities. Proceedings of the Royal Society B: Biological Sciences, 281: 2013–2701.
[13] Jaqcuet, JL. And D. Zeller. 2007a. National conflict and fisheries: Reconstructing marine fisheries catches for Mozambique. *In* Reconstruction of marine fisheries catches for key countries and regions (1950-2005). *Ed.* Zeller and Pauly.
[14] Jaqcuet, JL. And D. Zeller. 2007b. Putting the úníted'in The United Republic of Tanzania: Reconstructing Marine Fisheries Catches. *In* Reconstruction of marine fisheries catches for key countries and regions (1950-2005). *Ed.* Zeller and Pauly.
[15] Law, R., Plank, M. J., and Kolding, J. 2014. Balanced exploitation and coexistence of interacting, size-structured, fish species. Fish and Fisheries, doi:10.1111/faf.12098.
[16] Neis, B., Schneider, D.C., Felt, L., Haedrich, R.L., Fischer, J., and Hutchings, J.A. 1999. Fisheries assessment: what can be learned from interviewing resource users? Can. J. Fish. Aquat. Sci. 56: 1949–1963.
[17] Pauly, D. 1997. Small-scale fisheries in the tropics: marginality, marginalization and some implications for fisheries management. Pages 40-49 *in* E.K. Pikitch, D.D.Huppert, and M.P. Sissenwine, editors. Global trends: fisheries management, American Fisheries Society, Symposium 20, Bethesda, Maryland.
[18] Retnowati, S.D. and Satria, F. 2013. Fisheries data collection system in Indonesia. IOTC–2013–WPDCS09–17_rev1. 5 p.
[19] Saeger, J. 1993. The Samar sea: a decade of devastation. NAGA, the ICLARM (International Centre for Living Aquatic Resources management) Quarterly 16(4): 4-6, Penang, Malaysia.
[20] The World bank, United Nation Development Programme, Commission of the European Communities, and Food Agriculture Organization. 1991. Small-scale fisheries research needs. World bank technical Paper Number no 152. Fisheries Series.
[21] UNDP, 2005. Human development report: International cooperation at a crossroads. United Nation Development Programme. New York, 372p.
[22] Vakily, M. 1993. Dynamite fishing in Sierra leone. NAGA, the ICLARM (International Centre for Living Aquatic Resources management) Quarterly 16(4): 7-9, Penang, Malaysia.
[23] Yamamoto, T. 1980. A standard statistical system for current fishery statistics in Indonesia. A report prepared for the Fisheries Development and Management Project, Indonesia. Rome. FI:DP/INS/72/064, Field Document 7. 79 p.
[24] World Bank, FAO, and WorldFish Center. 2010b. Fisheries' hidden harvest, case-studies of data collection for subsistence and household-level fisheries. WorldBank Agriculture and Rural Development Department Sustainable Development Network, Washington, USA. 64 pp. http://siteresources.worldbank.org/EXTARD/Resources/336681 24775570533/TheHiddenHarvestsConferenceEdition.pdf.
[25] Zeller, D., Booth, S., Craig,P., and Pauly, D. 2005. Reconstruction of coral reef fisheries catches in American Samoa, 1950-2002. *Coral Reef*
[26] Zeller, D., S. Booth, P. Craig, and D. Pauly. 2006. Reconstruction of coral reef fisheries catches in American Samoa, 1950–2002 . *Coral Reefs* 25:144–152.
[27] Zeller, D. Booth, P., Craig, P., and Pauly, D. 2007. Re-estimation of small-scale fisheries catches for US flag island areas in the Western Pacific: the last 50 years. *Fisheries Bulletin* 105: 266-277.

[28] Zeller, D., Darcy. M., Booth, S., Lowe, MK, and Martell S.J. 2008. What about recreational catch? Potential impact on stock assessment for Hawaii's bottomfish fisheries. *Fisheries research* 91:88-97.

[29] Zeller, D., Rossing, P., Harper, S., Persson, L., Booth, S., and Pauly, D. 2011. The Baltic Sea: Estimates of total fisheries removals 1950-2007. *Fisheries Research* 108: 356-363.

[30] Kolding, J., Jacobsen, N. S., Andersen, K. H., & van Zwieten, P. A. (2015). Maximizing fisheries yields while maintaining community structure. *Canadian Journal of Fisheries and Aquatic Sciences*, 73(4), 644-655.

IOP Conf. Series: Earth and Environmental Science **89** (2017) 012004 doi:10.1088/1755-1315/89/1/012004

Exploitation Status of Devis'Anchovy in Kei Island Waters: Based on Total Length Data

E.A Supeni[1], J Dobo[1]

[1]Tual State Fisheries Polytechnic, Indonesia

Email: ekaantosupeni@gmail.com

Abstract. Devis' anchovy (*Encrasicholina devisi*) are a group of small pelagic fishes, which is thought to be one of the most abundant fishery resources in Indonesian waters. The high utilized of this resources will certainly affect to population status. This research aimed to know the exploitation status of anchovy in Kei Island waters. Sampling was carried out by using the data collected from lift net catches during April to July 2016 at Sathean, Selayar and Lebetawi, Kei Islands. A total of 3,524 anchovies were analyzed based on total length that ranged from 23 to 87 mm with a mean length of 55.4 $\pm$ 11.8 mm. The von Bertalanffy growth parameters estimated were L_∞ = 89.25 mm and K = 0.74 year^{-1}, with the growth models equation is $L_t = 89.25\ (1 - e^{-0.74(t+0.161)})$. The result of mortality analysis showed that Z = 2.97, M = 1.098, F= 1.87 year^{-1} and E = 0.63. This study reveals that exploitation status of anchovy in Kei Island is on fully exploited and leads to the over exploited condition.

1. Introduction

Devis' anchovy (*Encrasicholina devisi*) are a group of small pelagic fishes, which is thought to be on of the most abundant fishery resources in Indonesian waters. This resource is a neritic whose spreading in the near coastal. In areas where is a process of upwelling, these resources can form large biomass [1].

The potential of anchovy resources in southeast Maluku is quite high and important resource to advancing coastal community [5]. In Kei Island waters anchovy is caught mainly by lift nets. The preliminary study reveals that fishing activity is carried out throughout the year with the peak season during August to October. Until 2015 the lift net in Kei islands approximately 81 units, and the production reaches 1,016 tons. The fishing ground of anchovy in this region spread on the west coast of Kei Kecil Island, Nerong strait and north coast of Dullah island.

The high utilized of these resources will certainly affect to population status. Therefore it is necessary to steps that can support the sustainable uses of anchovy related to population structure information.

According to Smith *in* [15], determining stock status can be made on biological reference points, such as mortality rate, spawning, biomass or age groups. It is used to obtain diagnosis that warns of diminishing stocks. Some countries are trying to develop and implement of fish resource management policy based on study of biological and population dynamics (9, 17, 8]. [2] Describes six indicators of sustainable fishing: mean length of catch, length-at-50% maturity, total mortality, exploitation rate, ratio of bycatch, and center of gravity of commercial catches. In this study, we emphasize on three indicators of Fairweather, namely length of catch, total mortality and exploitation rate.

Fishing effort, ideally supported by more information about biology, economics and stock assessment. Information of stock includes the total catch data, a number of fishing effort and the catch per units effort (CPUE), while a biological aspect includes length and weight size, gonad maturity level, sex ratio and others according to Gulland *in* [12].

The sustainable uses of devices' anchovy in Kei islands influenced by how these resources were exploited. Therefore, the responsible and sustainable of fishing management needs supported by data and information of stock status. Today, information on various parameters of anchovy in this region is very limited; therefore it is necessary to study the status of the exploitation of these resources.

2. Methodology

2.1. Data collection

Monthly samples of anchovies were collected from the landings of lift net in Sathean, Selayar and Lebetawi during April to July 2016. A total of 3,524 specimens of anchovy, total length was measured to the nearest mm.

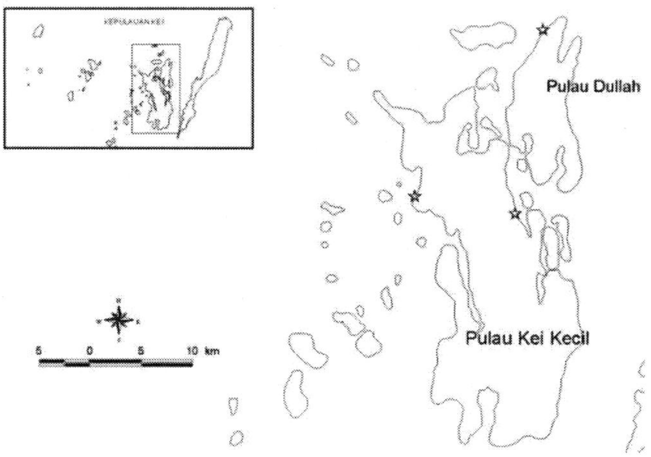

Figure 1. Map of the sampling station

2.2. Data Analysis

Ford-Walford plot was used for estimating growth parameters L∞ and K from the von Bertalanffy equation with the same sampling time interval [7]. The theoretical age at length zero was estimated separately using an empirical equation as described in [11], then the natural mortality coefficient (M) was estimated using Pauly's equation, while total mortality rate (Z) estimated with curve of length-converted catch [13] and the fishing mortality coefficient (F) were estimated [3].
The exploitation rate (E) was calculated according to Gulland *in* [14], to judge whether the stock is overfished or not. Exploitation ratio was estimated:

$$E = \frac{F}{F+M} = \frac{F}{Z} \quad (1)$$

The exploitation rate is compared to the optimum rate according to Gulland in [11] which is 0.5.

3. Result and Discussions

3.1. Length frequency distribution

As shown in Table 1, a total sample 3,452 anchovies with the maximum and minimum total length was 23 and 87 mm, respectively. The mean total length of anchovy was 54.4±11.8 mm. The size frequency distribution of anchovy in presented in Figure 1, in Kei Island water, anchovy were caught in the range of length of 23 mm to 87 mm and were dominated by anchovy the size of 58-62 mm.

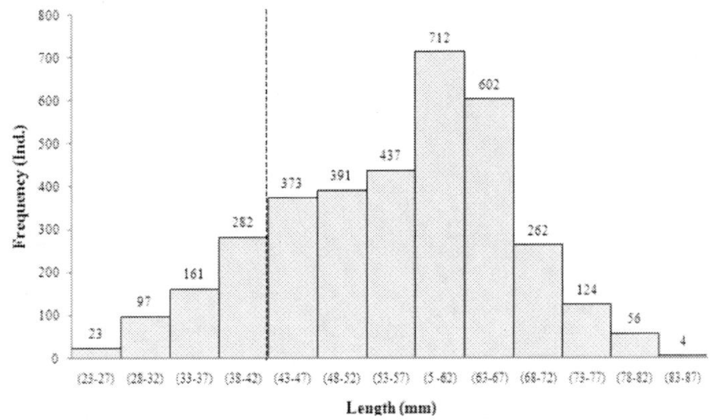

Figure 2. Length frequency distribution of devis' anchovy

Catches of anchovy during study greater than 43 mm of 84.02 percent, while smaller than 43 mm of 15.98 present.

Table 1. Statistics of length size of devis' anchovy during the study.`

Month	N	Total length (mm)		
		Maximum	Minimum	Mean
April 2016	537	30	87	52.2
May	1,695	24	77	56.4
June	820	23	73	47.7
July	472	53	81	68.7
Total	3,524	23	87	55.4

The normal separation method used in the separation of age groups also produced a normal curve that describes the number of cohorts of length frequency distributions. From the figure it is also seen that April and July there was a shift mode which tended to the right, but in May there was a slight shift mode to the left. The growth showed by the shift curve to the right shows and the shift to the left shows the of recruitment. Based on Figure 3 showed that the results of the grouping of size contained two age groups of anchovies caught during the study.

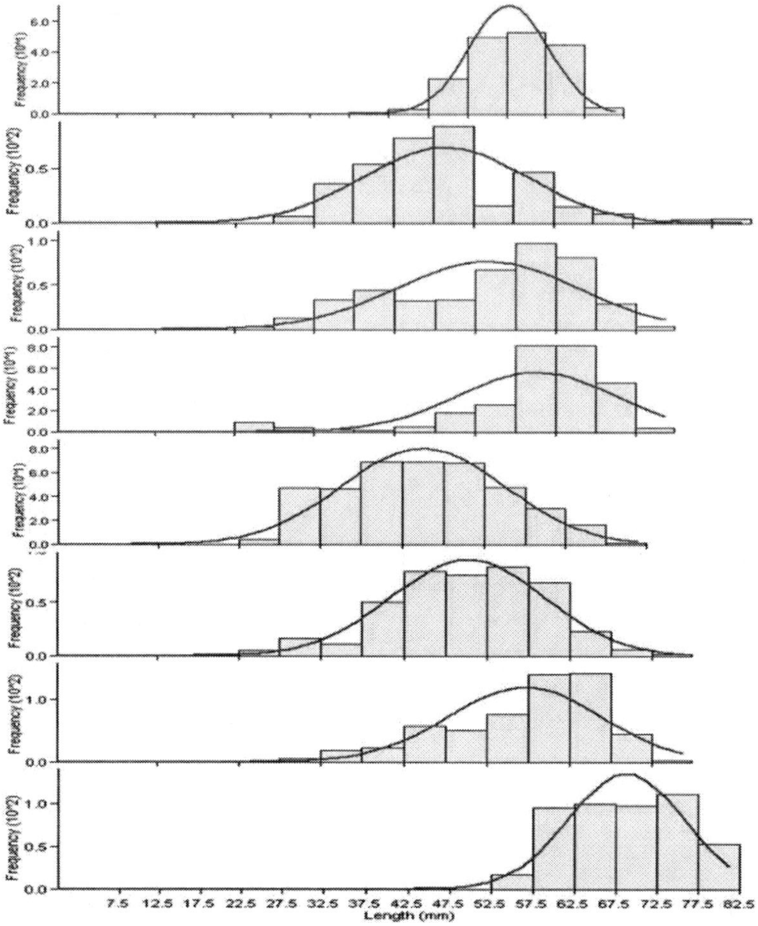

Figure 3. Shifting length frequency mode of devis' anchovy

3.2. Growth

Growth parameter analysis by using von Bertalanffy models (13, 7] with FISAT program, Elefan 1 sub program obtained value of asymptotic length (L_∞) anchovy in Kei island waters is 89.25mm, while growth coefficient (K) is 0,74 yield^{-1}. The analysis result based on (Figure 4) showed one age group of anchovies population in Kei island waters occurs from August to September, in other words of spawning season of anchovy from August to September.

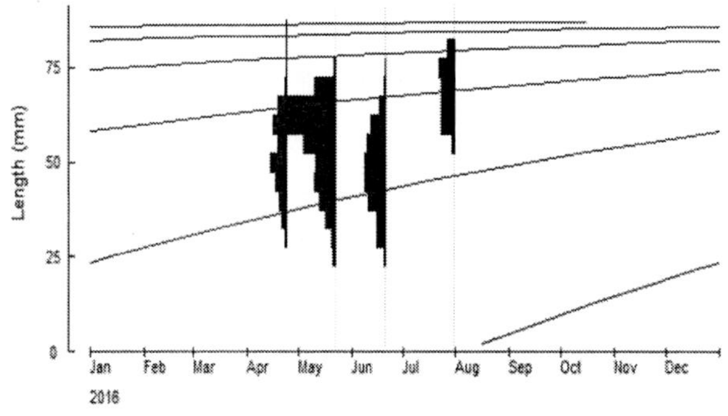

Figure 4. Distribution of length frequency and growth curve for devis' anchovy

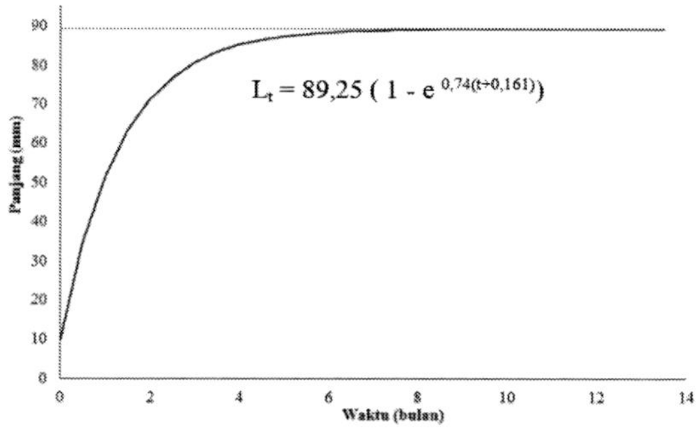

Figure 5. Curva of alleged growth based on total length

The resulting analysis used Pauly approach for anchovies to be t_0 = -0.161. Then from growth parameter value above, with using Von Bertalanffy obtained by the growth of anchovy as follows:

$$L_t = 89{,}25\ (1 - e^{-0{,}74(t+0{,}161})$$

Figure 5 showed that the anchovies found in the Kei island waters could reach a maximum length of 89.25mm, while the actual length obtained 87mm during of research. The analysis result showed anchovies reach infinity length (L_∞) in months 13.5. In figure 5 above shows that the growth of anchovy is significant in the 1st until the 6th month, then experiencing a slowdown until it reaches the asymptotic length.

In general, fish will experience rapid growth in young fish, because most of the energy obtained from food is used for growth. While the old fish, in addition to growth is also used to self-reliance and replace damaged cells [6].

3.3. Mortality

Measured mortality is the total mortality rate (Z), natural mortality rate (M), and fishing mortality rate (F). Total mortality rate to estimate with curve of length converted catch based on total length [14] and natural mortality rate used empirical formula Pauly with mean of sea surface temperature 29^{0}C. While fishing mortality rate (F) is total mortality rate minus the natural mortality rate.

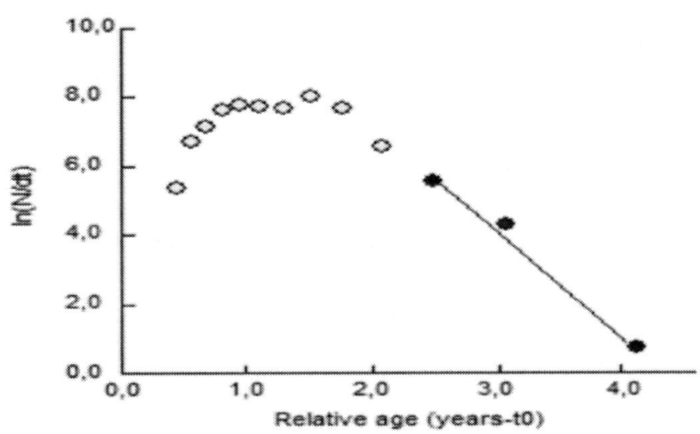

Figure 6. Curva of length-converted catch

The analysis result showed that the total mortality rate (Z) = 2.97 year^{-1}, natural mortality rate (M) = 1.098 year^{-1}, and the fishing mortality rate (F) = 1.87 year^{-1}.

Table 2. Mortality and exploitation rate of devis' anchovy

Parameter	Value (year^{-1})
Total mortality (Z)	2.97
Natural mortality (M)	1.098
Fishing mortality (F)	1.87
Exploitation rate (E)	0.63

From the analysis shows that mortality rate is dominated by fishing factors compared with natural factors. This is due to the fishing of anchovy resources in the Kei island waters throughout the year. Anchovy is also one of the small pelagic species that has important economic value in the research location.

3.4. Exploitation status

Exploitation rate (E) is obtained from the values of Z and F with the equation E = F/Z. Based on criteria from [11], the value of exploited rate of the rational and sustainable in water is at value of <0.5 or the highest E = 0.5. In such conditions a will be obtained of maximum sustainable yield/MSY. Exploitation rate of anchovy in Kei island waters E = 0.63, that meaning exploitation status is fully exploited and leads to the over exploited condition.

Table 3. Comparison of the exploitation status of devis' anchovy in some location

Location	E = exploitation's pace	Status exploitation	Sumber
Belawan	E = 0.34	Under exploitation	[16]
Belawan	E = 0.68	Fully exploitation	[4]
Ambon	E = 0.55	Fully exploitation	[10]
Kei islands	E = 0.63	Fully exploitation	This study (2016)

According to Jones *in* [14], if the value of the natural mortality rate (M) and the fishing mortality rate (F) is equal, then the rate of exploitation can be assumed to be F/Z= 0.5. If the value of E is greater than 0.5, it can be categorized more biologically. This condition is occurring in the devis' anchovy resources in the Kei island water where exploitation rate of the value greater than 0.5 indicates that there has been a biological overfishing.

4. Conclusion

A total of 3,524 devis' anchovies were analyzed based on total length that ranged from 23 to 87 mm with a mean length of 55.4 $\pm$ 11.8 mm. The von Bertalanffy growth parameters estimated were L_∞ = 89.25 mm and K = 0.74 year^{-1}, with the growth models equation is $L_t = 89{,}25\ (1 - e^{-0{,}74(t+0{,}161})$. The result of mortality analysis showed that total mortality rate Z = 2.97, natural mortality rate M = 1.098, fishing mortality rate F= 1.87 year^{-1} and exploitation rate E = 0.63. This exploitation status of devis' anchovy in Kei Island is on state fully exploited and leads to the over exploited condition.

Acknowledgements

This paper is part of our research work funded by Ministry of Research, Technology, and Higher Education of Indonesia (Competitive Research Grant from Directorate of Research and Community Services).

References

[1] Csirke J 1988 *Small Shoalding Fish Stocks* in J A Gulland ed Fish Population Dyamic, 2nd John Willy and Sons Chechester 271-302

[2] Fairweather T P, C D van der Lingen, A J Booth, L Drapeau and J J van der Westhuizen 2006 *Indicators of sustainable fishing for South African sardine Sardinops sagax and anchovy Engraulis encrasicolus* African J of Marine Science 28(3&4): 661–680

[3] Gulland J A 1971 *Fish Resources of the Oceans* West Byfleet Surrey Fishing News (Books) Ltd 255p

[4] Gurukinayan Z A, Yunasfi and A Muhtadi 2014 *Kajian Pertumbuhan dan Laju Eksploitasi Ikan Teri Nasi (Stolephorus spp.) di Perairan Belawan Sumetera Utara* Jurusan Manajemen Sumberdaya Perairan Fakultas Pertanian Universitas Sumatera Utara Medan

[5] Hamid S K dan II D A R Susanty 2013 *Analisis kegiatan usaha penangkapan ikan teri Stolephorus sp. di Desa Selayar Kabupaten Maluku Tenggara Provinsi Maluku* J Neritic 4(2):82

[6] Jalil dan A Mallawa 2001 *Biologi Populasi Ikan Baronang Lingkis (S.canaliculatus) di Perairan Kecamatan Bua Kabupaten Luwu* UjungPandang

[7] King M 1995 *Fisheries biology, assessment, and management* United Kingdom: Fishing News Books 341p
[8] Manojkumar P P 2004 *Some aspects on the biology of Nemipterus japonicus (Bloch) from Veraval in Gujarat* Indian J Fish 51:185-191
[9] Murty V S R, T A Rao, M Srinath, E Vivekanandan, K C S Nair, S K Chakraborty, S G Raje & P U Zacharia 1992 *Stock assessment of threadfin breams (Nemipterus spp.) of India* Indian J Fish 39:9-41
[10] Ongkers O T S 2006 *Pemantauan terhadap Parameter Populasi Ikan Teri Merah (Encrasicholina heteroloba) di Teluk Ambon Bagian Dalam* Prosiding Seminar Ikan IV [Abstrak] http://repository.ipb.ac.id/handle/123456789/42159 diakses 22 Januari 2015
[11] Pauly D 1983 *Some simple methods for the assessment of tropical fish stocks* FAO Fish Tech Pap 234:52
[12] Prihartini A 2006 *Analisis tampilan biologis ikan layang (Decapterus spp.) hasil tangkapan purse seine yang didaratkan di PPN Pekalongan* Program Pascasarjana Universitas Diponegoro Semarang 90p.
[13] Ricker W E 1975 *Computation and interpretation of biological statistics of fish populations* Bull Fish Res Board Can 191:382
[14] Sparre P and Venema S C 1999 *Introduction to tropical fish stock assessment* Part 1-Manual. FAO Fish Tech. Pap. Rome 306.1, Rev.1: 376 p.
[15] Sutjipto D O, S Muhammad, Sumarno, dan Marsoedi 2013 *Dinamika Populasi Ikan Kurisi (Nemipterus hexodon) dari Selat Madura* J Ilmu Kelautan 18(3):165-171
[16] Yuanda D, M S Mulya and A Muhtadi 2015 *Pertumbuhan dan Laju Eksploitasi ikn Teri Pekto (Stolephorus waitei) di Perairan Belawan Kota Medan* Jurusan Manajemen Sumberdaya Peraiaran Fakultas Pertanian Universitas Sumatera Utara Medan
[17] Zacharia P U 1998 *Dynamics of the threadfin breams Nemipterus japonicus exploited off Karnataka* Indian J Fish 45(3):265-270

IOP Conf. Series: Earth and Environmental Science **89** (2017) 012005 doi:10.1088/1755-1315/89/1/012005

Quality Indexes and Feasibility Analysis of Traditional Dried Anchovy From Buru Island

Imelda Krisanta Enda Savitri[1], R.B.D. Sormin[1], Bernita Silaban[1]

[1]Department of Fish Processing Technology, Fishery and Marine Science Faculty, Pattimura University. Kampus Unpatti Jln. Mr. Chr. Soplanit Poka 97233 Ambon Maluku

Email: endahsavitri@gmail.com

Abstract. Dried anchovy is a prominent fishery product in Buru Island Maluku. It is produce traditionally in villages Batuboi and Siahoni at Kayeli bay. Drying process was done in open air under the sun directly during two days to produce dried anchovy with a prospective marketable. This preliminary research was aimed to investigated the quality indexes and business feasibility of dried anchovy produced by processor comunity. The quality index was found by laboratory analysis of sample dried anchovy referring to Indonesian National Standard. The data for feasibility analysis was found by interview and by using quesioner list. The average Quality indexes of dried anchovy as follows: Water 14.22 grams percent; Protein 55.68 grams percent; Fat 2.29 grams percent; Ash 9.02 grams percent; Total Bacteri (TPC) 3,5 x 10^2-; organoleptic value 7.79; *E. coli*, *Salmonella* and Mold were unidentified. Total cost of dried anchovy production is 37,050,720 Rupiahs/year with the income was 224,000,000 Rupiahs/year. Total profit is 217,824,880 Rupiahs/year and the Revenue Cost ratio was 6.05. Dried anchovy from Buru Island has a good quality and also has a good business feasibility.

1. Introduction

Dried anchovy is a prominent fishery product in Buru Island – Maluku, among other products such as dried cucumber, dried seaweed and smoked fish. Dried anchovy is produced in Batuboi and Siahoni villages at Kayeli bay. Anchovy can be harvested throughout the year, but the highest season is twice a year on Mart-April and October-December. Bagan (floating object equipped light source to attract fish get together) placed on the water then harvesting held on it by using anchovy net at night or in the early morning. On the highest season, sometimes the fishes are unhandled and buried at the beach, thereupon should optimize the utilization of anchovy for a better business prospect and development.

The drying process held in two days (12-15 hours effective) under the sun and put on the ground based on anchovy net to produce a prospective marketable dried anchovy. So far this product is not popular yet as well as dried anchovy from Medan. Moreover the quality indexes and feasibility analysis are not explored yet. According to Indonesia National Standardization (1992) , the quality indexes of dried (salted) anchovy should have minimum organoleptic value 7 without mold appearance; maximum Total Plate Count (TPC) 5.0 x 10^5 cfu/g; *Escherichia coli* maximum 3 cfu/g; *Salmonella*, *Staphylococcus aureus* and *Vibrio cholera* are negative; water content maximum 40% and Ash (not acid soluble) maximum 0.3%.

IOP Conf. Series: Earth and Environmental Science **89** (2017) 012005 doi:10.1088/1755-1315/89/1/012005

Quality indexes and feasibility analysis are essential for its development as a superior product that will contribute to local government revenues. Based on that, this preliminary exploration study was done to investigate the quality indexes and business feasibility of dried anchovy from Buru Island.

2. Methodology

2.1. Sampling Dried Anchovy and Collecting Feasibility Data

Dried anchovy was sampling from villages Batuboi and Siahoni at the coastal of Kayeli Bay, Buru Island and brought to the industry institution of research and standardization laboratory and Fish processing Technology laboratory in Ambon for quality evaluation. Business feasibility data was got by interview 30 respondents and also by using questioner list. The respondents have a range age around 36-56 years old.

2.2. Proximate Composition Analysis

Proximate analysis were done referring to the method of Association of Officials Analytical Chemistry (AOAC, 2000). The following parameters were measured crude protein (Kjeldahl method), crude fat (Soxhlet method), moisture (thermostatically controlled forced air oven) and ash content (muffle furnace).

2.3. Organoleptic test

This test aimed to know the quality of dried anchovy by observing the appearance, smell, consistency and presence or absence of mold growth. Organoleptic test was done according to Indonesia National Standardization (1991) involved 15 panelists by using score sheet organoleptic dried (salted) anchovy with the highest scale of 9 and the lowest 1.

2.4. Microbiology Test

2.4.1. Total Plate Count (TPC), according to Indonesia National Standardization (BSN) (SNI 01-2332-3-2006). Pipetting 1 ml homogenized 25 grams dried anchovy in 225 ml Butterfield Phosphate (BFP) into a tube filled 9 ml BFP (10^{-2} dilution) then pipetting 1 ml 10^{-2} dilution into other tube filled 9 ml BFP (10^{-3} dilution) and further up to dilution 10^{-5}. Pipetting 1 ml from every dilution tube into sterilized petri-disc then 12-15 ml Plate Count Agar (PCA) (T 45°C) poured into the petri-disc then shaken generously until homogenous and let it solidified. Furthermore all of the petri-disc put into incubator (T 37° C) for 24-48 hours. Counting colony on the petri-disc with 25-250 colony.

2.4.2. Escherichia coli, According to Indonesia National Standardization (SNI 01-2332.1-2006). 1 ml dilution dried anchovy (25 g) in BFP solution (225 ml) pipetted into 3 serial tubes of LTB in Durham and incubated it (*Coliform* prediction). Positive LTB tube inoculated into BGLB Broth by using ose and incubated it at 35° C for 48 hours. Each positive inoculated LTB inoculated into FC Broth tube by using ose and incubated it at 45° C in the water bath for 48 hours (*Coliform* confirmation). Determination of fecal *Coliform* MPN/G based on total *E. coli* tube. 1 ose positive *E. coli* solution was took and streak on LTMB agar and incubated it at 35° C for 24 hours. More than 1 colony *E. coli* was took and streak on slanting PCA and incubated it at 35°C for 24 hours.

2.4.3. Salmonella sp. According to Indonesia National Standardization (SNI 01-2332-4-2006. 25 mg dried anchovy homogenized in 225 ml Lactose Broth solution for 2 minutes. 0.1 ml sample solution was moved into 10 ml RV medium and 1 ml into 10 ml TTB, whisked thoroughly and incubated it at 35°C for 24 hours. The tube vortex and took by using loop (3 mm) and streak on TTD which is incubated into HE, XLD and BSA medium. Streak into RV Broth then incubated it at 35°C for 24 hours. Two or more *Salmonella* colony from each selective agar took by using sterilized needle and streak TSI agar in the scraping oblique and vertically piercing firstly the streak on oblique agar. Both of TSI and LIA incubated at 35°C for 24 hours then continued with biochemistry test.

2.4.4. Mold Test, according to Indonesia National Standardization (BSN) (SNI 2332.7:2009). Pipetting 1 ml homogenized 25 grams dried anchovy in 225 ml Butterfield Phosphate (BFP) into a tube filled 9 ml BFP (10^{-2} dilution) then pipetting 1 ml 10^{-2} dilution into other tube dilled 9 ml BFP (10^{-3} dilution) and further up to dilution 10^{-5}.15-20 ml media Potato Dextrin Agar (PDA) poured into sterilized petri-disc And let it solidified. Pipetting 0.1 ml from each dilution tube and put onto PDA and spread evenly. After 1 hour incubated PDA plate (37° C) for 5 days with the cup position upside down. Make control procedure without sample and mixing dilution solution with PCA.

3. Result and Discussion

3.1. Quality Indexes

Quality indexes of dried anchovy produced by the community from Batuboi and Siahoni villages can be seen on Table 1.

Tabel 1. Quality indexes (average) of dried anchovy from Buru Island

Proximate (gam percent)			Organoleptic			Microbiology (cfu/g)		
Protein	55.76	55.4	Organoleptic	7.7	7.5	TPC	$3.5x10^2$	$3.6x10^2$
Fat	2.98	3.6				E. coli	Negative	Negative
Water content	14.21	14.22				Salmonella	Negative	Negative
Ash	9.6	8.44				Mold	Negative	Negative

Dried anchovy produced by the community in Buru Island has a good organoleptic with an average value 7.6. The growth and activity of microorganism especially putrefactive bacterial produce substances which greatly affects appearance, smell and consistency of the dried anchovy. Highly protein content on fish is very good condition for growth and activity of bacterial. Bacterial decay of macromolecule especially protein produce simple metabolism substances such as hydrogen sulfide, organic acids and ammonia are a very bad influenced on appearance, smell and consistency.

The microbiology quality of dried anchovy describes a good quality of a product with a relatively small total bacterial (TPC) average 3.55 x 10^2 cfu/g and negative *Escherichia coli, Salmonella* and mold. 14.2 grams percent a relatively low water content of dried anchovy creating condition that inhibits bacterial growth and activity and also mold.

According to Indonesia National Standardization (1992), the quality indexes of dried (salted) anchovy should have minimum organoleptic value 7 without mold appearance; maximum Total Plate Count (TPC) 5.0 x 10^5 cfu/g; *Escherichia coli* maximum 3 cfu/g; *Salmonella*, *Staphylococcus aureus* and *Vibrio cholera* are negative; water content maximum 40%, the dried anchovy from Buru Island has a good quality.

3.2. Feasibility Analysis

3.2.1. Profile of Dried Anchovy Business in Buru. Dried anchovy business is a family business, managed by the family member involving 3 or 4 people with a length of time ranging from 5-10 years. They use their own money for capital venture, without supported or treated by local government or financial institution and also has not taken under legal institution. They have bagan (floating object equipped light resource for attracting fish to come and gather around it) and some of them have more than one. Bagan put on the water then catching fish on it by using anchovy net (called karoro) at night or in the early morning. The price of bagan is 20,000,000 – 25,000,000 rupiahs and the boats for fishing made of wood or fiber with or without driving machine. The highest season of catching anchovy twice a year, in Mart-April and October-December. Therefore there are 2 months effective time to catching anchovy a year and catching anchovy done 4-5 times a month during high season.

The drying process did traditionally, newest catching fish put on the ground based on anchovy net and drying under the sun for 2 days (12-15 effective hours) then winded aired under the roof and packed in a big plastic sack. When the season abundant coincides with the rainy day, there is a failed production where the fish are thrown into the sea or buried on the beach to avoid the stench of the untreated fish. Dried anchovy production range on 500 kg to 1 ton a week (average 700 kg/week). The selling price of dried anchovy at the Namlea market is 40,000 rupiahs/kg, 55,000 rupiahs/kg in Ambon and 20,000 rupiahs a kg at the producer level. The merchants come to the producer every 2 weeks.

3.2.2. Distribution and Marketing of Dried Anchovy

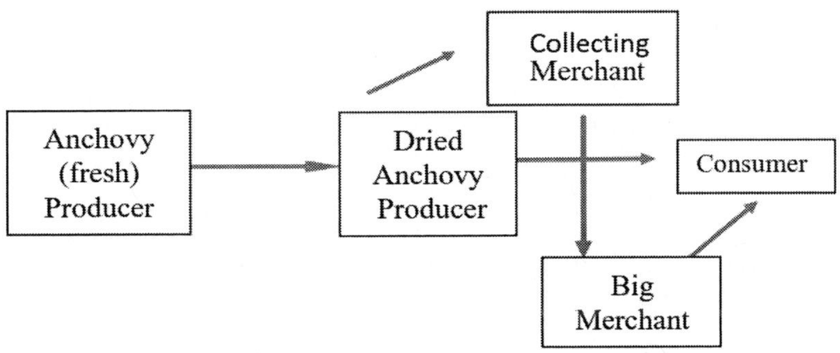

Figure 1. Marketing of Dried Anchovy in Buru

3.2.3. Cost of Dried Anchovy Production. The cost of dried anchovy production a week can be seen on Table 2.

Tabel 2. Cost of Dried Anchovy Production per week

Fixed Cost		Variable Cost	
Component	Amount (Rupiah)	Component	Amount (Rupiah)
Depreciation cost of bagan	95,890	Cigarette	50,000
Repair/maintenance of bagan	150,000	Fuel	200,000
Repair/maintenance of anchovy net	200,000	Food	50,000
Administration/Retribution/Task	1,000	Transportation	25,000
Total	446,890	Total	325,000

The total cost a week is Rp. 771,890, therefore total cost for a month is Rp. 3,087,560 and Rp. 37,050,720 spend for a year.

3.2.4. Total Revenue of Dried Anchovy Production. The revenue is selling price of product multiplied by the amount of dried anchovy production. The average production a week is 700 kg and the selling price at Namlea market is Rp. 40,000, therefore total revenue in a week is Rp. 28,000,000. Furthermore total revenue in a month is Rp. 112,000,000. In a year, there are 2 months effective time for catching fish, so the total revenue in a year is Rp. 224,000,000.

3.2.5. Profit and Loss Analysis. This analysis aims to know the profit or the loss of business, which is a profitable venture will have a greater revenue than total cost [7].

$$\text{Profit} = \text{revenue} - (\text{total fixed cost} + \text{total variable cost})$$

Total revenue a week is Rp. 28,000,000 and the total cost is Rp. 771,890, can be calculated the total profit is Rp. 27,228,110 a week. Therefore the total profit a month is Rp. 108,912,440 and Rp. 217,824,880 a year.

3.2.6. Feasibility Business analysis. This analysis is used to investigate the feasibility of a venture or a business and aims to know the profit level, return on investment or even a business break-even point. R/C analysis is used to investigate the profits relativity of the cost spend in a business. The feasibility is good when the R/C>1 [7].

Feasibility business of dried Anchovy production can be calculated by using formula Revenue Cost Ratio below:

$$^{R}/_{C} = \frac{\text{Total Revenue}}{\text{Total Fixed Cost} + \text{Total Variable Cost}}$$

Total revenue a year is Rp. 224,000,000 and the total fix cost plus variable cost are Rp. 37,050,720, therefore the feasibility business of dried anchovy production is 6.05. According to [7] the dried anchovy production in Buru has a good feasibility business and moreover very prospective to be developed in order to produce a featured product in Buru

4. Conclusion

Dried anchovy produced by the community from Kayeli bay Buru Island has a good quality and a good feasibility business

Acknowledgment

Authors would like to thank to Research-Technology and Highly Education for funding this study.

References

[1] AOAC (2000). Official Method of Analysis. In M.D. Gaithersburg (Ed.), 18th edition Association of Official Analytical Chemist, Maryland, USA, 1193.pp
[2] BSN 1991. Method of Organoleptic Test on Fishery Product (SNI 01-2345). Guidance Institution and Fishery Product Quality Testing. Directorate General of Fishery
[3] BSN 1992. Indonesia National Standardization Dried Salted Anchovy. Guidance Institution and Fishery Product Quality Testing. Directorate General of Fishery. Jakarta.
[4] BSN 2006. Microbiology Tests part 1. *Coliform* and *Escherichia coliform* testing on fishery product. SNI 01-2332-1-2006. Jakarta.
[5] BSN 2006. Microbiology Tests part 2. *Salmonella* Testing on fishery product. SNI 01-2332-4-2006. Jakarta.
[6] BSN 2009. Mold Test on Fishery Product. SNI 2332.7:2009. Jakarta
[7] Effendi, I and Oktariza (2006). Management Agribisnis Perikanan. Penerbit Swadaya. Jakarta

IOP Conf. Series: Earth and Environmental Science **89** (2017) 012006 doi:10.1088/1755-1315/89/1/012006

Correlation Between Existence of Reef Sharks with Abundance of Reef Fishes in South Waters of Morotai Island (North Moluccas)

Darmawan Ahmad Mukharror[1], Isnaini Tiara Baiti[2], Muhammad Ichsan[1], Niomi Pridina[1], Sanny Triutami[1]

[1]Shark Diving Indonesia Morotai, Kampung Yayasan, Daruba, Morotai
[2]Marine Science Department, FPIK, Universitas Padjdjaran, Jl. Raya Bandung-Sumedang KM 21, Bandung

Email: darmawan_ahmad_m@yahoo.com

Abstract. Despite increasing academic research citation on biology, abundance, and the behavior of the blacktip reef sharks , the influence of reef fish population on the density of reef sharks: *Carcharhinus melanopterus* and *Triaenodon obesus* population in its habitat were largely unassessed. This present study examined the correlation between abundance of reef fishes family/species with the population of reef sharks in Southern Waters of Morotai Island. The existence of reef sharks was measured with the Audible Stationary Count (ASC) methods and the abundance of reef fishes was surveyed using Underwater Visual Census (UVC) combined with Diver Operated Video (DOV) census. The coefficient of Determination (R^2) was used to investigate the degree of relationships between sharks and the specific reef fishes species. The research from 8th April to 4th June 2015 showed the strong positive correlations between the existence of reef sharks with abundance of reef fishes. The correlation values between *Carcharhinus melanopterus/Triaenodon obesus* with *Chaetodon auriga* was 0.9405, blacktip/whitetip reef sharks versus *Ctenochaetus striatus* was 0.9146, and *Carcharhinus melanopterus/Triaenodon obesus* to *Chaetodon kleinii* was 0.8440. As the shark can be worth more alive for shark diving tourism than dead in a fish market, the abundance of these reef fishes was important as an early indication parameter of shark existence in South Water of Morotai Island. In the long term, this highlights the importance of reef fishes abundance management in Morotai Island's Waters to enable the establishment of appropriate and effective reef sharks conservation.

1. Introduction

The reef sharks (family: Carcharhinidae) are predominantly found in Indo-Pacific waters with three of its species most commonly sighted near shores [25-26]: Blacktip reef shark, *Carcharhinus melanopterus*; grey reef shark, *Carcharhinus amblyrhinchos*; and whitetip reef shark, *Triaenodon obesus*. Human activity affects the on-going collapse of coral reef ecosystem which further domino effect to lesser the reef sharks population. Recent estimation suggested the shark population has declined regionally by 90% or more in 30 years [9]. The observation on Cleveland Bay, Australia, showed the decline of 71% of juvenile *Carcharhinus melanopterus* within almost two years [36], while other examination in Great Barrier Reef indicated the reduction of whitetip reef sharks (*Triaenodon obesus*) population by 7% annually and gray reef sharks (*Carcharhinus amblyrhyncos*) population by 17% yearly [26].

Coral reefs are used by variety of shark species [35] as their unique habitat. Blacktip reef shark (*Carcharhinus melanopterus*) is known for its relatively high site fidelity by living in their coral reef's home range (up to 21 km^2) for 12.61% to 84.57% of their life [30]. Given this high residency of blacktip reef sharks, the change in this predator population has caused trophic cascading effect in their coral reef's habitat [5]. Tropical lagoons or coral reefs have ten components of food chain [3]: macroalgae, planktons (zooplankton and phytoplankton), and coral reef at the base of food chain; omnivore (including corallivore), planktivore, and herbivore at the second level of trophic; piscivore and large carnivore at the trophic level three; mesocarnivore (reef sharks) at the trophic level four; and apex predator (pelagic sharks) at the top trophic level.

One indicator of a healthy coral reef ecosystem is the abundance of herbivore fishes - as well as omnivore and planktivore [21], as this means the reduced population of macroalgae [10]. The dynamic of the food chain in the coral reef ecosystem shows that the presence of reef sharks (mesopredator) may promote the abundance of herbivore [28], as this group are critical to the progress and outcome of coral recovery. In other study, herbivore abundance increases in the area protected from shark fishing were observed only after cyclone disturbance [27]. However, as a response, other study showed that during the recovery period, abundance of herbivorous fishes increased significantly on both fished and unfished reefs, but the magnitude of this increase where greater on reefs where shark was still present [28].

Morotai waters are located in the middle of world's coral triangle where three of the most common reef sharks (Blacktip reef shark, whitetip reef shark, and gray reef shark) are present [22]. Only 2% of the Morotai waters have been explored and these areas mostly lie in southern part of Morotai waters. There were fast declining of life coral reef coverage in Morotai Waters from 1996 to 2002 (41.1%) and 17.4% loss from 2002 to 2013 [33]. Other study on the research location showed that the life coral coverage in Niomi Point (flat), Blacktip Point (flat), Fariman (flat) and Eagle Ray Point (slope) were 48%, 52%, 22% and 45.95% respectively [13,22] Comparative study in south waters of Morotai Island shows the value of shark diving industry is 568 times higher than the value shark fishing/shark finning [11]. Thus, it is important to understand the correlation between the existence of reef sharks with the abundance of reef fishes in this area to establish effective conservation means as well as the tool for determining the reef shark existence.

2. Methodology

Sampling was conducted on 4 different coral reefs (stations) around south waters of Morotai Island in April – June 2015. Selection of the stations is based on previous research location with additional stations in Niomi Point and Eagle Ray Point are selected based on report of shark sighting records by Shark Diving Indonesia (Ichsan et al., 2015). Station A is located at N 01° 58' 29.1" E 128° 14' 13.9", Station B lies in N 01° 58' 13.4" E 128° 14' 03.6", Station C is located at N 01° 58' 29.5" E 128° 15' 01", and Station D is on N 02° 01' 35.5" E 128° 13' 12.2". Complete location of all observation stations on the Morotai Island's map is shown in Figure 1 below.

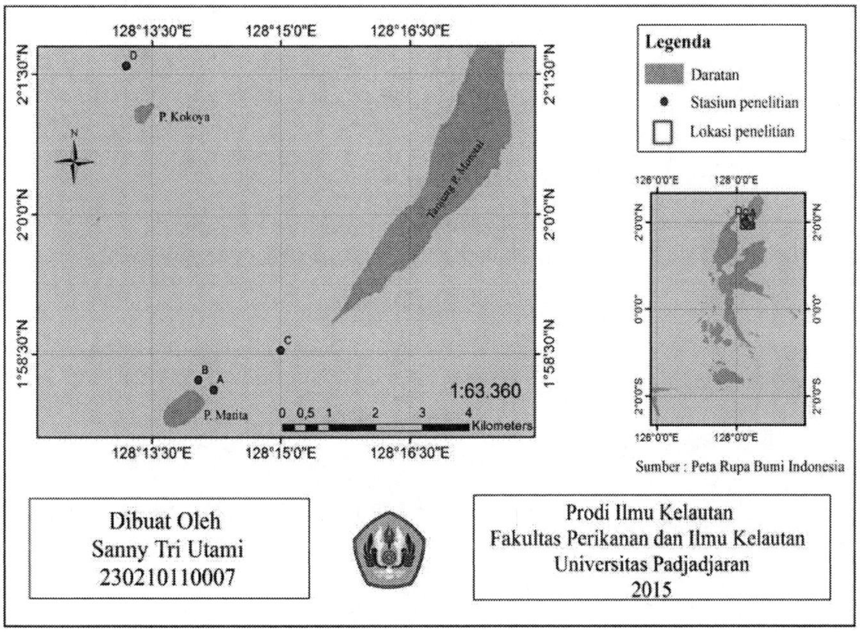

Figure 1. Observation stations in south waters of Morotai Island

Materials used for this research are: two (2) sets of basic SCUBA diving gear for one researcher and her buddy, one (1) underwater slate, one (1) transect meter, one (1) squeezable empty mineral water bottle, one (1) GPS locator, one (1) underwater camera. For analysis purposes, following tools were used: SPSS™ statistical software, Microsoft Excell™ 2007 version, Reef Fish Identification Books. Survey is conducted in each station every week between April to June 2015, eight (8) weeks in total, except in Station C which has one survey canceled out due to weather condition (safety reason).

We measured the existence of reef sharks with the Audible Stationary Count (ASC) method. ASC method is developed since 1970s using low-frequency sound (such as continuous squeezing of empty plastic bottle) to attract reef sharks to a stationary point [6]. It was demonstrated that ASC could be reliable to attract reef sharks at 80 m from the stationary point [16], with average response up to 250 m away [37]. Attracted reef sharks came by the stationary point were counted and recorded on underwater slate. Visibility during surveys was reliable enough (between 10 – 20 m) to reduce the counting error during census.

We surveyed the abundance of herbivore and carnivore reef fishes using Underwater Visual Census (UVC) surveys with a combination of Diver Operated Video (DOV) census. UVC has been used to estimate the densities of the reef fishes since 1950s [4]. There are two methods of UVC: transect count (an experienced diver swam at constant speed along the transect and noted the reef fish species observed on a slate) and point count (an experienced diver stayed at a vantage point and noted the surrounding reef fish species observed on a slate). UVC with transect and point count have relatively similar significance ($p = 0.054$ and 0.048 for 10 hectare count respectively), therefore both censuses is deemed reliable [32] to be selected for small area of survey. Novel method, combination of UVC and diver operated video (DOV) presents great potential to count the abundance of reef fishes in a research station (Tessier et al., 2013).

Identified reef fishes were then classified based on classification on www.fishbase.org. Set of data of the number of reef sharks sighted during ASC and abundance of reef fishes based on their diet category were then analyzed using R^2 coefficient of determination to examine the degree of relationship between two variables. The coefficient of determination must represent a goodness-of-fit and it must be unit free, should range from 0 to 1, where 1 represents a perfect fit. There are eight expressions for R^2 appears as [13]:

IOP Conf. Series: Earth and Environmental Science **89** (2017) 012006 doi:10.1088/1755-1315/89/1/012006

$$R_1^2 = 1 - \sum(y - \hat{y})^2 / \sum(y - \bar{y})^2$$

$$R_2^2 = \sum(\hat{y} - \bar{y})^2 / \sum(y - \bar{y})^2$$

$$R_3^2 = \sum(\hat{y} - \bar{\hat{y}})^2 / \sum(y - \bar{y})^2$$

$$R_4^2 = 1 - \sum(e - \bar{e})^2 / \sum(y - \bar{y})^2, \quad e = y - \hat{y},$$

R_5^2 = squared multiple correlation coefficient between the regressand and the regressors

R_6^2 = squared correlation coefficient between y and $\hat{y}$

$$R_7^2 = 1 - \sum(y - \hat{y})^2 / \sum y^2$$

$$R_8^2 = \sum \hat{y}^2 / \sum y^2$$

However, for the efficiency in analysis we used Microsoft Excell™ to obtain the R^2 value through linear regression between two variables, the interpretation of the correlation values are as follows [38]:

0.00 - 0.19	=	very low
0:20 - 0.39	=	low
0:40 - 059	=	moderate
0.60 - 079	=	strong
0.80 - 1.00	=	very strong

3. Results

3.1. Shark existence

There are two reef sharks species found during the survey: blacktip reef shark (*Carcharhinus melanopterus*) and whitetip reef shark (*Triaenodon obesus*), none of the grey reef shark is observed during the survey. Figure 2 and Figure 3 below shows the existence of blacktip reef sharks and whitetip reef sharks respectively each week in four observation stations. None of the grey reef shark (*Carcharhinus amblyrhinchos*) was found in all research stations during the period of observation. The maximum blacktip reef shark aggregation in one station is 11 individual recorded in station B (blacktip point), while station D (eagle ray point) showed that averagely low blacktip shark sighting.

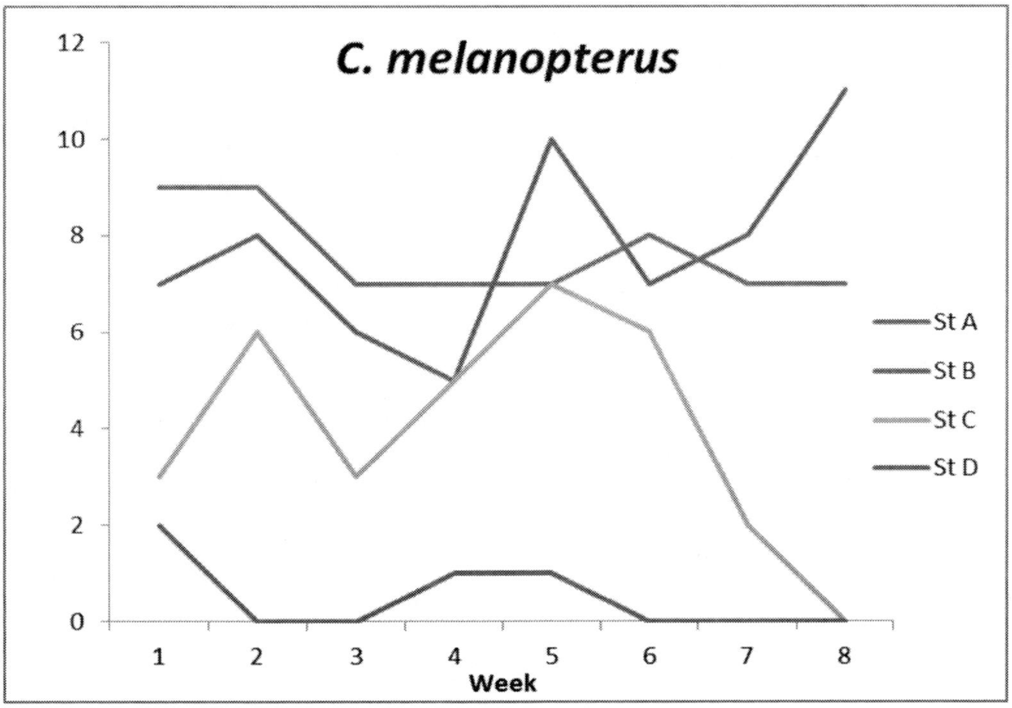

Figure 2 Existence of Blacktip Reef Shark in four observed stations

The frequency of sighting of white tip reef shark is between 0 – 3 sharks/dive, with the maximum aggregation is 3 at Niomi Point (Station A). No sighting of *Triaenodon obesus* in blacktip point (station B) was ever recorded.

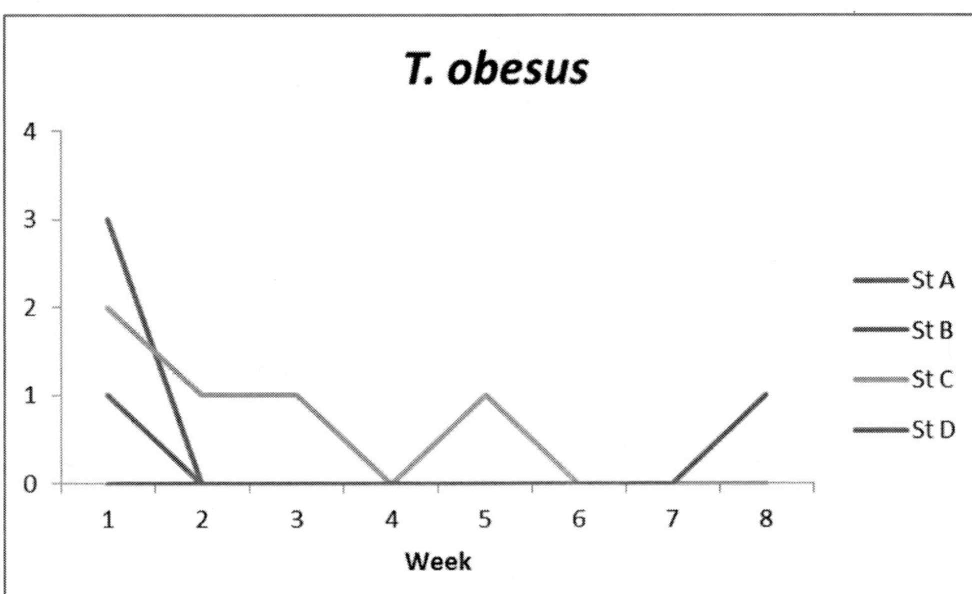

Figure 3. Existence of Whitetip Reef Sharks in four observed stations

IOP Conf. Series: Earth and Environmental Science **89** (2017) 012006 doi:10.1088/1755-1315/89/1/012006

3.2. Abundance of reef fishes in south waters of Morotai Island

There are 117 - 121 species from 32 families of reef fishes recorded in four observed stations in south waters of Morotai Island with abundance variety from 1.11 individuals/m2 in station D to 1.59 individuals/m2 in station C. *Pomacentrus auriventris* and *Chromis margaritifer* were two species of reef fishes that most abundant in Southern Waters of Morotai Islands. The most common reef fishes families found in Morotai Island's Southern Waters were pomacentridae, labridae, balistidae, and acanthuridae.

Table 1. Abundance of Reef Fishes in four observed station in south waters of Morotai Islands

Station ID	Total No. of Obsereved Fishes	No. of Species	Abundance (ind/m^2)	Abundance of Dominant Species (%)	Abundance of Reference Species (%)
A	11,455	119	1.39	Xanthichthys caeruleolineatus (14.84%), Chromis margaritifer (7.49%)	Ctenochaetus striatus (1.99%), Chaetodon kleinii (2.08%), Chaetodon auriga (0.22%)
B	13,737	119	1.57	Chromis margaritifer (8.23%), Pomacentrus burroughi (7.41%)	Ctenochaetus striatus (2.02%), Chaetodon kleinii (2.42%), Chaetodon auriga (0.17%)
C	13,025	117	1.59	Xanthichthys caeruleolineatus (15.15%), Pomacentrus auriventris (13.58%)	Ctenochaetus striatus (0.96%), Chaetodon kleinii (1.57%), Chaetodon auriga (0.12%)
D	10,012	121	1.11	Pterocaesio pisang (13.31%), Neoglyphidodon crossi (5.50%)	Ctenochaetus striatus (0.55%), Chaetodon kleinii (0.73%), Chaetodon auriga (0.11%)

3.3. Correlations between reef sharks and reef fishes

Analysis result of the observed data of the existence of sharks and abundance of reef fishes in four stations are shown in the Table 2 below.

Table 2. Correlation between reef sharks, herbivore and carnivore

Fish	Shark	C. auriga	C. kleinii	C. striatus
Shark	1	0.9405	0.8440	0.9146
C. auriga	0.9405	1	0.7430	0.9277
C. kleinii	0.8440	0.7430	1	0.9148
C. striatus	0.9146	0.9277	0.9148	1

3.4. Shark existence

The existence of reef sharks in south waters of Morotai Island was confirmed during eight (8) weeks observations. Two of the main reef sharks that were observed exist in Moratai Island's south water are: blacktip reef sharks (*Carcharhinus melanopterus*) and whitetip reef sharks (*Triaenodon obesus*) with the density of 0-11 individuals and 0-3 individuals respectively. As previous research [16] hypothesised that reef shark could reliably use their hearing sensory at 80 m (2 ha of area) to approach the low frequency emitted sound from ASC method, then the density of reef sharks could be assumed as follow in Table 3.

Table 3. Reef sharks density (individuals/hectare) in observed stations with ASC

Station ID	Blacktip reef shark	Whitetip reef shark	Combined
A	3.5 – 4.5	0 – 1.5	0 - 6
B	2.5 – 5.5	0	2.5 – 5.5
C	0 – 3.5	0 - 1	0 - 4
D	0 – 1.5	0 – 0.5	0 - 2

Comparative study between this research results with other reef sharks census [25] in Great Barrier Reef shows that the existence of sharks in Morotai Island's south water was considered as higher (up to 2.61 sharks/ha in GBR compared to 6 sharks/ha in Morotai Island).

3.5. Abundance of reef fishes

Comparing the density of reef fishes in south waters of Morotai Island to Koh Phangan and Koh Tao in Thailand (Pusch, 2011) shows that they were almost similar (1.59, 1.77, and 1,4 individuals/m^2 respectively). However, abundance of reef fishes from four observed stations in Morotai Island showed the very strong correlation with the abundance of the reef fishes in Pulau Langkawi [1] with the coefficient of determination of 0.9578. This very strong correlation between Morotai Island and Pulau Langkawi may be affected by the average life coral cover in both islands that is not too different (42% and 35%, respectively).

In most area, the damselfishes (*Pomacentridae spp.)* were the most abundant, with the percentage varies from 18.10% in Nha Trang Bay to 93.51% in western area of Biawak Island [39], as it shown in Table 4 below. Some fish family such as Balistidae spp. That were abundant in Morotai is not exist in all another area. This may relate with the geographical factor which shows that Morotai is further east compared to other area and it is affected directly by the Pacific Ocean current.

Table 4. Abundance of Reef Fishes Comparison between Morotai and Other Area in Indo-Pacific

Fish Family	Morotai (%)	Tapanuli Selatan (%)	Nha Trang Bay (%)	Langkawi (%)	West Biawak (%)
Pomacentridae	42.64	30.09	18.10	65.62	93.51
Labridae	12.90	3.52	19.10	8.82	2.88
Balistidae	11.44	N/A	N/A	N/A	N/A
Acanthuridae	9.31	0.998	4.40	N/A	N/A
Caesionidae	5.94	3.46	N/A	6.64	0.72
Chaetodontidae	4.59	3.18	5.90	3.47	1.08
Anthinnae	3.38	N/A	N/A	N/A	N/A
Holocentridae	1.49	N/A	N/A	0.58	N/A
Nemipteridae	1.47	N/A	N/A	0.98	1.08
Pomacanthidae	1.40	1.94	1.10	0.12	N/A
Mullidae	1.19	0.15	N/A	0.02	N/A
Scaridae	0.37	2.26	1.70	5.82	0.36
Siganidae	0.14	0.59	6.20	1.27	N/A
Apogonidae	0.02	52.85	N/A	1.61	N/A
Zanclidae	0.56	N/A	N/A	0.29	N/A

3.6. Correlation between reef shark and reef fishes

The correlation value between reef sharks vs *Chaetodon auriga, Chaetodon kleinii and Ctenochaetus striatus* in south waters of Morotai suggested that the relation between the existence of reef sharks as predator were very strong (value = 0.9405, 0.8440, 0,9146 respectively). All of the survey locations are categorized as outer reef flats except station D (Eagle Ray Point), which is slope. The previous study shows that *Ctenochaetus striatus* have been observed foraging on the outer reef flats [14, 20] and prefer areas of high water motion [19]. Another study also shows that *Ctenochaetus striatus* were more abundant on outer reef slope rather than inner atoll rim reef slopes [29]. These were consistent with the *Carcharhinus melanopterus* that was heavily relied on their feed (site fidelity of 67%) in the outer reef [18].

It has been long hypothesised that butterfly fishes (Chaetodontidae) could be used as bio-indicator of coral health [24]. Strong correlation between chaetodontidae and life coral coverage has been studied in Buton, Sulawesi Tenggara and resulted in coefficient of determination of 0.73 [40], while other examination in Nusa Penida waters showed strong correlation of 0.6628 [49] and surveyed in Batam, Riau Islands Province showed the coefficient of determination of 0.5625 [41]. In reverse, the extensive life coral coverage depletion was strongly decline the abundance of the butterfly fishes [21]. Hence, the strong relationship between reef shark in Morotai with the butterfly fishes (represented by *Chaetodon kleinii* and *Chaetodon auriga*) may indirectly indicate strong bonding between reef sharks and its coral habitat.

Several previous studies have suggested that the habitat use or site fidelity of predators may reflect a relationship to prey abundance in that area (e.g; [42-44]. The primary characteristic of the nursery area that included abundant food has been focused in the several shark nursery studies [45-47]. Earlier diet study showed that *Carcharhinus melanopterus* feed mostly on teleost (bony fish) with the percentage of 56.8% at Northern Australia [48], 99.7% at Queensland, Australia [50], and 44.44% at Palmyra Atoll, Polynesia [51], 52% at Great Barrier Reef [52], and 56% at Alaska [54]. Detailed diet analysis of *Carcharhinus melanopterus* and *Triaenodon obesus* has been examined by the research at Great Barrier Reef [52], while the Grey reef shark (*Carcharhinus amblyrhinchos*) was studied in the Hawaiian Island [34]. Composition of the blacktip reef shark and whitetip reef shark diet and Morotai's reef fish abundance shows in Table 5 below.

Triaenodon obesus is well suited for feeding on reef animals, as indicated by the study at Hawaii [53] which suggest that whitetip reef shark feed on surgeon fish (acanthuridae). Table 5 also shows that all of the common prey for the reef sharks (blacktip reef shark, whitetip reef shark and grey reef shark) mostly are available in Morotai. Dominant prey such as pomacentridae, acanthuridae, labridae, and caesionidae were also found as abundant in Morotai waters. This showed that Morotai waters are the suitable ecosystem for the reef sharks.

Table 5. Comparison of Blacktip/Whitetip Reef Fish Diet and Abundance of Reef Fishes in Morotai

Fish Family	Blacktip reef shark (%)(1)	Whitetip reef shark (%)(1)	Grey reef shark (%)(2)	Abundance in Morotai (%)(3)
Labridae	23.0	16.69	0	12.90
Fistulariidae	23.0	0	0	0.04
Atherinidae	15.38	0	0	0.00
Scaridae	7.69	50.00	6.15	0.37
Muraenidae	7.69	0	18.80	0.00
Pomacanthidae	7.69	0	0	1.40
Lutjanidae	7.69	0	0	0.38
Synodontidae	7.69	0	0	0.14
Acanthuridae	0	16.67	12.69	9.31
Caesionidae	0	16.67	0	5.94
Apogonidae	0	0	0	0.02
Holocentridae	0	0	18.80	1.49
Pomacentridae	0	0	6.15	42.64

[6, 34]

4. Conclusion

The existence of reef sharks represents the density of predator in south waters of Morotai Island was confirmed and even higher than that of another area (such as Great Barrier Reef, Australia). The correlation between reef sharks and bio-indicator fishes (*Chaetodon auriga, Ctenochaetus striatus*, and *Chaetodon kleinii*) suggested that they are strongly positive interrelated. Strong correlation between reef sharks and chaetodontidae may represent strong indirect effect of the live coral coverage with the reef sharks.

As the economic value of the recreational shark diving is 568 times higher than the extractive economic value of shark finning, the presence of *Chaetodon auriga, Ctenochaetus striatus*, and *Chaetodon kleinii* are important as the initial indicator of the shark abundance in the dive site. This highlights the importance of these reef fishes abundance management in south waters of Morotai Islands to enable the establishment of appropriate and effective reef sharks conservation.

Since there was very limited reef shark's diet study, especially in Indonesia, it is suggested that further studies should focus on the reef shark diet to examine the relationship between the stomach content and the prey (reef fishes) abundance.

Acknowledgements

We thank the research team in Morotai Island for their support during collecting the data as well as to Marine Science Department of Padjadjaran University for offering many suggestions and comments on the manuscript.

References

[1] Andersson, Karin. 2002 A Study of Coral Reef Fishes along a Gradient of Disturbance in the Langkawi Archipelago, Malaysia. Undergraduate Thesis, Uppsala University

[2] Bornatowski H., et al. 2014 Ecological Importance of Sharks and Rays in a Structural Foodweb Analysis in Southern Brazil. *ICES J. Of Mar. Sci.*. 71: 1586-1592.

[3] Bozec, YM, et al.., 2004 Trophic Model of Lagoonal Communities in Large Open Atoll. *Aquat. Liv. Res.* 17: 151-162

[4] Brock VE., 1954 A Preliminary Report on a Method of Estimating Reef Fish Populations. *Wildl. M'ment.* 18: 297-308

[5] Dulvy NK., et al. 2004 Coral Reef Cascades and the Indirect Effects of Predator Removal by Exploitation. *Ecol. Letters.* 7: 410-416.

[6] Frisch, Ashley J. 2013 The Dr. Dorothea Sandars and Irene Lee Churchill Fellowship to Develop Robust Methods for Measuring and Monitoring Reef Shark Populations. Fellowship Report No. 158, Winston Churchill Memorial Trust of Australia.

[7] Frisch, A., et.al. 2016 Reassessing the Trophic Role of Reef Sharks as Apex Predator on Coral Reef. *Coral Reefs.* 35: xxx-xxx

[8] Goldman, B and FH. Talbott., 1976 Aspect of the Ecology of Coral Reef FIshes. In: Jones, OA and R. Endean (eds) Biology and Geology of Coral Reefs VII. Academy Press, New York.

[9] Heithaus, MR. et al. 2007 Predicting Ecological Consequences of Marine Top Predator Declines. *TREE.* 913: 1-9.

[10] Hughes, TP. et al. 2007 Phase Shifts, Herbivory, and the Resilience of Coral Reefs to Climate Change. *Curr. Bio.* 17: 360-365.

[11] Ichsan, Muhammad, et al. 2015 Shark Diving Tourism in Morotai, North Maluku, Indonesia. Sharks and Rays Symposium Proceeding. WWF Indonesia, Jakarta, pp 183-188.

[12] Karimah, Ashma, et al. 2017 Status of Coral Reef Ecosystem and Indication of Coral Bleaching Event in Morotai Island Waters, North Mollucas. The 4th International Marine and Fisheries Symposium, Makassar.

[13] Kvalseth, Tarald O. 1985 Cautionary Notes About R^2. *The American Statistician.* 39:4: 279-285.

[14] Marshell, A., P.J. Mumby 2012 Revisiting Functional Roles of Surgeon Fishes *Acanthurus nigrofuscus* and *Ctenochaetus striatus*. *Coral Reefs.* 31: 1093-1011.

[15] Mourier, Johann and Serge Planes, 201. Direct Genetic Evidence for Reproductive Phylopatry and Associated Fine-Scale Migrations in Female Blacktip Reef Sharks in French Polynesia. *Mol. Ecol.* 22: 201-214.
[16] Myberg et al. 1976 Attraction of Free Ranging Sharks by Low Frequency Sound. In: Schuijf A, A.D. Hawkins (eds) Sound Reception in Fish. Elsevier, Amsterdam. Pp 205-228
[17] Nanami, Atsushi and Moritaka Nishihara. 2003. Effects of Habitat Connectivity on the Abundance and Secies Richness of Coral Reef Fishes. *Mol. Ecol.* 22: 201-214
[18] Mc Cauley, D.J., et.al.. 2012 Assessing the Effect of Large Mobile Predator on Ecosystem Connectivity. *Ecol. Application* 22(6): 1711-1717
[19] Nelson, S.G., S. DeC. Wilkins. 1988 Sediment Processing by Surgeon Fish *Ctenochaetus Striatus* at Moorea, French Polynesia. *J. Fish Biol.* 32: 817-824
[20] Polunin, N.V.C., D.W. Klump, 1989 Ecologial Correlates Foraging Periodicity In Herbivorous Reef Fishes in teh Coral Sea. *J. Exp. Mar. Biol. Ecol.* 126: 1-20
[21] Pratchett, MS. et al. 2014 Reef Degradation and the Loss of Critical Ecosystem Goods and Services Provided by Coral Reef Fishes. *Curr. Op. In Env. Sus..* 7: 37-43
[22] Pridina, Niomi. 2015 Distributions of Reef Shark based on the Characteristic of Habitat on Morotai Waters, North Maluku. Undergraduate Thesis, Universitas Padjadjaran.
[23] Pusch, Susanne. 2011 Effect of Tourism on Herbivore Community Compositions in Coral Reefs in the Gulf of Thailand. Diploma thesis, Universitat Muenchen
[24] Reese, E.S., 1981. Predation on Coral by Fishes of the Family of Chaetodontidae, Implication and Management of Coral Reef Ecosystem. *Bull. Mar. Sc.* 31: 594-604
[25] Rizzari, J.R. et al. 2014 Impact of Conservation Areas on Trophic Interactions between Apex Predators and Herbivores on Coral Reefs. *Cons. Bio.* 0: 1-12.
[26] Robbins, W.D. et al. 2006 Ongoing Collapse of Coral-Reef Shark Populations. *Curr. Bio.* 16: 2314-2319.
[27] Roff, George, et al. 2016 The Ecological Role of Sharks on Coral Reefs. *TREE.* 2075: 1-13.
[28] Ruppert, J.W.L., et al. 2013. Caught in the Middle: Combine Impacts of Shark Removal and Coral Loss on the Fish Communities of Coral Reefs. *PlosOne.* 8: 1-8.
[29] Sluka, R.D. and Margareth W/ Miller. 200. Herbivorous Fish Assemblages and Herbivory Pressure on Laamu Atoll, Republic of Maldives. *Coral Reefs.* 20: 255-262.
[30] Speed, JWL., et al. 2012 Trophic Ecology of Reef Sharks Determined using Stable Isotoph and Telemetry. *Coral Reefs.* 31: 357-367.
[31] Tull, Malcolm. 2009 The History of Shark Fishing in Indonesia. Working Paper No. 158, Murdoch University.
[32] Watson, RA and TJ. Quinn II., 1997 Performance of Transect and Point Count Underwater Visual Census Methods. *Ecol Model.* 104: 103-112.
[33] Wahidin, Nurhalis, 2015 Morotai Coral Reef Ecosystem Classification Based on Objects and Pixels. Postgraduate Thesis, Institut Pertanian Bogor.
[34] Wetherbee, BM., et al. 1997. Distribution, Reproduction and Diet of Gray Reef Shark Carcharhinus ambhlyrhyncos in Hawaii. *Mar Ecol Prog Ser.* 151: 181-189.
[35] White WT, Sommerville E. 2010 Elasmobranchs of Topical Marine Ecosystems. In: Carrier JC, Musick JA, Heithaus MR (eds) Sharks and Their Relatives II: Biodiversity, Adaptive Psychology and Conservation. CRC Press, Boca Raton, Florida.
[36] Chin, A., Tobin, A. J., Heupel, M. R., and Simpfendorfer, C. A. 2013 Population structure and residency patterns of the blacktip reef shark Carcharhinus melanopterus in turbid coastal environments. *Journal of Fish Biology*, 82(4), 1192-1210.
[37] Myrberg Jr, A. A 2001 The acoustical biology of elasmobranchs. In The behavior and sensory biology of elasmobranch fishes: an anthology in memory of Donald Richard Nelson (pp. 31-46). Springer Netherlands.
[38] Sugiyono, M. P. P. (2007). Pendekatan Kuantitatif. Kualitatif, dan R&D, Bandung: Alfabeta.
[39] Rasdiana, H 2010 Studies on Coral Reefs Condition and Reef Fishes Community at the Biawak Island and its Surrounding Marine Tourism and Conservation Area, Indramayu District, West Java Province.

[40] Utomo, S. P. (2010). Study the Correlation of Coral Reefs Condition with the Abundance Chaetodontidae's Fish in Local Marine Conservation Areas Liwutongkidi Island, Buton Regency.

[41] Maharbhakti, H. R. (2009). Hubungan kondisi terumbu karang dengan keberadaan ikan Chaetodontidae di perairan Pulau Abang, Batam.

[42] Boyd, I. L 1996 Temporal scales of foraging in a marine predator. Ecology, 77(2), 426-434.

[43] Howell, P. T., Molnar, D. R., and Harris, R. B. 1999 Juvenile winter flounder distribution by habitat type. *Estuaries* 22, 1090–5

[44] Gotceitas, V., and Colgan, P 1990 The effects of prey availability and predation risk on habitat selection by juvenile bluegill sunfish. *Copeia* 1990, 409–17.

[45] Branstetter, S 1990 Early life-history implications of selected carcharinoid and lamnoid sharks of the northwest Atlantic. NOAA Technical Reports NMFS 90, 17–28

[46] Castro, J. I 1993 The shark nursery of Bulls Bay, South Carolina, with a review of the shark nurseries of the south-western coast of the United States. Environmental Biology of Fishes 38, 37–48.

[47] Springer, S 1967 Social organization of shark populations. In 'Sharks, Skates and Rays'. (Eds P. W. Gilbert, R. F. Mathewson and D. P. Rall.) pp. 149–74. (Johns Hopkins University Press: Baltimore, MD).

[48] Lyle, J. M. 1987 Observations on the biology of Carcharhinus cautus (Whitley), C. melanopterus (Quoy & Gaimard) and C. fitzroyensis (Whitley) from northern Australia. Marine and Freshwater Research, 38(6), 701-710.

[49] Lazuardi, M. E. 2000 Struktur Komunitas Ikan Karang (Famili Chaetodontidae) dan Keterkaitannya dengan Persentase Penutupan Karang Hidup di Ekosistem Terumbu Karang Perairan Nusa Penida, Bali.

[50] Salini, J. P., Blaber, S. J. M., and Brewer, D. T 1992 Diets of sharks from estuaries and adjacent waters of the north-eastern Gulf of Carpentaria, Australia. Marine and Freshwater Research, 43(1), 87-96.

[51] Papastamatiou, Y. P., Lowe, C. G., Caselle, J. E., and Friedlander, A. M 2009 Scale-dependent effects of habitat on movements and path structure of reef sharks at a predator-dominated atoll. *Ecology*, 90(4), 996-1008.

[52] Frisch, A. J., Ireland, M., Rizzari, J. R., Lönnstedt, O. M., Magnenat, K. A., Mirbach, C. E., and Hobbs, J. P. A 2016 Reassessing the trophic role of reef sharks as apex predators on coral reefs. *Coral Reefs*, 35(2), 459-472.

[53] Randall, J. E. 1977 Contribution to the biology of the whitetip reef shark (Triaenodon obesus).

[54] Cortés, E. 1999 A stochastic stage-based population model of the sandbar shark in the western North Atlantic. In American Fisheries Society Symposium (Vol. 23, pp. 115-136).

IOP Conf. Series: Earth and Environmental Science **89** (2017) 012007 doi:10.1088/1755-1315/89/1/012007

Technical Suitability and Static Stability of *Sungkur* Fishing Boats for Fish and Shrimp Catching

Rusmilyansari[1], E Rosadi[1], Iriansyah[1]

[1]Fisheries and Marine Faculty, Lambung Mangkurat University, Banjarmasin, Indonesia.

Email: r_melyan@unlam.ac.id

Abstract. Sungkur fishing gear is operated actively on one the side of fishing boat, which requires technical suitability and fishing gear stability to ensure success in fish catching. This is a case study which aimed to analyze some technical issues related to the boat, boat's hydrostatic parameters, and the boat's stability. The data were collected though observation, measuring the boat to obtain the offset table. The data were analyzed numerically and descriptively. The data were processed with technical formula, Microsoft Office's Excel software, graphic display, minitab, statistical data processing, and maxsurf program. The research results showed that: (1) the sungkur fishing boat dimensional ratio L/B (6.47 – 7.00); L/D (10.90 – 11.20) and B/D (1.60 – 1.668) is within the range value of Indonesian fishing boats suitable to operate the fishing gear by towing or dragging. However, during fish catching operation, there have been problems in a hydrodynamic force due to the fishing gear movement, which affect the fish catching efficiency. (2) The boat's coefficient of fineness is in the fine type shape; the displacement on each waterline has increased; the loads of the boat are getting larger following the increase of waterline from one to five; this is also shown from the increasing midship area value. Ton per centimeter immersion to change wl 1 by 1 cm needs 0.04 tons of weight. (3) Sungkur fishing boat have a good static stability, which is proven by the positive value of angle of maximum GZ by 79.1 – 83.6. In other words, the boat has the ability to return to its original position after tilting; however, stability dynamics happens because fishing gear operation are located on just one side of boat.

1. Introduction

1.1. Research Background

The Sungkur fishermen community in Tanah Laut make their own design for their fishing gear. Judging by the material, the fishing gear can only be classified as traditional fishing boats, which may not be found in regular commercial market. These Sungkur boats are more suitable for fishing in shallow coast with sandy mud and relatively flat surface. These boats are traditionally manufactured by Barito Kuala Regency [10]

Sungkur as a fishery unit, which consists of fishing boat, fishing gear, fisherman, and fishery resources, requires a balanced unity of mutual support for ensuring a successful catch. As a supporting facility, Sungkur boats should fulfill particular prerequisite to operate well. The Sungkur Fisheries are Active fishing gears are operated actively to catch fish as the object [10]. The Sungkur fisheries located in Tanah Laut Regency have an abundance of tools and properties which are feared to be disturbing the hydrostatic pressure. Therefore, it is important to calculate the hydrostatic parameter.

During the operation, Sungkur fish catcher is put on one side of the boat. The inside part of the fish catcher consist of shoes, tanjuran, net, and ballast. With these properties, the boat should fulfill certain criteria to operate well. The boats built in South Kalimantan usually are still traditional; the bottom part of the body still depends on the supply of woods cut in half, heated, and made into the boat main foundation, which they call it as kell. The boat manufacture has so far only considered the work load and funding from the buyer. Therefore, the boats are often times transversally not symmetrical when it is launched to the water surface. The boat commonly makers anticipated it by adding ballast on one side of the boat to make it stable. This action, however, increases the weight of the boat and thus increases the boat movement.

Along with the development and modernization of fishing boats, the structure design has been done through some approaches. The first approach in boat designing is measuring the technical suitability based on the characteristics suggested by [5]; that is by comparing the boat's main dimensions. [18] suggested that ratio of boats' main dimensions of traditional fishing boats in Indonesia has several parameter value differences in the body compared to the boats produced by Japan. Japanese boats' parameter value is bigger than that of Indonesian. According to [11] the smaller the value of L/B is, the more the ratio value will affect the ship speed; the bigger L/D value is, the weaker the boat power to longitudinal stretch s. Meanwhile, the bigger the value of B/D is, the more stabile the boat is; however, the worse the propulsive ability gets.

As one of sea ships/boats, fishing boat should also fulfill the requirement to be a sea boat. [17] explained that general requirements for a ship can sail safely in any sea circumstance, having stable shape and enough buoyancy, and efficient regarding size, power, cost, manufacture, and activities. The requirements should be fulfilled to make sure that the boat is ready to sail. Sungkur specialty, in this case, requires a specific boat that can accommodate the fish catcher used; it is, therefore, necessary to learn the technical suitability and static stability of Sungkur boat.

1.2. Research Purpose

This research aims in specific to analyze: (1) the technical suitability of Sungkur boat which consists of general arrangement, lines plan, and main dimension ratio, (2) hydrostatic parameter and (3) static suitability of Sungkur boat. This research is expected to be able to give comprehensive definition of the problems and weaknesses of Sungkur boat and the refinement of its design. Practically, this research can be a consideration to perfect the structure design of the fishing boat and increase its productivity and economical profit in the long terms.

2. Methodology

2.1. Study Sites

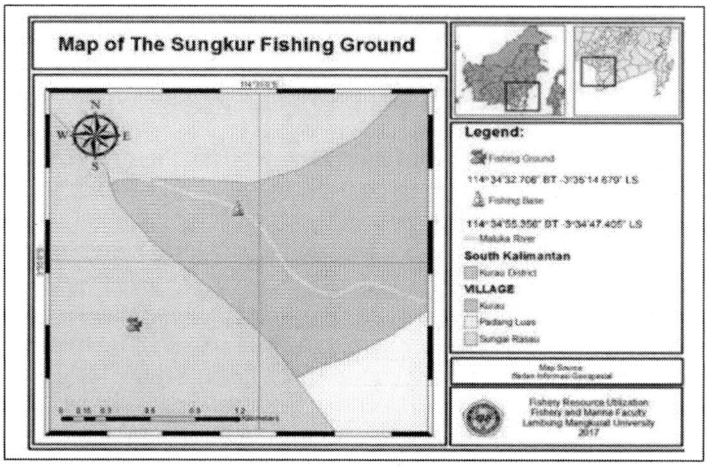

Figure 1. Map of Fishing Ground for Fish and Shrimp Catching with Sungkur Fishery in South Kalimantan Waters Scale Bar 1: 15000

This research was conducted from January to June 2017. This research was done at two sites: at the dry-dock of Sewangi, where Sungkur boats were made, and at the sea waterwork, where Sungkur fish catcher was operating in Tanah Laut regency, South Kalimantan province. Below is the Fishing Ground map of Sungkur catching (see Figure 1)

2.2. *Method*

The method of this research is descriptive and numerical case study. This research is conducted by learning certain particular case and limited object [15]. The research objects are the boats and Sungkur fish catcher. The data are collected through direct observation of the symptoms of the subjects observed during the catching operation, either in real situation or with special aid [1]. The ship is measured directly both at the dry-dock or at the quay in Tanah Laut regency, South Kalimantan province.
The tools used are meter roll, water pass, mast ruler, flexible curve, pendulum, small nail, and drawing table. Next, the tools are put in the field offset table. The soft tools used are Microsoft Excel, Software Maxsurf, and Notepad.

2.3. *Data Analysis*

Data from offset table are then made into picture in the form of general arrangement and Lines plan. The picture of boat line plan is the picture of each water line and ordinate manifested in three pictures: profile plan, half breadth plan, and body plan. The hydrostatic boat parameter is analyzed numerically based on the data resulted from geometrical measurement of boat shape using naval architecture calculation [5]
Newton Law II is used to assume the relationship between hydrodynamic (water resistance) and boat speed. The resistance calculation of Sungkur fish catcher is obtained by recalculating all resistances and wing opener of Sungkur which show the number of resistances from all net body and net pocket/bag as well as the components of fishing aid tools (shoes, surungan, all kinds of rope)

Suitability Analysis: L/B Ratio = Ratio between the length and width affects the boat resistance; L/H Ratio = Ratio between the length and the width affects the longitudinal strength of the ship; B/H Ratio= Ratio between the width and the length affects the boat stability. As a comparison, the standard ratio estimation in Nomura and Yamazaki (1977) is also used.

The Simpson Formula is used to count the midship area (A□), water area (Aw), and displacement volume, Ton Per Centimeter immersion (TPC), moment to change trim 1 cm (MCT 1cm). For calculating coefficient, these formula are used: Coefficient block (Cb), Coefficient prismatic (Cp), Coefficient vertical prismatic (Cvp), Coefficient midship (Cm), and Coefficient waterplan (Cw). The hydrostatic parameter is obtained by using the naval architecture formula [6][28].
To measure the ship stability, several formula is used to calculate the boat metacenter location: floating center location above basic line (KB), metacenter above floating center (BM), longitudinal metacenter above floating center (BM_L).

The static stability parameter is analyzed through static stability curve GZ using Atwood formula method [7]. This method analyzes static boat stability at an angle of heel 0^0-90^0. The calculation is below: GZ = BR – BT in which BR is the horizontal buoyancy center. At this moment, at shade area:

$$v \times hh_1 = BR \times \nabla$$

$$BR = \frac{v \times hh1}{\nabla}$$

in which: v is shade volume.
Hh_l is horizontal change of shade area.
∇ is boat displacement volume.

Therefore, $GZ = \frac{v \times hh1}{\nabla} - BG \sin \theta$

This method analyzes the static boat stability at angle of heel 0^0-90^0. The static stability curve GZ describes the height of Righting arm GZ at angle of heel 0^0-80^0. Based on GZ curve, an analysis then is conducted on several heeling angles. The stability calculation result is then compared with the standard boat stability released by kingdom regulation the fishing vessels
(Safety Provisional) [7] at Torremolinos International Convention For the Fishing Vessels Regulation 28 (1977) through curve GZ [5]

3. Results

3.1. Description of Fishing Unit

3.1.1. Fishing gear

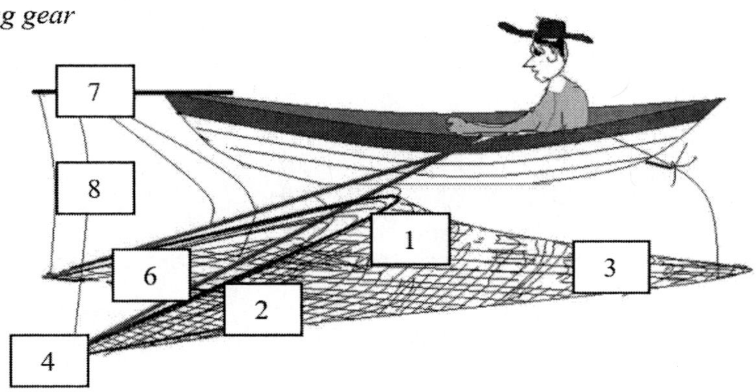

Figure 2. Components of Sungkur Design Model Information

No	Component	Information	Material	Resistance (Kgf)
1	: Webbing	Locking up fish/shrimp. 6 m, mesh size 3 m	PE	11,8426
2	: Wing	Shrimp leader net. 7 m length, mesh size 3 cm	PE	16,3437
3	: Cod end	Shrimp/Fish. 1, 10 m length and 2 cm	PE	1,3282
4	: Shoes	Preventing surungan from embedding at the sea bottom. 45, 0 cm length and 14,0 cm width inside or 2,5 cm thickness.	Wood	3,2421
5	: Surungan	Wing opener of net. 10, 63 m length	Bamboo	18,0406
6	: Sinker	Maintaining the position of wing at the bottom position, 12 item/meter; 3 cm length, 1,2 cm diameter	Rock	438,8204
7	: Tanjuran	A tool to hung ropes to determine the deepness of Sungkur position in the water	Bamboo	0
8	: Ropes	To open net, lift the fish cached, and manage the deepness during the operation	See Table 1	0

Sungkur is a fish catching tool with transversal wing opener forming triangle with a bag at the back. Its bottom part does not have any hull. For further details, the Sungkur fish catching tool is displayed below. The technological technical review on Sungkur fish catcher has been done by applying resistance calculation formula on all components. The biggest resistance is on sinker; the second biggest resistance is on surungan, and the third is on wing. Tanjuran is located at the upper part of the boat; therefore, it does not affect the boat resistance. Beside tanjuran, the ropes also do not affect the resistance. Several kinds of ropes used for Sungkur fish catcher can be viewed in Table 1.

Table 1. Several Rope Sizes at Sungkur Boat

Name of the Rope	Material	Length (m)	Diameter (mm)
Right Tanjuran	PE	12	12
Left Tanjuran	PE	7,3	7,3
Right Surungan	PE	3	20
Opener	PE	24	24
Kamong	PE	5,7	5,7
Upper Ris	PE	5,7	5,7
Lower Ris	PE	3,18	7,37
Sinker	PE	7,21	3,45

The bigger the fish catcher resistance is, the slower the boat speed is. The decrease of Sungkur speed is due to the resistance which is caused by the Sungkur fish catcher or the water resistance toward Sungkur fish catcher during operation.

3.1.2. Operation Method. Sungkur Operation uses motor boat (Figure 3). Fishing operation is started by determining the fishing ground. The sea water bottom for fishing ground is required to have 5 meter depth. Sungkur fish catcher is an evolution of push net. The position of the fish catcher is at the side of fisherman and the position of cod end is in front of the back wheel; otherwise, it may be enlaced by the ships' screw. This position also eases the catches lifting since Sungkur is only operated by one or two fishermen only.

Figure 3. Sungkur Boat Operation

The water depth can be reached by Sungkur fish catcher depending on the ropes attached at Sungkur. To arrange the reach of the depth, the rope could be extended or pulled. The depth also affects the boat speed because the deeper the water passed by the fish catcher, the bigger the Sungkur resistance is. The length of operation time is 1 hour/trip; in one day, it is possible to perform 7 operations. During operation, the pitch movement can happen when pushing the fish catcher. To neutralize it, one, two or three ballast tanks should be attached to the back part of the boat with the capacity 75 kg/tank. For boat impetus, the fishermen use these machines: Dongfeng SI 100 A, Diesel Engine, 1 Rating output 12 kw/2200r/min 16,5 Engine, S/2200 RPM 16 HP, Net Weight 155 kg, engine no 122299 dates 1990, Changzhou Diesel Engine Works The Republic of China.

3.2. Characteristics of Sungkur Boat Technic

3.2.1. General Arrangement of Sungkur Fishing Boat. The general arrangement is a technical image which shows general display of the boat room arrangement viewed from upper side and left/right side. The general arrangement of Sungkur boats at Tanah Laut regency consists of machine room, fish catcher room, ballast tank room, and catch room which are located in one deck. General arrangement (Figure 4) is planned for a platform which consists of catching goals and catch storing in one day trip.

IOP Conf. Series: Earth and Environmental Science **89** (2017) 012007 doi:10.1088/1755-1315/89/1/012007

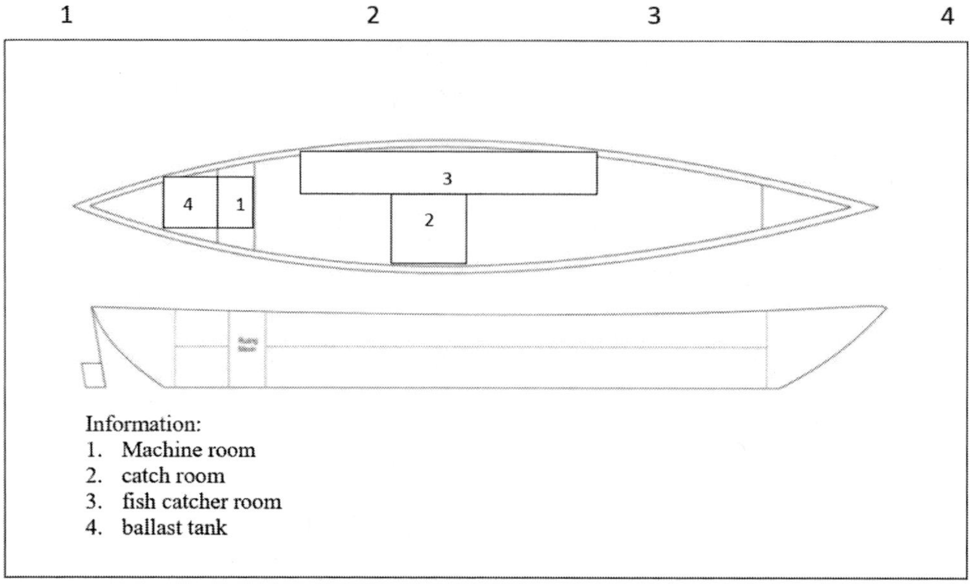

Figure 4. General Arrangement of Sungkur Boat

Machine room is an activator machine along with as propeller which connects it with the propeller at the back. Beside machine room, there is a spare fuel tank to ease the refill. The nature of the boat is still traditional; the machine room is not soundproof yet causing the machine sound roaring. The catch room is in the middle of the boat, which is the biggest part of it. Due to the circumstance, the biggest volume is in midship part. Partition between catch room and machine room is made so that the temperature of machine room does not disturb the temperature at hatch.

The catch room is put at one side of the boat to ease the catching operation. Tank ballast room is room to give pressure to the boat to make it balance and to reduce the movement of boat pitching during the operation.

3.2.2. Lines Plan of Sungkur Boat. Lines plan is an image that displays something in the form of line plan drawn at water lines and ordinates. Lines plan of Sungkur boat is divided into several ordinates stretching out along the boat with 1-meter distance among ordinates. The boat being observed is also divided into the five similar water lines starting from baseline until the highest draft (d) or load water line. The result of boat measurement based on the line plan is put on the offset table (table 2) which is needed for naval architect calculation.

Table 2. Offset of Sungkur Boat Table

Ordinal	½ B	Half Breath Plan				
		WI 1	WI 2	WI 3	WI 4	WI 5
0	0	0	0	0	0	0
1	37	32	28	20	11	3
2	67	60	55	39	24	3
3	75	70	63	48	32	3
4	80	77	68	56	40	3
5	83	78	69	61	45	3
6	81	75	66	55	43	3
7	76	70	62	48	36	3
8	71	65	56	43	29	3
9	62	55	47	36	23	3
10	36	32	26	16	3	0
11	0	0	0	0	0	0

Table 2 depicts the measurement result of Sungkur boat; that is, along the full load line of the boat, an vertical line is made dividing 11 parts into similar shape. The vertical waterline is given number 1-11 starting from after perpendicular (AP) until fore perpendicular (FP). This line is a useful ordinate to make half breadth plan and body plan pictures. Based on the data on the Table offset, a picture of plan line is obtained, manifested in three pictures: profile plan, half breadth plan, and body plan. Profile plan is shown by base line and 5 waterlines. WL 5 shows that the boat is full. Based line is considered as the start of waterline (0.0 wl). Waterline shows the boat position towards water surface when the boat is shrinking.

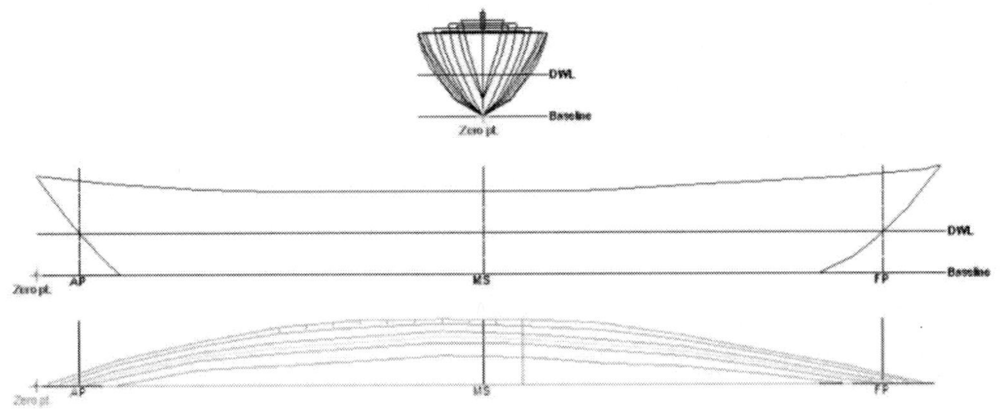

Figure 5. Lines Plan of Sungkur Boat

Half breadth plan is a picture that shows boat waterline viewed from the upper side of each buttock line. Buttock line shows vertical distance between side of the boat and flat surface which is in line with baseline, that is, a line cutting wl and in line with the centerline. Body plan shows the length of front line plan. In this picture, it is shown from the boat body on each ordinate. What is drawn here is half of the whole body. Ordinate 0-5 on the left side is a form from after perpendicular (AP) to midship. Ordinate 5-11 on the right side is a form from midship to fore perpendicular (FP). Based on Figure 5, lines plan of Sungkur boat on the bow and the stern of the boat forms Raked Bow (V). The form of boat is double pointed and the hull is Akatsuki bottom. This is related with the boat work system which focuses on the bow, starting from net placement to catching operation.

3.2.3. Ship Main Size/ The ship main size is the measurement consisting of Length (L), Breadth (B), Depth (D), and Draft (D). the boat characteristics can be viewed from main dimension ration of the boat. The ratio value of main size of each L/B, L/D, and B/D can be viewed at Table 3.

Table 3. Main Dimension Ratio of Sungkur Boat Sample

Boat dimension	Boat 1	Boat 2	Boat 3	Boat 4	Boat 5
L (Length)	10.90	11.20	10,5	12,05	11,55
B (Breadth)	1.68	1.60	1,36	1,85	1,50
D (Depth)	1.00	1.00	0,60	0,75	0,55
d (Draft)	0.70	0.50	0,40	0,50	0,40
Towed/dragged gear					
Ratio L/B (2,86 – 8,30)	6.49	7.00	7,70	6,51	7,70
Ratio L/D (7,20 – 15,12)	10.90	11.20	17,50	16,06	21,00
Ratio B/D (1,25 – 4,41)	1.68	1.60	2,66	2,40	2,72

IOP Conf. Series: Earth and Environmental Science **89** (2017) 012007 doi:10.1088/1755-1315/89/1/012007

Based on Table 3, the value obtained for the main dimension of Sungkur boat has mostly fulfilled the standard value determined, but still on the upper threshold limit. The L/B value is relatively closer to the upper threshold then the lower one. It shows that the movement resistance experienced is small, which affects the boat speed. B/D ratio affects the boat stability and impetus. The value relatively tends to the upper threshold even passes it; the stability is good but the propulsive ability could get worse if the resistance of fish catcher used is big. Based on the calculation, L/D ratio has not fulfilled the standard; the longitudinal strength affects the propulsive ability badly. The boat, therefore, experiences difficulty when veering left or right during the catching operation.

3.3. Sungkur *Boat Hydrostatic Parameter*

During a boat expedience test, one of phases to do is calculating hydrostatic parameter. The value of this parameter indicates the boat body statically. The parameter is displayed in table 4.
Displacement value of hydrostatic table range from 0,0999 -5,76 tons. Displacement value of sample boat on each waterline increases, which indicates that there are more stuffing on the boat as well as the draft. The fishermen can estimate the load weight the boat can carry, referring the maximum displacement value, which is 15, 75 tons.

Water plan area value (Aw) is the breadth of flat surface of the boat on certain water line slice. This value shows the breadth of the area on the boat for each waterline. There has been an increase in the value, which shows that the area breadth gets bigger. In other words, the closer the load placement to the deck is, the freer it gets.

Table 4. Hydrostatic Parameter of Sungkur

No	Parameter	Unit	Boat 1			Boat 2		
			WL 1	WL3	WL 5	WL 1	WL 3	WL 5
1	Ton displacement	ton	0.2751	2.146	5.02	0.0999	0.8648	2.159
2	Water area (Aw)	m^2	4.603	11.724	18.094	2.702	7.664	12.231
3	Midship area (Ao)	m^2	3.858	8.45	11.344	1.952	5.124	7.444
4	Ton Per Centimeter Immersion (TPC)		0.04	0.087	0.116	0.04	0.087	0.116
5	Coefficient block (Cb)		0.281	0.372	0.435	0.249	0.294	0.325
6	Coefficient prismatic (Cp)		0.563	0.593	0.638	0.499	0.511	0.539
7	Coefficient waterplane (Cw)		0.566	0.631	0.705	0.5	0.535	0.574
8	Coefficient midship (Cm)		0.5	0.628	0.681	0.5	0.575	0.603

The midship area value is the breadth of longitudinal section of the slice athwart the boat in the middle. This value shows that the number gets bigger and the midship breadth area gets bigger too following the waterline movement. Placing catch room in the midship, therefore, is correct since it can carry maximum load. Ton per centimeter immersion (TPC) is a value that indicates that the weight needed to change draft is 1 cm. Based on hydrostatic table, the two samples show similar value; changing 1 cm wl needs 0,04 tons load, wl 2 needs 0, 097 tons, and ll 5 needs 0, 116 tons. The coefficient value of waterplane shows the comparison between breadth value of LWL (length waterline) and BWL (breadth waterline).

The expedience value of the boat can be viewed from its coefficient value of the fineness, which consists of Cb, Cp, Cvp, Cw, Cb and Cm. The boat's coefficient value of fineness is displayed below, in Table 5.

Table 5. Coefficient Value of Sungkur Ship Fineness

Coefficient of fineness	Boat 1	Boat 2	Towed/dragged Gear Standard Value
Coefficient block (Cb)	0,435	0.325	0,40 – 0,60
Coefficient prismatic (Cp)	0.638	0.539	0,51 – 0,62
Coefficient water plan (Cw)	0.705	0.574	0,66 – 0,77

The value usually used in determining the boat overweight is Coefficient block (Cb) value. The closer the number to 1, the weightier the boat gets. Based on the hydrostatic value, cb value which ranges between 0,249 – 0,435 shows that the boat is in fine type category. Coefficient midship shows the diameter breadth of the boat midship area; the bigger the coefficient midship gets, the broader the ship diameter is. The Cm ranges between 0,60 - 0,68 % of the boat diameter breadth athwart the boat.

BM distance (radius metacenter) is metacenter vertical radius and BML is metacenter longitudinal radius. KM distance is a virtual distance from metacenter longitudinal point to the base line and KML is a virtual distance of metacenter longitudinal point. This parameter shows both the position of buoyancy point (B) and metacenter point (M) which affects the boat stability. The closer the distance between K point and B point towards M point, the more negative its influence on the boat stability.

3.4. Sungkur Boat Stability

3.4.1. Sungkur Boat Initial Stability. Stability is the most important thing on the ship, especially fish boat, considering that it has to do fish catching operation in any weather. Stability is the boat ability to return to its original position after heeling which is caused by external force. The analysis on the Sungkur boat static stability is conducted by assuming that Sungkur boat has intact stability and at static condition it does not experience trim; draft difference between bow and stern by then is not significant. Therefore, the static boat stability is tested until vanishing angle. As long as the heeling angle is still below the vanishing angle, the righting arm GZ still has positive value.

Sungkur boat static stability occurs at vertical gravity point (KG) and the boat design draft is calculated based on the position of three points: gravity point (G), buoyancy point (B), and metacenter point (M). All these points are calculated based on baseline point reference K, which then is called KG, KB, and KM. An object floating on the water is balanced when the gravity point (G) and buoyancy point (B) is on the same vertical line. The parameter value of the boats' static stability is shown below on Table 7.

Longitudinal Centre Buoyancy (LCB) is a virtual distance where vertical buoyancy center point exists. Negative CLB point shows that the position of buoyancy point (B) is in behind the midship, heading to the stern. The LCB point which decreases along with the increase of boat draft shows that the boat buoyancy longitudinally moves to the stern. This is due to the position of machine room and ballast which is in the stern.

KB value increases in each WL, with the biggest point 0, 44 for Sungkur boat 2; it shows that when the boat increases its load, the distance between buoyancy point above base line will increase. The distance gets even less until maximum draft. BM, a vertical metacenter radius, is positive. It shows that the buoyancy point is below metacenter point (M), stable equilibrium boat. The positive value shows that when the boat heels because of the wave or load distribution, the boat can quickly return to its original position.

Table 6. Parameter Value of Sungkur Boat Static Stability

No	Parameter	Unit	Boat 1			Boat 2		
			WL1	WL3	WL5	WL1	WL3	WL5
1	Longitudinal Centre Buoyancy (LCB)	m	-0.075	-0.046	-0.024	-0.198	-0.185	-0.189
2	KB distance (m)	m	0.067	0.197	0.322	0.093	0.269	0.44
3	BM distance (m)	m	0.122	0.23	0.251	0.301	0.335	0.309
4	KM distance (m)	m	0.189	0.428	0.574	0.395	0.604	0.748
5	BML distance (m)	m	81.259	27.249	17.794	65.723	22.11	14.209
6	KML distance (m)	m	81.326	27.446	18.116	65.817	22.379	14.648

When KM value is low, the boat heel slowly, not stiffly. The advantage of KM low value is that the tension is small. Crew will feel comfortable. The KML distance shows its longitudinal position from the midship. KM distance shows the distance from metacentre point towards K point vertically. BML distance shows the longitudinal position of BM from the midship.

3.4.2. Righting Arm of Sungkur Boat. The analysis on the sample boat static stability shows the higher value of the ship based on IMO. It means the sample boat used by the fishermen has good stability and high buoyancy. It can therefore support the operation well. GZ value and area value below curve GZ will change when load distribution change. Maximum GZ value is when the load is 0, 605 on boat 1 and 0, 675 on boat 2. The stability of sample boat displayed above shows that Righting arm GZ has bigger value compared to minimum value determined by IMO so it has good stability in the operation. To obtain good stability, the breadth of curve area GZ should be 30^0, not less than 0.005 m-rad; the breadth of the curve area should be 40^0, not less than 0.090 m-rad, and the breadth of the curve area should be 30^0 to 40^0, not less than 0.030 m-rad. In addition, the boat should not have heeling angle for upholder arm GX_{max} less than 25^0; the value of upholder arm GX_{max} should not be less than 0.2 m, and the metacenter height should not be less than 0.15 m.

Table 7. Comparison of Breadth under Righting Arm Curve GZ at Sungkur Boat in IMO Standard

Indicator	IMO Standard (Minimum Value)	Breadth under curve GZ	
		Boat 1	Boat 2
Area 0 - 30	0.050 m-rad	0.101 m-rad (+)	0.0791 m-rad (+)
Area 0 - 40	0.090 m-rad	0.1742 m-rad (+)	0.143 m-rad (+)
Area 30 - 40	0.030 m-rad	0.0732 m-rad (+)	0.064 m-rad (+)
Max GZ at 30 or greater	0.2 m	0.605 (+)	0.675 (+)
Angle of maximum GZ	25 deg	79.1°	83.6°
Initial GMt	0.15 m	0.744 m	0.566 m

3.4.3 Righting Arm Curve GZ of Sungkur Boat. Stability of two Sungkur boat samples was analyzed by calculating the upholder arm value (GZ) formed at curve GZ (figure 5). Those curves show that the upholder arm GZ resulted by sample boat is positive, which means the boat can return to its original position after heeling.

The curve shows the value of righting arm GZ on several heeling angles (0^0-90^0). The calculation of Sungkur righting arm GZ using GZ program was conducted when the boat during dead water. The value of righting arm GZ has a standard issued by International Maritime Organization (IMO).

Based on the curve at figure 6, it is known that GZ value of Sungkur boat tends to increase at angle of heel (0^0-60^0) with maximum performance, that is, 0.675 for boat 1 at heeling angle 80^0 and 0.605 for boat 2 at heeling angle 70^0. The boat have positive value, which means that the boats are capable to return back to its original position after heeling.

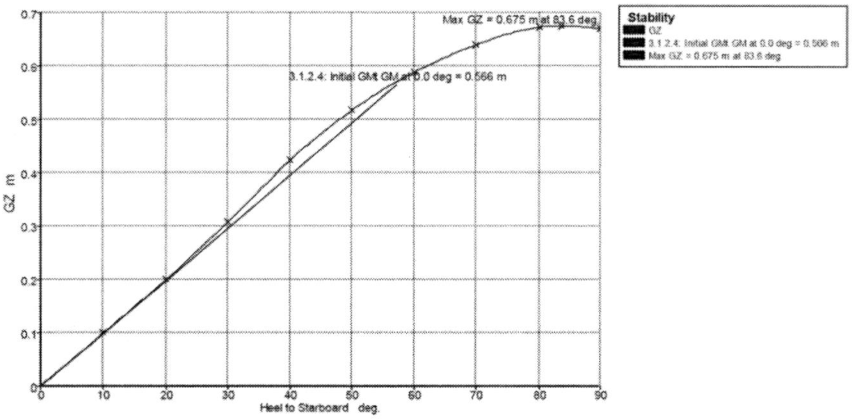

Figure 6. Righting Arm Curve GZ

Figure 6 shows a condition when the heeling point is 0^0 and the boat is in upright position because the righting arm GZ = 0 m. When the heeling point is 80^0, the returning arm GZ returns to upright position 0,6 m. Judging from the crossing points, the stability curve with heeling angle occurs at big angle. It shows that when the boat is heeling 80%, the boat will not turn upside down and can return to its original position because the upholder arm has not reached zero value.

3.5. Discussion

3.5.1. Technical Problem of Sungkur Boat. The value of boats' main dimension is a simple and easy approach to determine the ship's size. Main dimension ration of the boats is a parameter which very much influences the Sea Worthiness. There are several problems at Sungkur boat viewed from the boat dimension ratio: (a) L/B relative value which is closer to upper limit than the bottom one; (b) L/D ratio has not yet fulfilled the standard; the longitudinal strength is poor because the boat is not resistant to the turning movement either left or right during the catching operation; (c) movement resistance experienced is small, which positively influences the boat speed; (d) the Sungkur boats' propulsive ability is hardly good because the fish catcher resistance is big; and (e) the shape of Sungkur boat is in fine type; it can heel on the wave.

L/B value of Sungkur boat is relatively closer to the upper limit than the bottom limit. The resistance value of fish catcher which is big very much affects the catching efficiency. Therefore, it is suggested not to exceed the impetus of the boat because it forces the machine to work harder; it requires more power from the boat to give balancing impetus during operation. According to [2], the bigger the value of L/B, the smaller the resistance is. According to [22] the L / B ratio has a great influence on hull resistance and maneuverability.

Longitudinal strength of Sungkur boats is bad, affecting the heeling period. Therefore, redesigning is necessary by adding breadth. To determine the most optimal boat size, [16] state that the broader the boat is, the broader the heeling period is. The shape of Sungkur boat is in fine type. [24] say that the boat which is in fine type shape is vulnerable to heeling which reduces the comfort on the boat. [4] Argued that the slender body of the boat and quick heeling period are the weakness of the boat which can be overcome if the boat manufacture follows the manufacturing procedure of modern boat. Changes in the boat's breadth and depth can give the boat better technical parameter on the same length.

Based on Lines Plan of Sungkur boat, the bow and stern of the boat form Raked Bow (V) which causes the boat to move forward pushing the Sungkur fish catcher. It is supported by [12] who says that

V bottom enables the boat in certain speed to cleave the water mass with small resistance. It is related to the boat work system focusing on the bow part, starting from placing the net until catching process. According to [20], the general requirement should be owned by the fish boat, such as: the bottom part of the boat is in V (V bottom) shape to enable the boat to move faster.
Sungkur boat which is double pointed with hull is akatsuki bottom. According to [25], hull boat happens because the boat is manufactured traditionally or manufactured in the dry-dock and relies on the boat maker's habit. The decision to produce hull form of a boat is not based on technical consideration or operation expedience.

Based on the General Arrangement, Sungkur boat has very efficient space. [21] Argued that in determining general arrangement of a boat, we rely on several things: length value, which is closely related to the placement of machine room and fish catcher; breadth value, which is closely related to the catch placement, depth value, which is closely related to the luggage placement and stability.

During the operation of fish catching, pitch happened a lot. However, it was solved by occupying stern of the boat by putting away ballast tank; it yet increases the boat resistance. This condition has to be considered before manufacturing a boat. Boat making starts from design. In the design, the boat maker should consider the design requirement which is also called as technical data. Technical data needed consists of five main components: boat type and kind, load, speed, route, sailors, and water limit [23]

3.5.2. *Hydrostatic Parameter*. Beside main dimension, fish also has hydrostatic parameter. Hydrostatic parameter shows the initial boat condition. [5] Explains several important hydrostatic parameters are: displacement, water line area, position per centimeter immersion (TPC) and coefficient of fineness (Cb, Cw, Cp, Cvp and Co). The boat displacement value gets bigger following the draft so that the fishermen can estimate that maximum displacement is 15,78 tons. Based on the coefficient of fineness values, Sungkur boat has fulfilled connection among the coefficients as it is stated by [3] that coefficient represents the body of the boat called block coefficient (Cb), prismatic coefficient (Cp), and midship coefficient (Cm). The connection among the coefficients is Cb = Cp x Cm, and the size of the fishing boats follows the following order: Cb < Cp < Cm.

Cp value works not only as coefficient of fineness and works to ensure the boat stability; it is also useful to determine the smallest value of the boat side endurance [30]. Cb value of Sungkur boat is low; it means the diameter of the boat both on the bow or stern is slender. [13] also say that the bigger Cb value is, the closer the diameter length across the bow and stern to the one across the midship. According to [29] when the coefficient of fineness of a boat is determined, the boat resistance value depends on several things: (1) weight distribution along the boat, indicated by CLB value, (2) water area shape, especially in the bow (shape of crossing cut and stern type). [29] also suggest that one of ways to reduce the total resistance of the boat is by shifting LBC direction to the bow.

3.5.3. Static Stability of Sungkur Boat. According [5], fish boat stability is the ability of the boat to return to its original position after getting external influence. The external influence can come from the wind or wave. Fish catcher in the water can also increase the boat inclination and reduce the boat ability to return to its original position and reduce the righting moment of the boat. The boat originally has its own ability to return to transverse stability, static stability or longitudinal stability, which is based on the rule that fish boat should have initial stability not less than 0.6 meters.

Equilibrium terminology on stability matters needs to be understood. There are three kinds of equilibriums: stable equilibrium, unstable equilibrium and neutral equilibrium [26][6][8][4]. Stable equilibrium is a condition where boat can return to its original position (initial stability/equilibrium) after a force on the boat causes heeling. Unstable equilibrium is a condition in which the boat heels because of the working force on the boat which makes it cannot return to its original position, and keep heeling to one side. Neutral equilibrium is a condition in which the boat heels due to the working force on the boat and it stays permanently.

[26] Elaborates more about it; he says that maximum condition can result from the positive righting arm GZ which can return the ship into its original position. Pitch movement on Sungkur boat usually

happens when Sungkur was operated by pushing. [14] Stated that on beam seas, pitch movement can always be seen on several types of wave, usually very small, so is yaw movement.

When a stabile boat gets external force and causes the boat to heel, the buoyancy center will shift to the lowest position. When the boat heels even more, the righting arm and the distance between weight force and buoyancy force will decrease until it reaches zero points or even negative. In that condition, sea water will enter the ship through the opening on the ship [8]. Maximum stability is maximum GZ value the boat can reach on certain angle and condition while stability estimation is the biggest heel angle of the boat without negative GZ [27].

4. Conclusions

Sungkur boat design is suitable to operate fish catcher by towing or dragging. However, there are several problems: movement resistance tends to be high and the longitudinal strength is not very good to make a turning either left or right. The general arrangement of Sungkur boat consists of machine room, catch room, fish catcher room on the side of the boat to ease the catching operation, ballast tank room to give pressure to the boat to balance it and reduce pitching movement during operation.

Lines plan of the boat consist of Body plan: Akatsuki, half breadth plan : Double pointed and profile plan : bow and stern forming Raked Bow (V) which is very good and suitable to cleave water, and suitable for a boat of which the work focuses on the bow, starting from net placement and catching process.

The boat's coefficient of fineness is Cb, which is in fine type shape, causing the boat to heel easily when there is external force and reducing the comfort on the boat. Displacement on each waterline experiences an increase and stuffing along with the waterline. The biggest load value is located at midship. Ton per centimeter immersion used to change wl 1 for 1cm needs 0.04 tons load.

The boat stability overall is in good condition with relatively high buoyancy capacity. The angle of maximum GZ is positive and the boat can return back to its original position even after getting external force.

Sungkur boat perfection is necessary especially in terms of load balance and boat resistance during fish catching operation by redesigning to ensure the safety and security of Crew on the boat. Therefore, a further research on the boat's resistance and dynamic stability is needed, either during heading to the fishing ground, fish catching operation, or when returning back to the fishing base.

Acknowledgements.
The authors would like to thank The International Conference On Applied Marine Science And Fisheries Technology (MSFT 2017), The Ministry of Technology and Higher Education for funding support through the Fundamentals scheme in 2017, Contract No:119 / UN8.2 / PL / 2017.

References

[1] Arikunto S 1998 Prosedur Penelitian Suatu Pendekatan Praktik. Ed. Rev. IV. (Jakarta: Penerbit PT. Rineka Cipta). P 378.
[2] Aydin M dan Salci A 2008 Resistance Characteristics of Fishing Boat Series of ITU. J. *Marine Technology*. 45 2: 194-210.
[3] Ayodhyoa, A. U. 1972. Suatu Pengenalan Kapal Ikan. (Bogor: Fakultas Perikanan. IPB.
[4] Derrett DR. 1981. Ship Stability for Masters and Mates. 4 Ed. Revised Great (Britain: B.H Newnes)
[5] Fyson J 1985 Design of Small Fishing Vessels. Senior Fishery Industry Officer (Vessel) Fisheries Industries Division FAO. (Roma Italy. Published by Arrangement With the FAO of United Nation by Farnham – Surrey England). p267.
[6] Gillmer TC and Johnson B 1982 Introduction to Naval Architecture. (Naval Institute Press. Annapolis. Maryland).
[7] Hind JA 1982 Stability and Trim of Fishing Vessels. 2 ed. Fishing News Books Ltd. (England: Farnham. Surrey) p130

[8] Hutauruk RM and Rengi P 2014 Respons Gerakan Kapal Perikanan Hasil Optimasi Terhadap Gelombang. J. Perikanan dan Kelautan. 19 1: 13-22.
[9] International Maritime Organization (IMO) 1983. International Confrence on Safety Fishing Vessels 1977. IMO. London.
[10] Iriansyah, Rusmilyansari 2011 Teknologi Alat Tangkap "Sungkur" dan Analisis Hidrodinamik. (Banjamasin: Alhaka Publishing). p92.
[11] Iskandar BH, Novita Y 2000 Tingkat Teknologi Pembangunan Kapal Ikan Kayu Tradisional di Indonesia. J. Buletin PSP 9 2: 53-67.
[12] Iskandar BH 2007 Stabilitas Statis dan Dinamis Kapal Latih Srella Maris. *J. Buletin PSP* 16 1 : 31-49
[13] Kantu L P N. Kalangi I dan Polii J. F 2013 Desain dan Parameter Hidrostatis Kasko Kapal Fiberglass Tipe Pukat Cincin 30 GT di Galangan Kapal CV Cipta Bahari Nusantara Minahasa Sulawesi Utara. J. Ilmu dan Teknologi Perikanan Tangkap 1 3: 81-86.
[14] Lloyd ARJM 1989 Seakeeping: Ship Behaviour In Rough Weather (England: Ellis Horwood Limited)
[15] Mantjoro E, Pohtoh O dan Wasak M 1989 Filsafat ilmu. (Manado: Fakultas Perikanan UNSTRAT)
[16] Marjoni, B. H. Iskandar, dan M. Imron. 2010. Stabiliats Statis dan Dinamis Kapal Purse Seine di Pelabuhan Pantai Lampulo Kota Banda Aceh Nanggroe Aceh Darussalam. *J. Marine Fisheries*. 1 2: 113-122.
[17] Novita Y 2003 Konsep Pengembangan Sektor Perikanan dan Kelautan Di Indonesia. (Bogor: Departemen PSP. IPB)
[18] Novita Y and Iskandar BH 2008 Hubungan anttara bentuk Kasko Model Kapal Ikan dengan Tahanan Kapal. *J. Buletin PSP* 17 2: 315 -324.
[19] Nomura M dan Yamazaki T 1977 Fishing Techniques I. (Tokyo: Seafdec. Japan International Cooperation Agency) pp 85-86.
[20] Rahman, A. dan Y. Novita. 2006. Studi Tentang Bentuk Kasko Kapal Ikan di Beberapa Daerah di Indonesia. J. Torani. 16 4 : 240-249.
[21] Rusmilyansari dan Rosadi E 2011 Kapal Perikanan Sarana Perikanan Tangkap di Perairan Kalimantan Selatan (Suatu Pendekatan Survey) (Banjarmasin Penerbit Alhaka Publishing)
[22] Saha GK. and Sarker AK. 2010. Optimization of Ship Hull Parameter Of Inland Vessel With Respect To Regression Based Resistance Analysis. (Proceedings Technology MARTEC December 10-11-2010, Dhaka, Bangladesh)
[23] Sahlan. Samudro. Wibowo HN, Arifin. and Ahmad SM 2012 Proc. InSINas. Kajian Disain Kapal Cepat Berbahan Aluminium Sebagai Sarana Transportasi Sungai dan Laut yang Aman, Nyaman dan Ramah Lingkungan (Indonesia) pp 81-86.
[24] Susanto A, Iskandar BH, Imron M 2011. Fishing Vessel Design and Stability Evaluation in Palabuhanratu (Case Study off PSP 01 Training-Fishing Vessel).
[25] Tompo S A 1990 Teori Merancang Kapal I. (Makasar: Jurusan Perkapalan Fakultas Teknik UNHAS).
[26] Taylor, L. G. 1997. The Principle of Ship Stability. (Brown and Son Publisher Ltd. Nautical Publisher. 52 Darnley Street.Glasgow)
[27] Tumiwa J H, Masengi KWA, Pamikiran RDCh 2012. Stabilitas Dinamis Kapal Pukat Cincin di Sulawesi Utara. J. Perikanan dan Kelautan Tropis. 8 3,76-79.
[28] Tupper EC. 2004. Intoduction to Naval Architecture. 4th. Edition. (England: Elsevier Butterworth-Heinemann. Pp30-37.
[29] Yaakob O, Lee TE, Way LY, Koh Kho King. 2005. Design of Malaysian Fishing vessel for minimum Resistance. *J. Technology*. 42 2 : 1-12
[30] Yoshimura Y, Kosanagi Y. 2004. Design of Small Fisheries Research Vessel with Low Level of Underwater-Radiated Noise. *J. Marine Acoustic. Soc. Jpn*. 31 3 1-9.

IOP Conf. Series: Earth and Environmental Science **89** (2017) 012008 doi:10.1088/1755-1315/89/1/012008

The growth and population dynamics of seagrass Thalassia hemprichii in Suli Waters, Ambon Island

C I Tupan[1], Pr A Uneputty[1]

Fishery and Marine Science Faculty, Pattimura University. Jl. Mr. Chr. Soplanit, Kampus Unpatti, Poka, Ambon, Indonesia

Email: lotjetupan@yahoo.com

Abstract. The objectives of the research were to determined growth of rhizome, age structure, recruitment rate, and mortality rate of *Thalassia hemprichii.* Data were collected by using reconstruction technique which the measurements were based on past growth history. The age of seagrass was based on plastochrone interval. The recruitment rate was estimated by age structure of living shoots while mortality rate was estimated by age structure of dead shoots. The research was conducted on coastal waters of Suli where divided into two stations with different substrates, namely mixed substrates of sand and mud (S1) and mixed substrates of sand and coral fragment (S2). The growth rate of horizontal rhizome ranged from 4.15-8.68 cm.year^{-1} whereas the growth rate of vertical rhizome was 1.11-1.16 cm.year^{-1}. The average age of *T. hemprichii* varied between 3.22-4.15 years. The youngest shoots were found at age 0.38 years and the oldest shoots were 7.82 years. Distribution of age was polymodal which reflecting cohort. The recruitment rate ranged from 0.23-0.54 year^{-1}. Otherwise, the mortality rate ranged from 0.21-0.26 year^{-1}.Seagrass population of T. hemprichii in Suli Waters indicated an increasing condition which shown by higher recruitment rate than mortality rate.

1. Introduction

One of the coastal ecosystems that can be found on Ambon island is seagrass ecosystem. The seagrass ecosystem is found to flourish in tidal areas, coastal waters or lagoons that essentially are mud, sand, gravel and dead coral fragments at up to 4 meters depth. In clean waters, seagrass can be found growing up to 8-15 meters even up to 40 meters depth [1]. These ecosystems have important functions as sediment traps, stabilizing basic sediments, reducing wave and current energy (physical functions), and ecological functions such as providing important habitats and shelter for some animal species and habitats that rich in nutrients substances for flora and fauna diversity.

Suli coastal waters which are part of Ambon Island waters have a fairly dense seagrass ecosystem with healthy status [2]. One of the species found abundantly in this waters is *Talassia hemprichii.* This species is found to grow and spread on muddy, sandy substrates up to coral beds.

Recently, the utilization of coastal area has been higher along with the increasing of establishment. This has resulted in greater changes to the environment and the resources in it. The utilization activities such as solid and liquid waste disposal to the sea, sand and stone mining, boat docking and *bameti* conducted by people of Suli coastal area are feared could impact on the degradation of seagrass ecosystem. This problem can occur due to lack of public awareness on the role of seagrass ecosystem and the minimum information about the biology of seagrass itself. The biological aspect that still lacks of information is about dynamics population including growth of rhizome, age structure, recruitment and mortality. To date, the research on seagrass growth is mostly conducted on the leaves. The growth of seagrass other than from the length of the leaves can also be determined by the length of rhizome [3]. Seagrass can reproduce asexually (vegetative) through rhizome elongation [4]. From rhizome, the growth, age structure, recruitment rate and mortality rate can be determined. Thus, this research was aimed to determine the population dynamics of seagrass *T. hemprichii* which includes rhyzome growth, age structure, recruitment rate and mortality rate.

2. Method

2.1. Sample collection method

This research was conducted on 2 observation stations based on substrate difference located in Suli Waters. Station 1 represents muddy sand substrate and Station 2 represents sand substrate mixed with coral fragments. Seagrass samples of *T. hemprichii* were collected in squares of 1 x 1 m size, as many as 10 observation squares covering minimum 200 stands [5]. The seagrass samples were then calculated for dead shoot and live shoot, number of vertical nodes and internodes, number of horizontal nodes and internodes, number of leaves, number of living roots, as well as calculated the length and width of leaves, horizontal length of rhizome between two shoots and vertical length of rhizome.

2.2. Data analysis method

Age estimation is generated in interval plastochrone unit which was then converted into an absolute time unit. Interval plastochrone is defined as the average time interval between the successive growth of two leaves in one shoot [6, 5, 7].

The growth rate of horizontal rhizome was estimated as the slope of linear regression equation between rhizome length between shoot connected along the rhizome cut (y) and the age difference of shoots. Similarly, the growth rate of vertical rhizome was estimated as the slope of linear regression between vertical rhizome length (y) and the age of the shoots

Recruitment rate (in unit per year) was estimated from the total number of shoots ($\sum_{t=0}^{\infty} N_t$) and the number of shoots in more than one year ($\sum_{t=1}^{\infty} N_t$) contained in the population [8, 5, 7] as follows: $R_{gross} = \ln \sum_{t=0}^{\infty} N_t - \ln \sum_{t=1}^{\infty} N_t$

Mortality rate was calculated based on exponential decrease from the number of dead stands with age and allows the estimation of mortality rate of exponential shoots (M, in units per time).

$$N_t = N_0 e^{-Mt}$$

N_0 is the number of dead shoots of the same age as the model, and N_t is the number of dead shoots at age interval in t time [8]. The growth rate of net population was calculated as the difference between annual recruitment and mortality.

Mortality (M) and recruitment (R) were used to evaluate the seagrass population status and to predict its growth [8].

If $R > M$ = Population increased
$R = M$ = Population balanced
$R < M$ = Population decreased

3. Result and discussion

3.1. Substrate of research station

Seagrass grows almost in all substrate types ranging from muddy substrates to coral fragments. In the Suli waters, seagrass is found to be dominant in muddy sand and mixed muddy with coral fragments. Substrate fraction composition in two observation station can be seen in Table 1. In Table 1, it is seen that in substrate station 1 is dominated by fine sand followed by very fine sand, while station 2 is dominated by medium sand. This type of substrate supports the growth of *T. hemprichii* seagrass. According to Tomascik *et al.*, [9] that *T. hemprichii* is an abundant species of seagrass in intertidal areas on coral reefs with sand substrate and coral fragments. Substrates play a role as a protector of plants from ocean currents and as a place of processing and suppliers of nutrients [10].

Table 1. Percentage of substrate fraction size at both Suli waters research stations

Station	Percentage of substrate fraction size (%)							
	G (4mm)	VCS (2mm)	CS (1mm)	MS (0.50mm)	FS (0.25mm)	VFS (0.125mm)	S (0.063mm)	C (0.0038mm)
1	0	0	6	14	44	23	12	0
2	8	11	10	46	25	0	0	0

Note: G = gravel; VCS = very coarse sand; CS = coarse sand; MS = medium sand; FS = find sand; VFS = very fine sand; S = silt; C = clay

3.2. Rhizome growth

Rhizome in monocots such as seagrass *T. hemprichii* is a modification of the stem which also contained two or more vascular bundles and numerous fiber bundles to distribute nutrient needs. Rhizome develops extensively and has a major role in vegetative propagation. This vegetative propagation is more important than generative propagation because it is more profitable for seagrass dispersal. The average length of horizontal rhizome *T. hemprichii* between two shoots in the Suli waters ranged from 4.51 ± 1.99 - 4.56 ± 1.88 cm, where the longest size was 12.50 cm and the shortest size was 1.00 cm (Table 2). Its growth rate ranges from 4.15 - 8.68 cm.year^{-1} (Table 2). This value is smaller when compared to the growth rate of horizontal rhizoma *T. hemprichii* in the waters of Tanjung Tiram, Poka [11]. Also it is also smaller when compared to other species of seagrasses such as *Cymodocea nodosa* [7]. Morphologically the horizontal rhizome of the genus *Cymodocea* is smaller but longer than the genus *Thalassia* [12, 13].

The average length of vertical rhizomes ranges from 1.80 ± 1.72 - 2.43 ± 1.89 cm, where the longest length was 9.90 cm and the shortest size was 0.30 cm (Table 2). The growth rate of vertical rhizomes ranges from 1.11 - 1.16 cm.year^{-1} (Table 2). The vertical rhizome consists of some nodes and internodes. The longer the vertical rhizome, the more nodes and internodes, which characterize the older or longer life of the seagrass shoot, on the contrary, the shorter the vertical rhizome, the less the nodes and the internodes that characterize the younger or new shoot.

In one shoot consist of 2 - 5 leaves with an average length of 12.45 ± 4.82 - 13.87 ± 5.31cm, where the longest size was 28.30 cm and the shortest size was 1.20 cm (Table 2). The average leaf width was 0.76 ± 0.18 - 0.91 ± 0.18 cm, the largest leaf width was 2.60 cm, and the smallest leaf width was 0.30 cm. The size of this seagrass leaves was much larger than the *T. hemprichii* seagrass found in Tanjung Tiram, Poka with an average length of 5.68 ± 2.17 and the average width of 0.59 ± 0.12 [11]. The average production of leaves of both stations was the same, that is 13.04 leaves.year^{-1} (Table 2.). Production of this leaf was also greater than the production of *T. hemprichii* leaves in the waters of Tanjung Tiram, Poka as much as 12.71 [11]. If the results of this study were compared with the results of the study on the species *C. nodosa*, then the species of *C. nodosa* has more varied leaf production in one year as many as 12-14 leaves.year^{-1} [5].

Table 2. Summary of rhizome growth and morphometric of *T. hemprichii* (Mean ± standard errors) at the two stations

	St 1	St2
Average length of horizontal rhizomes (cm)	4.56±1.88	4.51±1.99
Longest horizontal rhizome (cm)	12.50	12.00
Shortest horizontal rhizome (cm)	1.20	1.00
Horizontal rhizome elongation cm.year^{-1}	4.15	8.68
Average length of vertical rhizomes (cm)	2.43±1.89	1.80±1.72
Longest vertical rhizome (cm)	9.90	9.60
Shortest vertical rhizome (cm)	0.30	0.30
Vertical rhizome elongation (cm.year^{-1})	1.16	1.11
Average production of leaf (leaves.year^{-1})	13.04	13.04
Average leaf length (cm)	12.45±4.82	13.87±5.31
Longest leaf (cm)	24.00	28.30
Shortest leaf (cm)	1.20	3.00
Average leaf width (cm)	0.91±0.18	0.76±0.18
Biggest leaf width (cm)	2.60	2.40
Smallest leaf width (cm)	0.30	0.40
Average number of leaves (shoots^{-1})	2.86±0.83	3.10±0.69

3.3. Age Estimation

The seagrass age structure of the Suli Coastal Waters is shown in Table 3 and Figure 1. The age of *T. hemprichii* ranges from 0.38 to 7.82 years. *T. hemprichii* in Suli Coastal Waters is mostly concentrated at the age of 0.50 - 1.99 years, that is 12 - 39%. This indicates that most *T. hemprichii* population at the two stations in the Suli Coastal Waters are new shoot or a young shoot. New shoot or young shoots are characterized with shorter vertical rhizomes, fewer vertical internodes, fewer leaves and more life roots. The average age of *T.hemprichii* ranges from 3.22 to 4.15 years (Table 3). The youngest shoots were found to be 0.38 years old and the oldest was 7.82 years old (Table 3). Tupan *et al.*, [11] found *T. hemprichii* on the Tanjung Tiram, Poka having an age range of 0.38 - 7.36 years. Compared to this research, *T. hemprichii* of Suli waters have a slightly longer lifetime than seagrass of Tanjung Tiram Poka. Cunha and Duarte [5] found the oldest *C. nodosa* in Ria Formosa, southern Portugal with the oldest age of 7.6 years, and not much different from this study. The old seagrass is characterized by more vertical internodes, longer vertical rhizome length and more leaves.

3.4. Mortality and Recruitment

Figure 1. shows the distribution of shoots age is polymodal which reflects the presence of cohort, and looks very oblique indicating a high mortality rate. This annual cohort is due to an annual cycle in the rhizomes growth [7]. This figure also shows the more the age; the less *T. hemprichii* shoots survive, which indicating high mortality rates of shoot.

The age structure of dead shoots shows the number of dead shoots ranging from 0.00 – 4.99 years old, and most shoots die at less than 2 years old (Figure 3). This condition indicates the high mortality occurred in the first years (Figure 3). The mortality rate found ranging between 0.21 – 0.26 year^{-1} (Table 4). The mortality rate in this research was smaller than the mortality rate in Tanjung Tiram Waters, Poka which found between 0.69 – 0.89 years^{-1} [11], and also smaller that mortality rate of *C nodosa* which ranging between 0.99 – 3.70 year^{-1} [5].

Recruitment rate describes the population growth rate obtained ranging between 0.23 – 0.54 year^{-1} (Table 4). This value is smaller than the recruitment rate of *T. hemprichii* population in Tanjung Tiram Poka with range value of 0.29 – 0.72 year^{-1} [11] and *C. nodosa* seagrass population with range of 0.7 –

3.6.year^{-1} [5]. The net growth rate of population varies considerably from non-growth to growth in the range of 0,02 – 0.33 years^{-1} (Table 4) . The low net growth rate of population is in line with the low growth of rhizomes. This limited growth of rhizomes reduces the recruitment of the seagrass shoots.

Table 3. Age shoots of *T. hemprichii* at the two station

No	Age range (year)	percentage of life shoots		percentage of dead shoots	
		St 1	St 2	St 1	St 2
1	0.00 - 0.49	0.97	1.93	15.63	31.25
2	0.50 - 0.99	19.81	39.61	31.25	50.00
3	1.00 - 1.49	23.67	26.57	21.88	6.25
4	1.50 - 1.99	18.84	12.08	3.13	3.13
5	2.00 - 2.49	8.70	3.86	3.13	0.00
6	2.50 - 2.99	6.76	3.86	6.25	3.13
7	3.00 - 3.49	7.73	2.90	3.13	0.00
8	3.50 - 3.99	5.80	2.90	9.38	3.13
9	4.00 - 4.49	1.93	0.97	0.00	3.13
10	4.50 - 4.99	1.45	0.97	6.25	0.00
11	5.00 - 5.49	0.48	0.97	0.00	0.00
12	5.50 - 5.99	0.48	1.45	0.00	0.00
13	6.00 - 6.49	0.48	0.48	0.00	0.00
14	6.50 - 6.99	0.48	0.48	0.00	0.00
15	7.00 - 7.49	0.48	0.48	0.00	0.00
16	7.50 - 7.99	1.93	0.48	0.00	0.00
	Total	100.00	100.00	100.00	100.00

3.5. Population Status.

To evaluate the population status of *T. hemprichii* seagrass in Suli coastal waters, both of recruitment and mortality value analysis were conducted. Based on this analysis, the average recruitment value of both stations were 0.38 year^{-1} and the average mortality value was 0.23 year^{-1}. With reference to these values, it is concluded that the *T. hemprichii* population in Suli waters has increased because the recruitment value was higher than mortality value. The results obtained in this research were different from the research results in Tanjung Tiram, Poka. The population status of *T. hemprichii* in Tanjung Tiram, Poka waters are decreasing [11]. Although the growth value of horizontal rhizomes and the recruitment rate of *T. hemprichii* in the Suli waters is smaller than Tanjung Tiram, Poka, however its population status is increasing. This is supported by a smaller mortality rate. This condition is different from the waters of Tanjung Tiram Poka which has high growth rate of horizontal rhizome, high recruitment rate but also high mortality rate causing the population to decline.

The research on the population dynamics of the seagrass based on past historical growth that is reflected in the current morphological structure. The morphological structure of seagrass is influenced by the environmental quality of the surrounding waters. According Ambo Rappe [13], heavy metal pollution can affect the morphological structure of the seagrass such as shortening the rhizomes and reduce the number of nodes and change the size structure of seagrass *Halophila ovalis*. Although the current population status of *T. hemprichii* is increasing, it is necessary to monitor the condition of waters regularly due to the discovery of various utilization activities around the Suli Coastal waters. Various utilization activities can produce waste that potentially polluting waters, so it is feared to affect the growth and population dynamics of seagrass *T. hempricii* population in these waters.

IOP Conf. Series: Earth and Environmental Science **89** (2017) 012008 doi:10.1088/1755-1315/89/1/012008

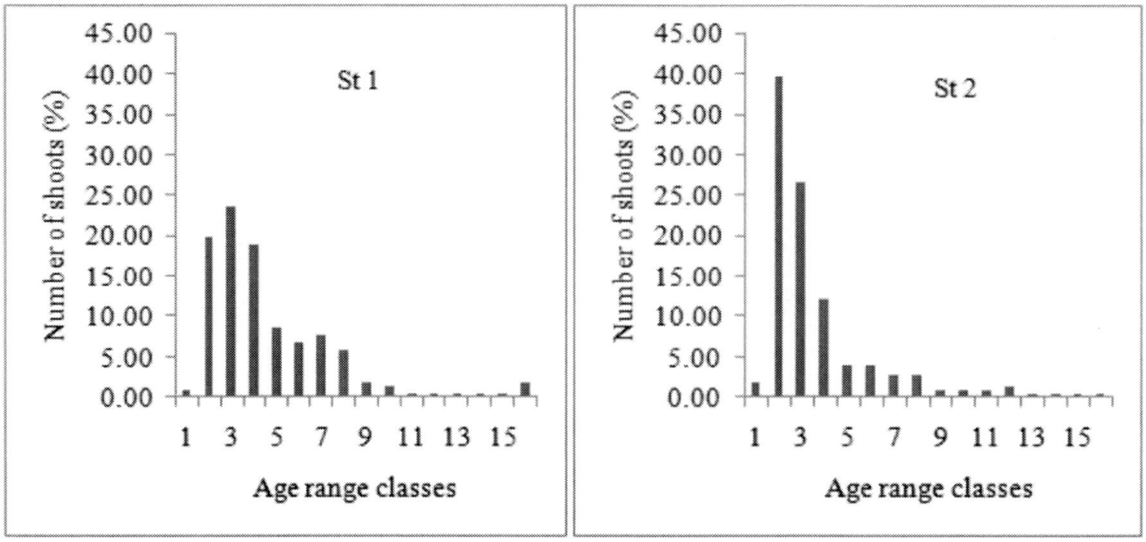

Figure 1. Age distribution of *T. hemprichii* life shoots

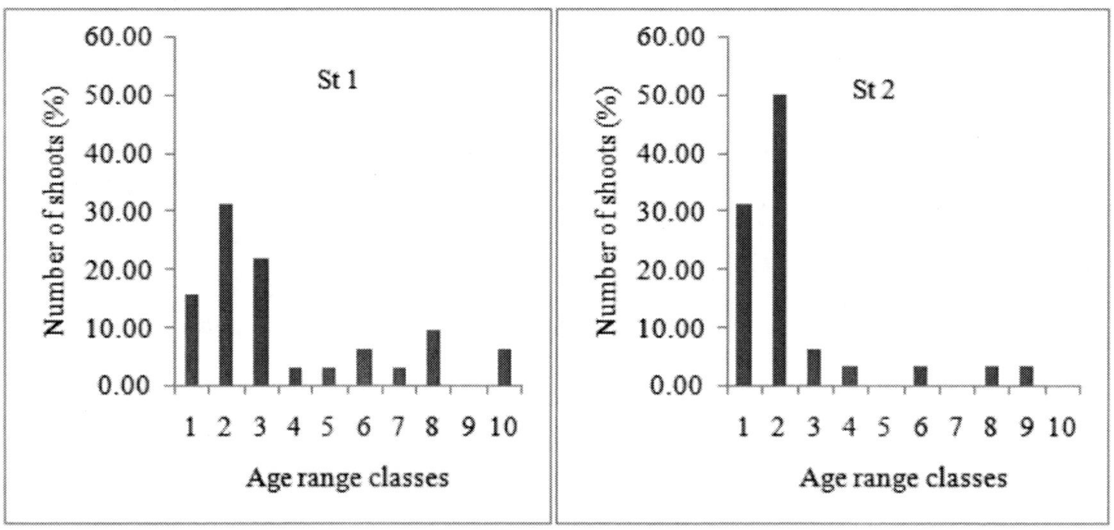

Figure 2. Age distribution of *T. hemprichii* dead shoots

Table 4. Summary of population dynamic of *T. hemprichii*

	St 1	St 2
Average shoot age (years)	4.15	3.22
Youngest shoot (years)	0.38	0.46
Oldest shoot (Years)	7.82	7.52
Recruitment rate(years^{-1})	0.23	0.54
Mortality rate (years^{-1})	0.26	0.21
Net population growth rate (years^{-1})	-0.02	0.33

4. Conclusion

The population of *T. hemprichii* in Suli waters have limited rhizomes growth, but have higher recruitment rate than the mortality rate which causes the seagrass population to increase.

Acknowledgements
We are grateful to Director of Research and Community Service, the Director General of Strengthening Research and Development, Ministry of Research, Technology and Higher Education, Republic of Indonesia through Research University of Pattimura, Ambon, Maluku, Indonesia who have funded this research through Fundamental Research Program.

References

[1] Erftemeijer P A L 1993 *Factors Limiting Growth and Production of Tropical Seagrass: Nutrients Dynamics in Indonesia Seagrass Bed* Thes (Nijmegen Catholic University: Nijmegen the Netherland)
[2] Sinmiasa V C 2015 *Status keberlanjutan komunitas lamun di perairan pesisir Negeri Suli, Kecamatan Salahutu, Maluku Tengah*. Thes (PS Manajemen Sumberdaya Kelautan dan PP Kecil, Pascasarjana Unpatti: Ambon) 123 p
[3] Short F T and Coles R G 2001 *Global Seagrass Research Methods* (Amsterdam: Elsevier Science B V)
[4] Tomlinson P B 1974 Vegetative morphology and meristem dependence: the foundation of productivity in seagrass. *Aqua* 4 107-130
[5] Cunha A H and Duarte C M 2005 Population age structure and rhizome growth of *Cymodocea nodosa* in the Ria Formosa (southern Portugal) *Mar. Biol.* 146 841 – 47
[6] Brouns J M 1985 The plastochrone interval method for study of the productivity of seagrass: possibilities and limitation. *Aquat. Bot.* 21 71 – 88
[7] Cabaco S, Ferrira O and Santos R 2010 Population dynamics of the seagrass *Cymodocea nodosa* in Ria Formosa Lagoon following inlet artificial relocation. *Estua, Coast and Shelf Sci* 87 510 – 16
[8] Duarte C M, Marbã N, Cébrian J, Enriquez S, Fortes M D, Gallegos M E, Merino M, Olesen B, San-Jensen K, Uri J and Vermaat J 1994 Reconstruction of seagrass dynamics: age determination and associated tools for the seagrass ecologist. *Mar. Ecol. Prog. Ser.* 107 195 – 209
[9] Tomascik T A J, Nontji M A and Moosa M K 1997 *The ecology of the Indonesian seas* (Periplus Eds: Singapore)
[10] Dahuri R 2003 *Keanekaragaman Hayati Laut. Aset Pembangunan Berkelanjutan Indonesia.* (Gramedia Pustaka Utama: Jakarta) pp 38-52
[11] Tupan C I, Pentury R and Uneputty Pr A 2016 Population dynamics of seagrass *Thalassia hemprichii* in Tanjung Tiram Waters, Poka, Ambon Island, Indonesia *AACL Bio* 9(6) 1286 – 93

[12] Lanyon J 1986 *Guide to the identification of seagrass in the Great Barrier Reef Region* (Great Barrier Reef Marine Park Authority Special Publication Series/ Townsville, Queensland) 54p
[13] Kuo J den Hartog C 2001 Seagrass taxonomy and identification key In Global *seagrass research methods*. Short F T and Coles R G (eds) (Amsterdam: Elsevier Science B. V) pp 31-58
[14] Ambo Rappe R, Lajus D L and Schreider M J 2011 Heavy metal impact on growth and leaf asymmetry of seagrass, *Halophila ovalis J. of Environ Chem and Ecotox* 3(6) 149-159

IOP Conf. Series: Earth and Environmental Science **89** (2017) 012009 doi:10.1088/1755-1315/89/1/012009

Impact of tidal phenomenon "Met Ef" on the exploitation of benthos at inshore shoals in Kei islands, Indonesia

Eugenius Alfred Renjaan[1], Marvin Mario Makailipessy[1]

[1]Tual State Fisheries Polytechnic, Langgur - Sathean Ring Road, Km. 6. P.O. Box 1001, Southeast Maluku, Indonesia

Email: earenjaan@polikant.ac.id

Abstract A study on benthos exploitation at five shoals located in Rosenberg and Nerong Straits, the kei islands had been conducted during period of the lowest ebb tide phenomenon which locally termed Met Ef on 2016. The purpose of the study is to know the impact of the Met Ef on benthos exploitation at the shoals by local communities during the Met Ef. Data of tidal amplitudes were obtained from the Tide Charts Mobile Applications which confirmed the observations on tide pole during October, November, and December 2013 until 2016. Data of benthos exploited during periods of Met Ef at the shoals was obtained through direct observation on benthos exploited by the local communities, also by interviewing them using questionnaires. The results showed that the lowest ebb tide of the Met Ef occurred in November, i.e., at 2 to 5 days after the full moon and/or new moon, with an average tidal range of 2.66 m and even have ever one reached 2.80 m. The most exploited benthos at the shoals is Giant clam (*Tridacna* sp.), Spider conch (*Lambis* sp.), Hammer oyster (*Malleus* sp.), Octopus (*Octopus* spp.). The intensity of benthos exploited at the shoals increased during the period of Met Ef especially in October because at that time the sea was very calm and clear due to relatively lower wind speed and the rain fall was relatively lower. This has promoted an easier accessibility of the communities to exploit benthos at shoals and, therefore October is considered by the local communities as the peak of Met Ef, instead of November. During November and, December the availability of benthos in shoals has been reduced due to it has been exploited intensely in October.

1. Introduction

Met Ef, a local term in kei language, but popularly known as Meti Kei. Met Ef is the lowest ebb tide phenomenon with maximum tidal amplitude on kei islands causing the tidal plains to become very wide, as well as the appearance of rock shoals and sandbars at sea level around the kei islands, especially kei kecil islands. The phenomenon occurs every year in October, November, and December. The Met Ef phenomenon is always accompanied by high air temperatures, low rainfall, fewer rain day frequency, and low wind speed. The phenomenon of Met Ef improves people's accessibility to exploit benthos maximally at inshore shoals located in the Rosenberg and Nerong Straits since, certain species of benthos that are now rarely found in the intertidal zones due to increasing exploitation rates from years to years. Exploitation of benthos in the inshore shoals is conducted by people from several villages. On the other hand, most of benthos species taken from the shoals have long reproductive cycles that are vulnerable to extinction. Exploitation of the shoals has been conducted from generation to generation by local people, and according to them the current number of particular benthos such as Giant clam (*Tridacna* sp.) or locally called Hanoat is more difficult to find in the shoals and tend to have smaller size. The aim of this study is to know the impact of the Met Ef Phenomenon on the intensity of benthos exploitation in several inshore shoals, located in Rosenberg and Nerong Straits, the kei kecil islands, Indonesia. The results of this study are expected to provide basic information about the pattern of benthos exploitation in the inshore shoals, as well as the preliminary information for further studies on the sustainable use of benthos in the inshore shoals.

2. Methodology

Materials and methods used in this study include; plastic samples, tape ruler, questionnaires, Satellite Image, Tide charts mobile Applications (Copyright @ 7th gear, LLC), SAS software, Planet software ArcGIS 10.X, Software Surfer 15. While Tools used include; Tide pole, Android, laptop, GPS, and Camera. Data of tidal range and tidal time was obtained from Tide charts mobile Applications (Copyright @ 7th gear, LLC), which confirmed the observations on tide pole installed at a jetty located at Rosenberg Strait. Field tide observation only conducted at Lower Low Water (LLW) condition during October, November, and December 2013 until 2016, to confirm the results of tide chart mobile application. The survey method used to obtain benthos data is the Creel Method, which is to observe directly the benthos species taken by local people who utilized five inshore shoals located in Rosenberg Strait and Nerong Strait, the kei islands of Indonesia during Met Ef in 2016. Besides also interviews the local people who exploit the five shoals. Local names of the inshore shoals they are Wama, Tibun, Vurlilan located on the Rosenberg Strait, while North Samar, South Samar, and Wid which located in the Nerong Strait. The shoal's location maps are presented in Figure 1. The climate data that always associated with Met Ef, such as air temperature, rainfall, rainy day frequency, and wind speed from 2013 to 2016 were obtained from the Indonesian Agency for Meteorology, Climatology, and Geophysics in Langgur, kei islands, Indonesia. Monthly data from the four-year average are used to describe the climatic conditions associated with Met Ef which also enable accessibility of the people to exploit the inshore shoals.

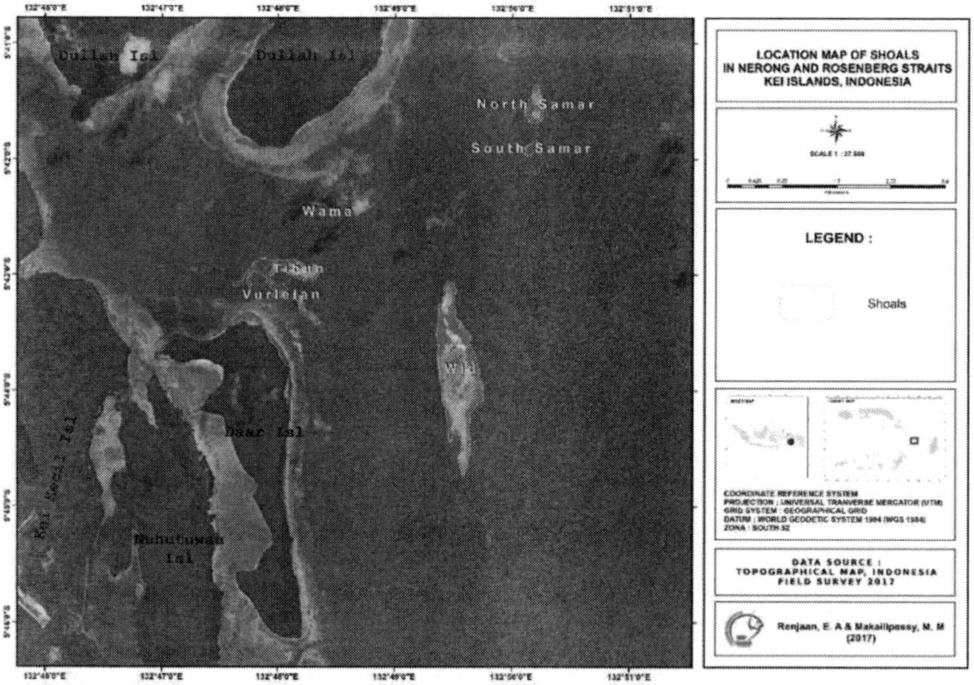

Figure 1. Map of the five inshore shoals in the Rosenberg and Nerong Straits, Kei Islands, Indonesia

IOP Conf. Series: Earth and Environmental Science **89** (2017) 012009 doi:10.1088/1755-1315/89/1/012009

Acquisition of bathymetry of the shoals' site was conducted by analyzing satellite image which described as follows.

a. Acquisition of bathymetry Data
   The bathymetry data of the study sites in from Image Marine Maps Navionics Sonar Chart (TM). The Image consists of depth contour line map that downloaded according to areas of interest by using software SAS Planet.
b. Data Processing (editing and lay outing data)
   i. Downloaded Image that is available then opened using ArcGIS 10.x software; the editing process is then conducted by creating new depth point following the contour lines that exist in satellite image.
   ii. After the process of making the point according to the depth is completed, the results are then exported to the Surfer software for the making of depth contour and distribution.
   iii. Contour and depth results are then exported back to ArcGIS for the making of water depth classes around AOI as well as map view.

3. Results

3.1. Description of the Inshore shoals

Shoals in kei local language are called "uran vat" which means pot-shaped stone, which describes the number of brain corals. In oceanography, geomorphology, and earth science, a shoal is a natural submerged ridge, a bank, or bar that consists of, or is covered by, sand or other unconsolidated material, and rises from the bed of a body of water to near the surface[19, 22). Inshore shoals Samar and Wid are located in the Nerong Strait waters that separate kei besar islands and kei kecil islands. While inshore shoals Wama, Tibun, and Vurlilan are located in the Rosenberg Strait that separates dullah islands and nuhuroa islands within the group of kei kecil islands. Some parts of the shoals are still above sea level during the daily flood tide. The seabed substrate of the shoals consists of gravel and sand with depth at flood tide ranging from 0 to 3 meters. The bathymetry map in the shoals and around the shoals as well as the geographical position and the size of the inshore shoals is presented in Figure 2 and Table 1, respectively.

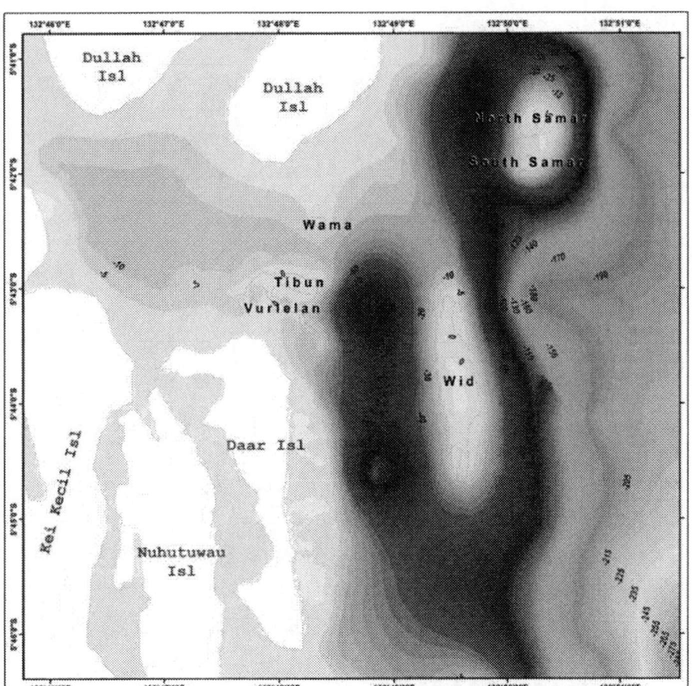

Figure 2. Bathymetry Map in and around the five inshore shoals

Table 1. Local names of shoals, geographical position, areas and position on sea level (SL) during Period of Met Ef.

No.	Names of Shoals	Latitude; Longitude	Area (Ha)	Remarks
1	North Samar	132° 50'12E; 5°41'29" S	12.5	Above SL
2	South Samar	132°50'10''E; 5°41'54" S	3.1	Above SL
3	Wid	132°49'34"E; 5°43'49"S	114.3	Above SL
4	Wama	132°48'25"E; 5o42'28"S	4.2	Below SL
5	Tibun	132°48'10"E; 5°42'28" S	12.8	Above SL
6	Vurlelan	132°48'02"E; 5°43'10"S	25.1	Above SL

The lowest depth level category on the Mef Ef period ranges from 0 to 5 meters and even partially above sea level except for the shoal Wama which ranges from 5 to 10 meters. Shoal Wid is the widest shoal compared to the other and the most accessed shoal by surrounding villagers, the second most accessed are Tibun and Vurlelan, and the third is Samar. While the Wama which is always below the sea level.

3.2. Characteristics of Met Ef as a tidal phenomenon

Type of tide in the waters of kei archipelago is categorized as a Mixed tide prevailing Semi-Diurnal; namely, tide occurs twice a day. Larger tides occur in the morning while smaller tides occur in the afternoon and evening. The peak of Met Ef in 2016 occurred on November 16, at that time the tidal amplitude reached 2.80 m. Tidal type curves, time and tidal amplitude are presented in Figure 3.

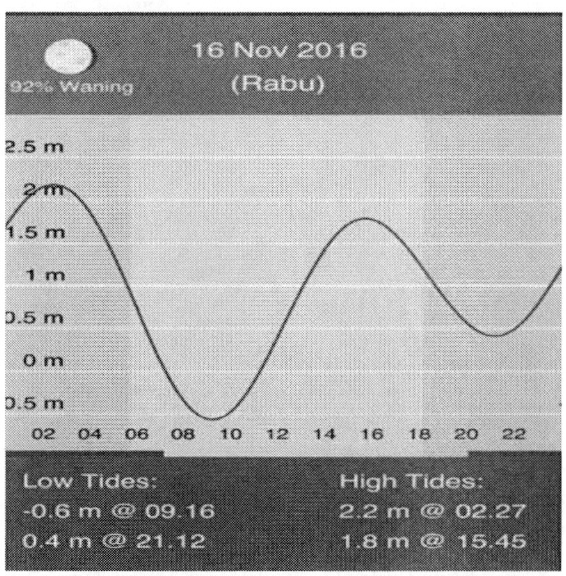

Figure 3. Tidal type curves, tidal amplitude, and time of tide on November 16, 2016 in kei islands (taken from tide charts mobile applications),

IOP Conf. Series: Earth and Environmental Science **89** (2017) 012009 doi:10.1088/1755-1315/89/1/012009

Maximum monthly average of tidal range in 2013 to 2016 increased from September to November but then back to decrease in December which had been confirmed on the tide pole is shown in Figure 4 below,

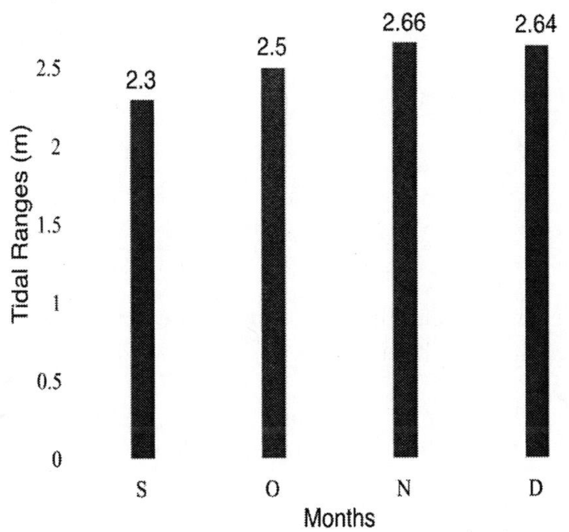

Figure 4. Maximum tidal range average during Met Ef that occurs around the full and the new moons periods in October (O), November (N), and December (D), 2013 to 2016.

The meaning of the term met ef, i.e., "met" means low tide, and "ef" means drought. The met ef term is intended for the largest tidal phenomenon occurring in October, November, and December. The local communities of the kei islands have long known that the tidal phenomenon always comes along with hot and dry climatic conditions and calm sea water surface due to low wind speed. These climatic conditions have become the indicator of the met ef period for the local communities. The average data of air temperature, rainfall, rainy day frequency and wind speed per month during 2013 to 2016 are presented in Figures 5, 6, 7 and 8.

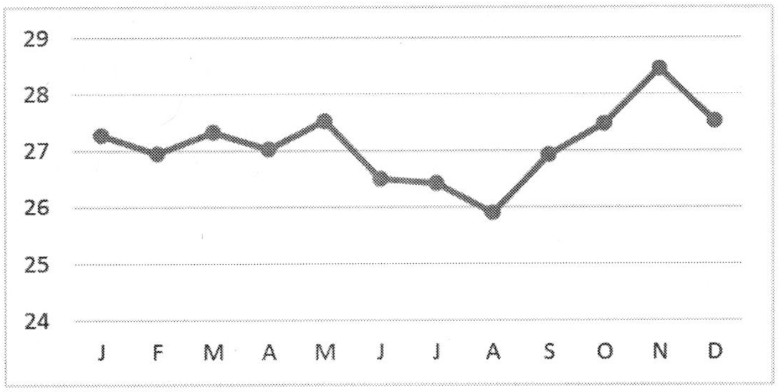

Figure 5. The average monthly (°C) air temperature from 2013 to 2016 in the kei islands

Figure 5 presents the average monthly air temperature conditions in the kei islands from 2013 to 2016. Namely, the lowest average air temperature occurred in August, (26 ° C) whereas the highest temperature occurred in November 28.5°C.

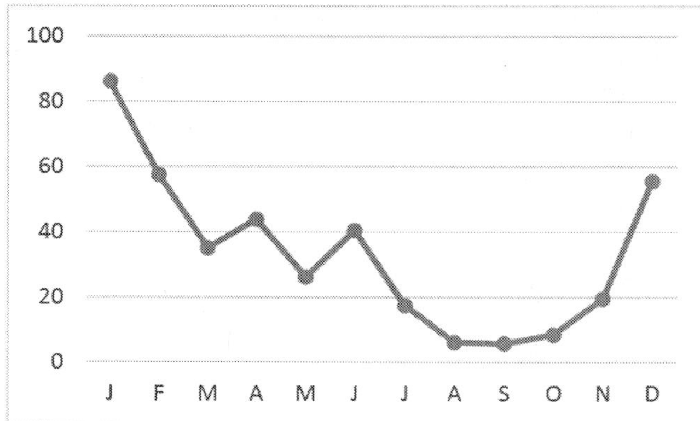

Figure 6. Rainfall (mm) monthly average from 2013 to 2016 in the kei islands

Figure 6 presents the relatively lower rainfall (mm) based on the average monthly rainfall from 2013 until 2016 occurred in August, September and October while the highest average rainfall occurred in January.

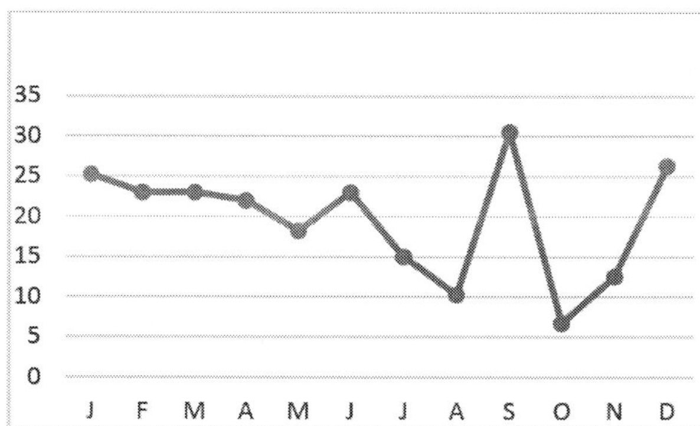

Figure 7. Frequency of average monthly rainy days from 2013 to 2016 in the kei islands

Figure 7 presents the frequency of the average monthly rainy days from 2013 to 2016 in the kei islands, i.e., the lowest rain frequency (day) occurs in October (less than 10 days) while the highest in September (30 days).

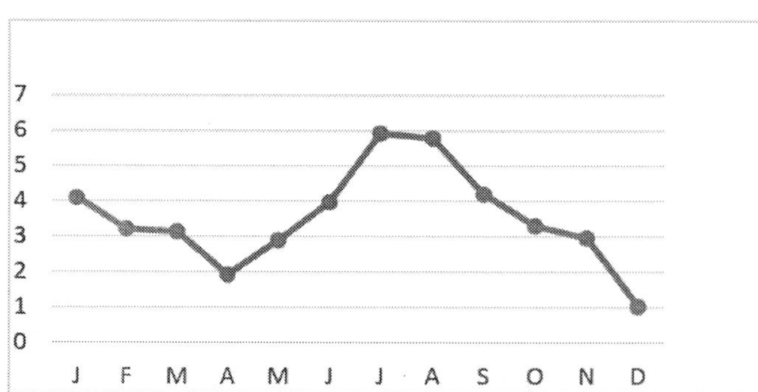

Figure 8. Monthly average wind speed (knots) from 2013 to 2016 in the kei islands

Figure 8 presents the average monthly wind speeds from 2013 to 2016 in the kei islands, i.e., the highest average speed occurred in July and the wind speed decreased gradually until reaching the lowest speed in December.

3.3. Exploitation of Benthos at the inshore shoals

The five inshore shoals are shared by several villages around shoals even though there are villages claiming to be their own. Intensive use happens only in the Met Ef period, especially October. The benthos collected by the people in the five shoals are generally Hanoat or Giant clam (*Tridacna* sp.), "Ngeng" or Spider conch (*Lambis* sp.), "Arut" or Hammer Oysters ("*Malleus* sp.") "Eb or Sea cucumber (*Stichopus* sp, and *Holothuria scabra*), and Krit or Octopus (*Octopus* spp,). The people who exploit the shoals consist of people from several, i.e., Ohoitel, Vatran, Taar, Faan, Wearlilir, Sathean, Ibra Iso, Disuk, Revav. But Sathean and Ibra are the villages that the most intensively exploit the benthos in the shoals compared to other villages. According to the fisherman, although the number of people who exploit the benthos in the shoals has decreased lately the yields are getting smaller and size are also getting smaller. Based on interviews that the decreasing of people who currently exploits benthos in the shoals compared to previous time due to local people turn to seaweed culture business which is more economical value. The peak of benthos exploitation during the Met Ef period, occurs at least 1 to 6 times per month. It starts from the full moon to 6 days after the full moon and 2 to 6 days after the new moon. Benthos commonly collected from the shoals are presented in Table 2. The peak of benthos exploitation during the Met Ef period, occurs at least 1 to 6 times per month. It starts from the full moon to 6 days after the full moon and 2 to 6 days after the new moon.

3.4. Discussions

Met Ef phenomenon characterized by low tide with a relatively large tidal range classified as a tidal meso (2.5 m and 4 m). In addition, due to the gently sloping coastal bed topography of the kei kecil islands so that during low tide in October, November, and December where the tidal range reaches 2.60 to 2.80 m, the drought tidal plains become very wide and even the coral banks or shoals that normally lies below the surface of the water popping up above sea level. This has led to the accessibility of local communities to exploit benthos that inhabits shoals more intensively than in other months. Between October, November, and December, benthos exploitation with the highest intensity was conducted in October, this was due to at that time, the rainfall and the number of rainy days were very low, as well as the low wind speed causes the sea to be clear and calm. Similar climatic conditions also occurred in November, with the highest tidal range but since intensive exploitation has occurred in October so that in November and December, the number of people who exploit the shoals became diminished due to the reduced of benthos in those shoals.

Table 2. Benthos which collected in shoals during period of Met Ef and its user villages

No.	Name of Shoals	Local common and Scientific Names of benthos	User Village of shoals
1	North 'Samar'	Arut (*Malleus* sp.), Ngeng (*Lambis* sp.), Hanoat (Tridacna sp.) Eb (*Stichopus* sp., *Holothuria* sp.,	Ohoitel, Vatran, Taar, Faan, Wearlilir, Sathean, Ibra
2	South 'Samar'	Hanoat (*Tridacna* sp.) Krit (*Octopus* spp.), Eb (*Stichopus* sp., *Holothuria* sp.),	Ohoitel, Vatran, Taar, Faan, Wearlilir, Sathean, Ibra
3	Wid	Hanoat (*Tridacna* sp.), Arut (*Malleus* sp.), Ngeng(*Lambis* sp.), Krit (*Octopus* spp.) Eb (*Stichopus* sp., *Holothuria* sp.,)	Sathean, Iso, Disuk, Ibra, Wain, Revav, Rumat
4	Wama	Hanoat (*Tridacna* sp.), Eb (*Stichopus* spp., *Holothuria* sp.)	Taar, Sathean, Faan, Wearlilir, Ohoitel, Vatraan

The most collected Benthos are *Tridacna* sp., *Malleus* sp., *Lambis* sp, and *Octopus* spp. The species of benthos are abundant in shoals because they are generally not exploited in other months except at Met Ef. Especially numbers of *Tridacna* sp., have been greatly reduced in the tidal plains because it has been intensively exploited for a long time to be marketed as a source of protein and other things for local communities as elsewhere in the Indo Pacific, including the kei islands. Giant clams are commercially important for the food market [16, 22), shell craft industry [10] and in the live marine aquarium trade [13]. Overexploitation of this valuable and multi-facetted resource has led to the decline of natural stocks throughout its natural Indo-Pacific range and the ecological extinction [20] in some areas of the larger species (*Tridacna gigas* Linne, 1758; *Tridacna derasa* [4, 15, 11]. The exploitation of the Giant clam is alarming because its availability in nature has been reduced, whereas the eight species of Giant clams have been defined as endangered species, and are included in the Convention of International Trade of Endangered Species, and are included in the Convention of International Trade of Endangered Species (CITES) list (25, 8]. In addition, Giant clams are also of distinct ecological significance as they contribute to reef development and are an important prey item and provide shelter for many other marine organisms [18]. The small giant clam T. maxima (max known size < 40 cm, but size modes usually between 5 and 15 cm; [6] can be found on sandy bottoms, but is typically embedded in hard substrate. With large clams being preferentially targeted by fishers given the better catch per unit effort, this species remained relatively untargeted by fishers until recently, in countries where larger species co-occur (the giant clam *T. gigas*, the fluted giant clam, the southern giant clam and the bear paw clam; [14]. The small giant clam (T. *maxima*) is currently classified as lower risk/conservation dependent (i.e. species of 'least concern') under the IUCN red list of threatened species (2004). However, with most larger species having been over-exploited, some studies report the unsustainable exploitation of this smaller species (26, 12] and recommend that the species' IUCN status be reconsidered [17].

Another benthos species also targeted for the exploitation by people in the five shoals during Met Ef is the spider conch. The common spider conch *Lambis lambis* is distributed in shallow waters of the Indo-Pacific. The conches are sought by shell collectors of Philippines, Solomon Islands, Indonesia and India [2]. It is harvested and consumed as food in Japan. The conches are found to occur mainly on sandy patches among rocks or coral reefs, from the intertidal region to depths up to 20 m [9]. Of the eleven known species of *Lambis* sp., from Indian waters [1], six are listed in Schedule IV of the Indian Wildlife Protection Act, 1972, banning commercial exploitation. Studies on breeding of *Lambis lambis* are limited, the only known published work being that of [7]. In India, studies on *Lambis* spp. are limited to its habitat and feeding [23]. The present study is an attempt to collect, maintain and breed the common spider conch *Lambis lambis* under captivity in order to develop a methodology for brood stock maintenance and to study their feeding behavior, mating, spawning and larval development. The common spider conch *Lambis lambis* is distributed in shallow waters of the Indo-Pacific. The conches are sought by shell collectors of Philippines, Solomon Islands, Indonesia and India [2]. It is harvested and consumed as food in Japan. The conches are found to occur mainly on sandy patches among rocks or coral reefs, from the intertidal region to depths up to 20 m. Of the eleven-known species of *Lambis* from Indian waters [1].

Black hammer oyster is found in seagrass meadows with the 'T' shape buried. They range in depths of 1 to 15 meters. It is also found to be growing in close colonies. Maximum shell height is about 25 cm. The outer shell is often encrusted with calcareous algae and other organisms. Black hammer oysters are locally collected in Indonesia and the Philippines for production of lime and shell craft. It has a widespread distribution – Indo-West Pacific from East Africa, including the Persian Gulf, to Melanesia; north to Japan and south to Queensland [5].

Octopuses are commonly found in the sea and subtropics around the Mediterranean region, the far eastern regions and the South Pacific. In Indonesia, allegedly present in the waters of Kalimantan, Sulawesi, Maluku and Banda. Octopus can live in shallow water and also present at the tidal boundary until somewhat deep with depth of 4000 meters to 5000 meters. Mostly swim and move together in large herds. Actually, the octopus is benthic or sticking [3], and usually forms a sanctuary within the crevices of rocks, rocks, seaweed in coastal waters. The most preferred residence is hollow rocks. Octopus is active at night or called a nocturnal animal [24, 27].

4. Conclusions

Met Ef occurs in October, November, and December. The lowest tide average during 2013 to 2016 occurs in November with an average tidal range of 2.66 m but the highest exploitation intensity tends to be in October so that in November and December the number of benthos in the shoals has decreased and the exploitation intensity also has decreased. Giant clam (*Tridacna* sp.), Spider conch (*Lambis* sp.), Hammer oyster (*Malleus* sp.), and Octopus (*Octopus* spp.) are benthos targeted for exploitation in the shoals. *Tridacna* sp. decreased in number and size compared to previous time

Acknowledgements

We are grateful to a large number of individuals who helped by answering and confirming our questions in the Sathean and Villages. We are also grateful to Dr. Serge Andréfouët from Institut de Recherche pour le Développement, France, for providing us some references. Finally, we acknowledge anonymous reviewers and the editors for their constructive comments and corrections.

References

[1] Apte, D 1998 The book of Indian shells. Oxford University Press, 115 pp.

[2] Carpenter, K. E., Niem, V. H 1998 FAO species identification guide for fishery purposes. The living marine resources of the Western Central Pacific. Volume 1. Seaweeds, corals, bivalves and gastropods. *FAO*, Rome, 686 pp.

[3] Barnes, R.D 1967 Invertebrate Zoology W.B. Saunders. Co, London : 632 pp.

[4] Brown, J. and M. Muskanofola 1985 An investigation of stocks of giant clams (Family Tridacnidae) in Java and of their utilization and potential. *Aqaucult. Fish. Manage.* 16(1):25-39.

[5] Deepak, S 2013 Black Hammer Oyster – *Malleus malleus* (Linnaeus, 1758), Marine Life India, October <https://marinelifeindia.wordpress.com/2013/10/26/black-hammer-oyster-malleus-malleus-linnaeus-1758/>

[6] Gilbert, A., Remoissenet, G., Yan, L. and Andrefouet, S 2006 Special traits and promises of the giant clam (*Tridacna maxima*) in French Polynesia. Fisheries Newsletter – South Pacific Commission 118, 44–52.

[7] Hamel, J. F. and Mercier, A 2006 Note on the spawning and development of the common spider conch *Lambis lambis*. *SPC Trochus Information Bulletin*, 12: 19-21.

[8] Hilton-Taylor, C 2000 IUCN Red List of Threatened Species. Gland, Switzerland and Cambridge, UK: IUCN.

[9] Jagadis, I., Shanmugasundaram, k., and J. Padmanathan 2012 Observations on broodstock maintenance, breeding and early larval development of the common spider conch *Lambis lambis* (Linnaeus, 1758) in captivity. *Indian J. Fish*., 59(2): 165-169.

[10] Juinio, A., R. Menez and C. Villanoy 1987 Use of giant clam resources in the Philippines. Naga 10:7-8.

[11] Juinio, A, R. Menez and C. Villanoy, and E. Gomez 1989 Status of giant clam resources of the Philippines. *J. Moll. Stud.* 55:431-440.

[12] Kava, V., Udagawa, K. and Ikenoue 1995 H. Fishing pressure on the resource of the elongated giant clam, Tridacna maxima, in Tonga. *Fisheries Research Bulletin of Tonga* 3, 11–17.

[13] Knop, D 2004 Giant Clam Buyer's Guide. Coral - The Marine Aquarium Magazine 1(3):36. Ecosystem Publications, California 92705, USA.

[14] Lewis, A.D., Adams, T.J.H. and Ledua, E 1988 Fiji's giant clam stocks – A review of their distribution, abundance, exploitation and management. Giant Clams in Asia and the Pacific. *ACIRA Monographs* 9, 66–72.

[15] Lucas, J.S 1988 Giant clams: description, distribution and life history. In: Copland, J. & Lucas, J.S. (Eds.), Giant clams in Asia and the Pacific. *ACIAR*, Canberra, pp. 21–32.

[16] Murakoshi, M 1986 Farming of the boring clam, *Tridacna crocea* Lamarck. Galaxea 5: 239-254.

[17] Neo, M.L. and Todd, P.A 2013 Conservation status reassessment of giant clams (Mollusca: Bivalvia: Tridacnidae) in Singapore. Nature in Singapore 6, 125–133.

[18] Neo, M.L., Eckman, W., Vicentuan, K., Teo, S.L.-M. and Todd, P.A 2015 The ecological significance of giant clams in coral reef ecosystems. *Biological Conservation* 181, 111–123.
[19] Neuendorf, K.K.E., J.P. Mehl, Jr., and J.A. Jackson, eds 2005 Glossary of Geology (5th ed.).
[20] Pandolfi, J., R. Bradbury, E. Sala, T. Hughes, K. Bjorndal, R. Cooke, D. McArdle, L. Mc Clenachan, M. Newman, G. Paredes, R. Warner and J. Jackson 2003 Global trajectories of the long-term decline of coral reef ecosystems. *Science Wash.* 301(5635):955-958.
[21] Rutecki, D, Nestler E, Dellapenna T, and Pembroke A 2014 Understanding habitat value and function of shoal/Ridge/Trough Complexes to fish and fisheries on the Atlantic and Gulf of Mexico Outter continental shelf. Draft literature synthesis for the U.S. Dept. of the interior, Bureau of ocean energy management. Contract # M12PS00031. Bureau of Ocean Energy Management. US. Department of the Interior. 116pp.
[22] Shang, Y., L. Ping Sun and J. Brown 2014 Test marketing of giant clams for food in Honolulu, Guam and Saipan and observations in Okinawa and Taiwan. Pages 69-87 in Economics of commercial giant clam marine culture.
[23] Siraimeetan, P., Ameer Hamsa, K. M. S. and Satyanarayanarao, K 1988 On the habitat, habits and food of *Lambis lambis* and *Hemifusus cochlidium*. Proceedings of the National Seminar on shellfish resources and farming. *Bull. Cent. Mar. Fish. Res. Inst.*, 42 (1): 111-116.
[24] Welss, M.J 1962 Brain and Behavior in Cephalopoda, Stanford University Press, Stanford, California: 198 pp.
[25] Wells, S., R. Pyle and N. Collins 1983 Giant clams. Pages 97-107 in The IUCN Invertebrate Red Data Book, Gland, Switzerland.
[26] Zann, L.P. and Ayling, A.M 1988 Status of giant clams in Vanuatu. Giant Clams in Asia and the Pacific: *ACIAR Monographs* 9, 60–64.
[27] Roper, C.F.E., M.J. Sweeney and C.E. Nauer 1984 An Annotated and Ilustrated Catalogue of Species of Interest to Fisheries. In P.R. Boyle (Ed.). Cephalopods of The World. *FAO Fisheries Synopsis* No. 125 (3): 191-257.

The Quality and Food Safety of Dry Smoke Garfish (*Hemirhamphus far*) Product From Maluku

Alfonsina Marthina Tapotubun[1], Fredrik Reiuwpassa[1], Yolanda M.T.N. Apituley[2], Hellen Nanlohy[2], Theodora E.A.A.Matrutty[1]

[1]Fisheries Technology Study Program, Faculty of Fisheries and Marine Science, Pattimura University
[2]Fisheries Agrobusiness Study Program, Faculty of Fisheries and Marine Science, Pattimura University

Email: am.tapotubun@gmail.com

Abstract. Dry garfish is product of smoked process of "ikan julung" (*Hemirhamphus far*) and slowly the product getting dry, stiff and its colour become gold yellow-brown. The aim of this study is to find out quality and food safety of dry smoked "julung" from Maluku. The sample of this study is taken from production Keffing village, East Seram Regency, Maluku. Parameters to be analyzed are degrees of protein, fat, water, ash, TPC, *Escherichia coli, Salmonella, Vibrio* and *total Staphylococcus aureus* used standard analysis method for proximate (AOAC. 2005), sensosy parameters (BSN.2009) and food safety (BSN. 2006). Spreadsheet Ms Excel (Microsoft Inc., USA) is used for data processing; data is being analyzed descriptively to be interpreted in the research report. Dry smoked "julung" Keffing village, Maluku meet the good quality and food safety, that are ingredient degrees of water content 12.43%, protein 61.55%, fat 12.58%, ash 9.3%, TPC [6,8]x10^1 CFU, total *Staphylococcus* sp [1,7] x 10^2, total *E.coli* 6.4 APM/g. and negatively for *Salmonella* and *Vibrio*.

1. Introduction

Julung fish (*Hemirhamphus far*) is a small pelagic live in groups around the coral coast to offshore area, mainly while spawning. Julung can be found in East Indonesia Sea wiith high degrees of salt and tend to be oceanic [20]. The product of julung (*Hemirhamphus far*) in Maluku increase annually. In 2013 the production was 7.305 tonnes and in 2015 it was raised to 11.619.6 tonnes, with the highest production from the East Seram Regency, which was 8.739,8 tonnes [5].

Fisherman community in each area has capability to use their sea product by process the product to become various processing product based on their knowledge and local wisdom. Maluku as one of huge fishing area, has significant property of traditional fish processing products, one of them is Julung product. Traditional fish processing product is fish product that is processed, based on habitual heredity, using simple tools with simple performance but has a unique taste, usually unpacked and has not well preserved.

Smoke is a traditional processing method, combined with salting, drying and pouring natural chemical compound through the burning output which results in heat and composes smoke compound in the form of steam. This process, result in adhering smoke compound to the fish and fused in water layer of the surface body of fish. As a result the unique taste and aroma are formed, as well as the goldy brown. The basically smoked process of fish is a unification of salting, drying and smoked processing fish. The aim of these processes is to abolish bacteria and to support the adhering smoke particles of the natural burning [1].

The smoke process not only being a processing method to raise well-preserved of the fish product but also to get the taste, aroma, and color that fit consumers desire [17, 20]. Smoke can increase the long lasting of fish, as a combine of dehydration influence, activity of anti-microbe and antioxidant in several smoke constituent, especially: formaldehyde, carboxyl acid, and fennel. Fenol and acid are compound of anti-microbe.

Dry Julung is an output of smoke process of Julung fish (*Hemirhamphus far*), by doing it slowly until it dry, hard and goldy yellow. Dry julung fish is a traditional processing of fisherman community in East Seram Regency [21, 25]. In Manado, the smoke julung fish is known as "ikan roa" or "julung-julung asap" named *Galavea* , which is one of the traditional fish processing product consumed by people for generation [19].

The sale of this product is held in traditional market, unpacked, full body condition using waya (20 pieces each) or being tied less (5 pieces each). During distribution and temporary storage, the product is only packed in a sack or cardboard box, so it will not be protected from environment influence. The product that being storage in opened condition will be potentially contaminated result in quality downgrade or be an unhealthy product. So far,the study of food quality and safety of traditional dry julung is less and dominated by fish from Manado [9,19, 25]. The aim of this study is to know the quality and safety of dry smokes julung, as one of traditional food processing in Maluku.

2. Methodology

2.1. Material and Tool

The main material used in this study is dry smoke julung fish, produced in Keffing Village, East Seram Regency. The analysis materials are filter papers (Whatman 42), HCL (Merck), NaOH (Merck), H2SO4 (Merck), petroleum benzene (Merck), fenol (Merck), ethanol (Merck), active carbon (Merck), litmus papers (Mercherey-negel). The tools are scales (Apel-PD-3000UV), oven (Memmert), ash oven (Heraeus), hot plate (Cimarec), desiccator, reflux tools, distillation dan destruction (Pyrex).

2.2. Research Methods

This study used descriptive research method, in two (2) steps. The first step, sample collection, and direct observation at Keffing Village. The second step is quality analysis and safety of food.

2.3. Analytic Parameters

Parameters for quality analysis, using standard methods are water content (oven method), protein content (Kejdhal method), fat content (Soxhlet method), dust content (dusting method) [3]. Valuation of organoleptic quality is organize using 30 half trained panelist using smoke fish sensory score sheet SNI [6]. Safety food analysis, that are Total Plate Count, *Escherichia coli,* and *Salmonella, Vibrio* sp and *Staphylococcus aureus* using SNI method [7].

2.4. Data Analysis

Data is processed using spreadsheet *Microsoft office Excel* (*Microsoft Inc*., USA), and being analyzed. Descriptively to evaluate proximate, sensory parameter and food safety.

IOP Conf. Series: Earth and Environmental Science **89** (2017) 012010 doi:10.1088/1755-1315/89/1/012010

3. Result and Discussion

3.1. Production process of dry smoke julung fish

Dry garfish or dry julung, is a traditional processing of julung, using cold smoke process to the julung fish (*Hemirhamphus far*), which is done with the combination of hot and cold smoke process. Hot smokes to ripe the fish meat, and the cold one to dry the smoke fish. Smoke proceed and drying the fish, hard and change the colour into goldy brown. Production area of this product in Maluku is in Keffing, East Seram Regency. The process of dry smoke julung fish is done hereditary, which is clamped using pieces of bamboo, and then being smoked in a rack called *para-para*, and being combination of hot smoked and cold smoked.

Julung fish has a long slim body. Therefore a single smoke process is difficult to be held, there is why bamboo clamp is used to ease the process. Generally the traditional smoke process in East Seram Regency is as follows: Julung fish is washed using clean sea water and put it in *waya* made of bamboo and being clamped (20 pieces of julung in each *waya*) then being leaked for a view minutes before smoking process. Fish clamped in *waya* is then put in rack and the smoke process is started in lowest rack near the heat resource. After it ripe, *waya* is moved to the top level rack of the fireplace and covered with gunny sack and smoked using small fire until it dry. This cold smoked process is held until it was delivered to the buyer or to distribute to town in the Province of Maluku and another city. Space between first rack and fire resource in the smoke fireplace is 1 meter, while in the smoke fireplace there is 2 rack, with 1 meter space between them. The upper rack is used for cold smoke process and temporary storage until the products are distributed.

Figure 1. Smoked Dry Julung Fish of Keffing, East Seram Regency, Maluku

The dry smoked julung produce traditionally in Maluku, generally sold in its original form and without proper packing, while distribution and temporary storage are done in a sack or reuse carboard box. [12] Before distributed and delivered to the market, the product was in the storage which is potentially contaminated with external factors that cause quality decreasing, damage during storage (humidity, storage, packing, etc).

Chemical Quality Composition

In general, product of smoke julung, processed traditionally in Keffing, East Ceram, Maluku, could be said well in quality. Analysis results show that quality composition of dry smoke ruling from the two village does not show any difference (Table 1). The reason is that the methods of handling and smoke process on those two villages are similar. This result is better compared to the same dry smoke julung at the different times and places [9, 19, 25].

The quality of smoke fish is influenced by several factors: smoke process, kind of fuel for the fireplace, smoke composition, temperature, humidity, velocity and density of smoke [24]. [2] mentioned that processing method, fresh fish quality, and storage, also influence the proximate of the smoke fish.

Table 1. Quality Profile of Dry Smoke Julung Fish Product

Quality Parameter	Content
Water (%)	12.43
Protein (%)	61.55
Fat (%)	12.58
Ash (%)	9,30
pH	6,05

This result shows that water content of traditional smoke julung is low so that it can be categorized as dry product. This result is similar to the same product sales in traditional market in Manado that is 12.1 % [19], but lower than dry smoke tandipang fish from Sulawesi 15.2 % [14], and three kinds of traditional smoke fish from Sokoto Nigeria i.e. 17.15 -19.13 % [17]. Low content of water, is a result of high-rise smoke process of the product: firstly hot smoke (temperature 80 – 100°C) for less than one hour, then continued with cold smoke treatment (temperature 29 – 35°C) for about 24 hours. This cold smoke was held at the top level the rack far from the fireplace, to dry the smoke julung, until it delivered to the buyer. The temperature of the cold smoke process is about 30 - 40°C [10] and hot smoke temperature is more than 100°C and until temperature of the fish flesh becomes 60°C [27].

One of the factors related to the long lasting of the processing product is water content. The water content in the smoke fish will affect the storage period of this product [19]. The lowest water content, the slowest growing of the microbe, reversely the higher water content will accelerate the grow of the microbe because water is a good media for the microbe to grow so that the spoiled process will be faster [25].

Protein content related to water content. The lower water content, protein becomes concentrated. The protein content of the dry smoke julung in this study is 61.56%. This value is higher than the previous study by Sormin *et al.* [25], protein content of dry smoke julung is 56.1%; and protein content of smoke fish of three kinds of fish (*Clarias gariepinus, Mormyrus rume, Heterobranchus longifilis*) is 53.54 – 58,05 [17]. The protein content of fish after smoke process is higher than the fresh fish [4, 11, 17].

Ash content of three kinds of smoked being observed by [17] is between 7.18 – 7.41%. The increase of dust content after the smoked process is a result of the lost of humidity during the process [16]. The result of this study shows that the fat content of the smoke julung is quite high. Fat existence in the food gives the delicious taste because it influent the taste. Fish contains small amount of fat, compare to protein, but it contributes to the taste and aroma of the smoke fish [26].

3.2. Sensory Assessment

Smoke fish is a traditional processing product which is favor to the consumer for its unique taste and aroma, as a result of smoke processing treatment. The study shows that panelist gives the highest assessment to all of sensoric parameter, which is appearance, smell, and texture of the dry smoke julung. This high assessment is an appreciation to the perform of dry smoke julung of the Maluku. Sensoric specification is valued based on smoked fish score sheet.

Consumer likes of the smoke fish are related to the appetite in certain region, mainly in coastal area. This influence assessment decision toward the product. The result shows that the dry smoke julung from Maluku, get the high sensory assessment, that is 9 to 7, based on panelist assessment, and can be seen in the following table:

Table 2. Average assessment panelist toward sensory parameter of dry smoke julung

Sensoric parameters	Average
Appearance	8.4
Odour	8.6
Taste	8.4
Texture	7.6

Most of the panelist give the rate 9 and 8 to the appearance, odor and taste of the sample, and to the texture parameter,the greatest assessment are 7, because the score sheet used is one that usually used to the hot smoked fish to produce soft smoked fish. [8] revealed that fenolic compound combined with different smoked method will give direct effect to the sensory of smoked fish. Panelist assessment percentage to the sensory parameter of smoked julung fish is described in the following Table 3:

Table 3. Percentage to the Sencoric Parameter of Smoked Julung Fish

Scale	Appearance		Odour		Taste		Texture	
	Sum	%	Sum	%	Sum	%	Sum	%
9	15	46.7	17	56.7	12	40	6	56.7
8	13	43.3	13	43.3	16	53.3	7	23.3
7	3	10	0	0	2	6.7	17	20

Dry smoked julung fish is bright gold yellow-brown in color, whole flesh and compact production. Appearance specification of smoked julung fish according to the score sheet is, whole, clean, orderly (neat), glitter based on genre. Waya that used to put julung fish in stack is effective to assure the wholeness and orderly of fish during the smoked process. Volatile compound from the smoke like carbonite and phenol in the smoked fish will react to the protein component and form the specific color, taste and aroma [13].

The color of the smoked julung is goldy yellow-brown. [19] stated that appearance of the product usually related to the color and the wholeness of the product. Smoked component adhere to the fish during smoked process result in fish with the color mentioned above and shine. After the smoked process the color becomes goldy yellow-brown and shine [26]. The volatile compound from the smoke, like carbonyl and fennel at the smoked fish will react with protein component of the fish and generate specific color, taste and aroma [13]. Appearance is one of the determining factors for the consumer to make their decision toward smoked fish.

The dry smoked julung fish has aromatic smoke odor. Specification description of its odor, result of panelist assessment based on the score sheet of aromatic to the less aromatic, specifically without additional odor. A smoked process can eliminate the putrid odor of the fresh fish, and change it to the aromatic and smoked smelly. The smoked process results in specific smoked aroma of the fish [26]. Odour quality criteria of the smoked fish are, soft smoked smelly until strong smelly, no rancid, no spoiled, smell, no strange smell, no sour smell, and no sweaty smell [27]. Substances that dominate the odor existence are smoked components that are adhered to the product, like fennel

Smoked fish have a smoked delicious taste while being eaten. Dry smoked julung fish taste description in this study are: tasteful to the delicious. Kind specific with no additional taste. The smoked taste is in favor of consumers, especially those who are familiar with smoked production. The dry smoked julung which are produced traditionally using cold smoked with long period, enable absorption of smoked substances in the limb of the fish and result in unique smoked taste in favor of the consumer. The smoked process, not only is a method of long-lasting but also to improve the taste to meet the consumer's favor [23]. Fenol and Carbonyl content in smoked fish contribute to its taste [18, 26]. Its taste is influenced by the time of the smoke process and the fireplace.

Traditional dry smoke Julung fish from Maluku, have a hard texture and crispy, because of less water content during the cold smoked process. The texture assessment description of the dry smoke julung is dense, compact, dry enough to too hard, brittle, dry. The smoke fish texture is related to the water content. Cold smoked process treatment more than 24 hours will reduce water content significantly, and the fish flesh becomes harder. The lower water content, the harder texture of the smoked fish. The dry and hard texture of the fish according to water content of the product is 12.43%. [19] stated that the longer time of smoked process, the lower water content and product become dry and hard.

3.3. Food Safety

The results show that *Escherichia coli* bacteria assessment is beyond the top maximum margin of Indonesia National Standard (SNI). This condition shows the existency of *E. coli* contamination at the dry smoked julung, that possibly come from processor, intersecting contamination, fly or insect that infects the product. Contamination may occur easily because stack of fish is put in the rack on the fireplace and only the top part is covered by gunny sack.

Process person, tools, water for cleaning, and insect are source that causes contamination, so product should be protected with proper packing. In the food quality and safety requisite of smoked fish, Indonesia National Standard (SNI) establish SNI for maximum *E.coli* content should be less than 3 APM/g [6]. For this condition, the process of dry smoked julung fish before it consumed should be done properly with high temperature to reduce bacteria *E. coli* so that product can be consumed safely. State that in the temperature ≥ 70 °C is good enough to kill *E.coli* bacteria.

The maximum standard of TPC for food safety of smoked fish is $1{,}0 \times 10^5$ colony/g, *E.coli* less than 3 APM/g, *Salmonella* and *Vibrio cholerae* negative, *Staphylococcus aureus* maximum $1{,}0 \times 10^5$ [6]. Amount of microbe in the food influences its velocity of damage. The simple smoked fireplace and opened will make bacteria easier to contaminated smoked fish.

Table 4. Safety Profile of Dry Smoked Julung Fish from Maluku

Food safety parameter	Cantent
TPC (CFU/g)	$[6,7] \times 10^1$
E coli (APM/g)	6.4
Salmonella (per 25 g)	Negative
Vibrio (per 25 g)	Negative
Staphylococcus sp (colony/g)	$[6.1] \times 10^1$

Food safety becomes the need of modern community to decide kind of food would be consumed, to assure the health of foods. Bacteria contamination in product will accelerate the damage of the product, result in it become unhealthy. In general the result of this study shows that safety of dry smoked julung fish from Maluku is good enough, except at Escherichia *coli* parameter.

4. Conclusion

Dry smoked julung fish of Maluku has the good quality and safety, that are: water content 12.43%, protein 61.55%, fat 12.58%, ash 9.3%, TPC $[6,8] \times 10^1$ CFU/g, total *Staphylococcus* sp $[1,7] \times 10^2$ colony/gram, *Salmonella* bacteria, dan *Vibrio* negative but *Escherichia coli* positive.

Acknowledgements

Thanks to Government of Republic of Indonesia through the Directorate General of Higher Education on MP3EI funds for implementation of this research.

IOP Conf. Series: Earth and Environmental Science **89** (2017) 012010 doi:10.1088/1755-1315/89/1/012010

References

[1] Afrianto E and Liviawaty E 2010 *Pengawetan dan pengolahan ikan* (Yogyakarta - Kanisius press)

[2] Akinola OA, Akinyemi AA, Bolaji, BO. 2006. *Evaluation of Traditional and solar drying systems towards enhancing fish storage and preservation in Nigeria (Abeokuta Local Government as a case study)* Journal of Fisheries International. 1(2-4):44-49.

[3] AOAC 2005 *Official methods of analysis* 18th ed (Washington DC-Assosiation of analitical chemists)

[4] Bilgin S Ulusayin M Izci L and Gunlu A 2008 *The detemination of the shelf life and some nutritional components of gilthhead seabream* (Sparus aurata L 1758) *after cold and hot smoking* (*Turkish Journal of Veterinry and Animal Science*) 32: 49-56

[5] BPS 2016 *Maluku province in figures.* (Ambon - BPS Statistics of Maluku Province Press)

[6] BSN 2009 *Ikan asap - Bagian 1: Spesifikasi.* Standar Nasional Indonesia, SNI 2725.1:2009. (Jakarta - Badan Standardisasi Nasional Press)

[7] BSN 2006 *Cara uji mikrobiologi pada produk perikanan.* Standar Nasional Indonesia (SNI) 2332:2006 (Jakarta - Badan Standardisasi Nasional Press)

[8] Cardinal M Cornet J serot T Baron R 2006 *Effect of the smoking process on odour characteristics of smoked Herring* (*Clupea herengus*) *and relationship with phenolic compound content* (*Food Chemistry*) 96:137-146

[9] Damongilala LJ. 2009. *Kadar air dan total bakteri pada ikan roa (hemirhampus sp) asap dengan metode pencucian bahan baku berbeda. Jurnal Ilmiah Sains.* 9(2):190-198

[10] Eyo, A.A. 2011. *Fish processing technology in the tropics. National Institute for F. National Institute for Freshwater fisheries* Research, New Bussa, Nigeria, ISBN-B : 9781770457, 403 pp

[11] Goulas AE, Kontominas MG 2005 *Effect of salting and smoking method on the keeping quality of Clus Mackerel* (*Scumber Japonicus*). Biochemical and Sensory Attributes. Food Chem 95:511-520

[12] Heruwati, E.S., 2002. *Pengolahan Ikan Secara Tradisional: Prospek dan Peluang Pengembangan.* Jurnal Litbang Pertanian 21 (3).

[13] Izamu KT Purnomo H Yuwono SS 2012 *Physical, chemical, and organoleptic characteristic of smoked skipjack tuna* (*Katsuwonus pelamis*) *in Kendari city* (Jurnal Teknologi Pertanian) 13(2) p 105-110

[14] Kaparang R, Harikedua SD, Suwetja IK. 2013. *Penentuan mutu ikan Tandipang (Dussumeiria acuta C.V) asap kering selama penyimpanan suhu kamar. Jurnal Media Teknologi Hasil Perikanan.* 1(1): 1 – 6.

[15] Kawimbang E, Paransa IJ, Kayadoe ME. 2012. *An estimation of garfish stock and its fishing season Tagulandang waters, Siau Tagulandang Biaro Regency.* Jurnal Ilmu dan Teknologi Perikanan Tangkap. 1(1):10-17

[16] Kumolu-Johnson CA, Aladetohun NF, Ndimele PE 2010. *The effect of smoking nutitional qualities and shelf-life of Clarias gariepinus* (Burchell 1822). African Journal of Biotechnology. 9(1):073-076.

[17] Magawata, and Musa T. 2015. *Quality characteristics of three Hot-Smoked fish species using locally fabricated smoking kiln.* International Journal of Fisheries and Aquatic Studies. 2(5):88-92

[18] Martinez O, Salmeron J, Guillen MD, Casas S 2007. *Sensorial and physicochemical characteristics of salmon (Salmo salar) treated by different smoking process during storage.* Food Science and Technology International. 13(6):477-484.

[19] Patty NCh, Dotulong V, Suwetja IK. 2015. *Mutu Ikan Roa (Hemirhamphus sp) Asap yang Ada di Pasar Tradisional di Kota Manado yang Disimpan pada Suhu Ruang.* Jurnal Media Teknologi Hasil Perikanan. 3(2):45-54.

IOP Conf. Series: Earth and Environmental Science **89** (2017) 012010 doi:10.1088/1755-1315/89/1/012010

[20] Reppie E, Luasunaung A. 2011. *The status of roundscad net (talang) in Pahepa Island, Sangihe Talaud, North Sulawesi* (181-186). *In*: Carman O, Sulistiono, Purbayanto A, Suzuki T, Watanabe S, Arimoto T (ed). Proceedings of the 4th JSPS International Seminar on Fisheries in Tropical Area. Sustainable fisheries in Asia in the new millenium. Faculty of Fisheries and Marine Science, Bogor Agricultural Institute. TUF International JSPS Project vol.10

[21] Rieuwpassa F, Tapotubun AM, Matrutty ThEAA, and Sormin RBD 2015 *Opportunities of seafood processing development in Southeast Maluku District.* International Journal of Current Research 7(8): 19336-19339

[22] Rieuwpassa F, Apituley Y, Tapotubun AM, and Nanlohy H 2016. *Peningkatan daya saing produk ikan olahan tradisional melalui perbaikan mutu, keamanan pangan dan manajemen usaha.* Laporan Penelitian MP3EI. Universitas Pattimura

[23] Salan OE, Julinana AG, Marillia O 2006 *Use of smoking to add value to salmones trout*. Braz. Arch. Biol. Technol 49(1):57-62

[24] Simko P 2005. *Factors affecting elimination of polyciclic aromatic hydrocarbons from smoked meat foods and liquid smoke flavorongs: a review of melocular nutrition.* Food Research. 49:637-647

[25] Sormin RD, Latumahina MChA, Matruty THEA, Tapotubun AM, and Nanlohy H 2015. *Quality Profile and Economics Analisis of Dried Smoked Julung in Banggoi Village, District of Eastern Seram, Molucas.* International Journal of Current Research 7 (2) : 12775-12778.

[26] Swastawati F, Surti T, Agustini TW, Riyadi PH. 2013. *Karakteristik kualitas ikan asap yang diproses menggunakan metode dan jenis ikan berbeda*. Jurnal Aplikasi Teknologi Pangan. 2(3):126-132

[27] Wibowo S 2000 *Industri Pengasapan Ikan*. Panebar Swadaya. Jakarta

Fortification of seaweed (*Eucheuma cottonii*) flour on nutrition, iodine, and glycemic index of pasta

Muhamad Firdaus[1] Yahya[1], Galih Raditya Hardany Nugraha[1], Dyah Dwi Utari[1]

Department of Fish Processing Technology, Faculty of Fisheries and Marine Science, Brawijaya University, Malang – 65145, Indonesia

Email: muhamadfir@yahoo.com

Abstract. Pasta is a nutritious and energy product which produced from the dough of wheat flour and water. It contains less of iodine and high of glycemic index. *Euchema cottonii* belongs of red seaweed is food substance that contains much of iodine and dietary fiber. The objective of this study was to know the fortification effect of *E. cottonii* flour on the nutrition, iodine, and glycemic index of pasta. *E. cottonii* was collected from the culture farm of *E. cottonii* on the Wongsorejo beach, District of Banyuwangi, East Java on April-June 2015. Wheat flour and pasta ingredients were obtained locally at shops of Pasar Besar, Malang. Pasta was produced by weighing of components, mixing, dough, milling, steaming and drying. *E. cottonii* flour was added on mixing process at 0; 7; 14 and 21 % of ingredients. The parameter of this study was the level of water, lipid, protein, ash, and carbohydrate (by difference), iodine, crude fiber, the total of dietary fiber, soluble fiber, insoluble fiber, and glycemic index, respectively. Data were analyzed by variance and the least square difference used to determine the difference between treatments. The highest concentration group showed more nutritious than other treatments. The characters of its product were water 6.70%, lipid 2.26%, protein 23.09%, ash 14.11%, carbohydrate 53.84%, iodine 3.71 ppm, crude fiber 8.02%, the total of dietary fiber 20.88%, soluble fiber 11.69%, insoluble fiber 9.19%, and glycemic index 44.45, respectively. In conclusion, the fortification of *E. cottonii* flour enhances the nutrition value, iodine content, and glycemic index of pasta.

1. Introduction

Pasta is a food product that produced from wheat powder dough. Based on the type, this product can be divided to fresh and dried pasta, meanwhile, based on the form, it consists of long and short pasta. The long form of it is spaghetti and fettuccini whereas the short one is macaroni [1]. Pasta can be classified as the high glycemic food caused its carbohydrate content is relatively high [2].

Glycemic index is the indicator of a food product that can increase of blood glucose. The product has the high glycemic index if it consumed and then shortly increase of blood glucose. Food product with the high glycemic index is mainly produced from carbohydrate substances. This index value can be affected by many dietary factors, such as; fiber and iodine [3].

Iodine is one of essential micronutrient for the human body. The growth disturbance, intelligence, and insulin secretion associated to this mineral unavailability. The hyperglycemia phenomena can be related to the lack of iodine availability in the body [4]. Dietary fiber is food ingredient that can't be hydrolyzed by the digestion enzymes. Although this substance can't be digest and low in the nutrition value, its role in the obesity, cholesterol profile and hypoglycemic activity. Hypoglycemic effect of dietary fiber is their capability on the hindrance of glucose hydrolysis and absorption [5].

Eucheuma cottonii is carragenophyte that belonged to red seaweed. This seaweed contains nutrition substances, mineral, and dietary fiber [6]. This alga has been used, supplemented and fortified on bread, doughnut, and instant noodle. The enrichment of this alga on these products is addressed to enhance the nutritious and functional value of it [7, 8, 9].

Pasta is the high glycemic food that can be decreased this value by the fortification of iodine and dietary fiber. *Eucheuma cottonii* is seaweed that contains iodine and dietary fiber. The purpose of this research was to evaluate the fortification of *Eucheuma cottonii* flour on the nutrition, iodine, and glycemic index of pasta.

2. Material and methods

2.1. Material

Eucheuma cottonii were obtained from Wongsorejo Beach, Banyuwangi District, East Java Province, Indonesia on April-June 2015. Wheat flour and pasta ingredients were bought from the shops on Pasar Besar, Malang. Chemical materials for proximate analysis were obtained from Merck. Glucometer strips were found from GlucoDr. Thermamyl, pepsin, and pancreatin were obtained from Sigma-Aldrich.

2.2. Methods

The sample was washed by tap water, bleached by CaCO2 5% for a night, dried by the sun drying for 3 days, milled, and sieved on 80 mesh size and obtained seaweed flour. This study was experiment and treatment of this experiment was the percentage of *Eucheuma cottonii* flour, i.e., 0, 7, 14, and 21%, respectively, meanwhile, their parameter was proximate, iodine, dietary fiber, crude fiber, and glycemic index. Pasta was produced by mixing of seaweed flour and wheat flour (treatment portion), after that blending with salt and egg, and then mixing with water and finally cooking oil to obtained the dull dough. The dough was blended several times by blending machine to form dough sheets. These sheets were poured with wheat flour, after that the sheets were steamed for 10 minutes and after cooling at room temperature, pasta was dried for 7 hours at 55oC [10].

2.2.1. Proximate. Water content was assayed based on thermogravimetry method. The free water of sample was evaporated at 105oC for 3 hours to constant weight. Ash content was determined by burner system. The Sample was carbonized and burned in the burner at 500-600oC for several hours and the product was weighed and determined as ash content. Lipid content was assayed by Soxhlet method. The lipid of the sample was extracted by the organic solvent in the Soxhlet tube for several hours. Protein content was tested by the Kjeldahl method. Protein of the sample was assayed by the determination of nitrogen content of sample by destruction, distillation, and titration. Carbohydrate content was determined based on by difference method [11].

2.2.2. Iodine. Iodine content was assayed based on precipitation and titration method [12]. This mineral was determined based on the formation of amylum-iodine complex by iodate oxidation.

2.2.3. Dietary and crude fiber. The total, soluble and insoluble fiber were determined by enzymatic-gravimetric digestion method [13]. The crude fiber was evaluated by hydrolysis method [11]. The crude fiber was eliminated from the sample by extraction of the sample in acid and base solutions.

2.2.4. Glycemic Index. This index was determined by comparison of area under curve between the area of blood glucose after consuming sample and bread as standard [14]. Ten volunteers that health, non-diabetes and their blood glucose level between 70-120 mg mL-1 were used in this study.

2.3. Statistical analysis
Data of this study were stated as a mean. The difference among of treatments was evaluated by analysis of variance. The least square difference test was used to determine the difference among treatment groups. The significant level was set at $P < 0.05$.

3. Results and Discussion
The effect of seaweed flour percentage on proximate and iodine content of pasta can be seen in Table 1. The result showed that increasing of seaweed percentage decrease water, ash, protein, and lipid content but it enhances carbohydrate and iodine content of pasta.

Table 1. Proximate and iodine content of pasta on several *E. cottonii* flour percentage

Treatments (%)	Water (%)	Ash (%)	Protein (%)	Lipid (%)	Carbohydrate (%)	Iodine (ppm)
0	10.77^{d}	9.67^{a}	27.43^{b}	3.03^{ns}	49.10^{ns}	2.47^{a}
7	9.68^{c}	13.50^{b}	24.03^{a}	2.62^{ns}	50.17^{ns}	2.81^{a}
14	8.40^{b}	13.68^{b}	23.23^{a}	2.45^{ns}	52.24^{ns}	3.59^{b}
21	6.70^{a}	14.11^{b}	23.09^{a}	2.26^{ns}	53.84^{ns}	3.71^{b}

3.1. Proximate and *iodine*
Table 1 exposed that the water content of pasta is influenced by the increasing percentage of *E. cottonii* flour. The previous study [15] showed that supplementation of *E. cottonii* flour decreases water content of instant noodle. *E. cottonii* is foodstuff that contains hydrocolloid substance, such as, carrageenan. This constituent has hydroxyl compound that able to bind free water of substance.

Table 1 exhibited that the fortification of *E. cottonii* flour enhances the ash content of pasta. The former research [16] stated that *E. cottonii* is one of foodstuff that encloses mineral more than terrestrial food. Fortification of *E. cottonii* flour on the food product will consequently increase this mineral content in the product.

Table 1 displayed that the fortification of *E. cottonii* flour decreases the protein content of pasta. The earlier investigation [17] showed that protein content of pasta is related to the protein of wheat flour. The fortification of other flour on pasta, therefore, it will decrease protein content of pasta.

Table 1 revealed that the lipid content of pasta is not influenced by fortification of *E. cottonii* flour. The earlier research [18] showed that *E. cottonii* is less or not contain the lipid. It, therefore, does not influence on lipid content of product when the *E. cottonii* flour is fortified.

Table 1 showed that the carbohydrate content of pasta also not influenced by the increasing of *E. cottonii* flour concentration. The previous research [19] showed that supplementation of *E. cottonii* flour does not affect the carbohydrate content of pasta. This is possibly caused the mass equilibrium among the food component of the product. Pasta is a food product that produced mainly by carbohydrate compound of wheat flour. *E. cottonii* flour is a foodstuff that the containing generally carbohydrate. Supplementation of wheat flour by *E. cottonii* flour does not cause the changing of the carbohydrate content of this product.

Table 1 showed that fortification of *E cottonii* flour able to enhance the iodine concentration of pasta. A previous study [20] showed that fortification of *E cottonii* able to increase iodine content on an instant noodle. It is possibly caused *E. cottonii* is food substance that contains iodium. Fortification of *E cottonii* flour on pasta will affect to the increasing iodine content in this product.

3.2. Dietary and Crude Fiber
The effect of seaweed flour percentage on the dietary fiber content of pasta can be seen in Table 2. The result showed that increasing of *E. cottonii* flour percentage enhance total, soluble and insoluble dietary and crude fiber content of pasta.

Table 2. Total, Soluble and insoluble dietary and crude fiber content of pasta on several *E. cottonii* flour percentage

Treatments (%)	Total Dietary Fiber (%)	Soluble Fiber (%)	Insoluble Fiber (%)	Crude Fiber (%)
0	13.89[a]	8.90[a]	4.99[a]	5.26[a]
7	16.48[b]	9.27[b]	7.21[b]	5.53[a]
14	18.39[c]	9.90[c]	8.49[c]	7.70[b]
21	20.88[d]	11.69[d]	9.19[d]	8.02[b]

Table 2 showed that fortification of *E cottonii* flour able to increase the total, soluble and insoluble fiber concentration on pasta. A previous study [20] exhibited that the total, soluble, and insoluble dietary fiber of *E. cottonii* are 82.0; 38.8; and 43.2 % of dry weight, respectively. These compounds are huge contained in this seaweed, so the fortification of these compounds on foodstuff will subsequently affect the composition of the food product.

Table 2 showed that fortification of *E cottonii* flour able to enhance the crude fiber concentration of pasta. A previous study [21] that fortification of seaweed will increase the concentration of crude fiber of instant noodle. It is possibly caused the main component of *E. cottonii* is carbohydrate. This compound consists of polysaccharide polymer, and it is naturally composed of the fiber.

3.3. *Glycemic Index*

The effect of seaweed flour percentage on the glycemic index of pasta viewed on Figure 1. The result showed that the increasing of seaweed percentage decrease glycemic index value of pasta.

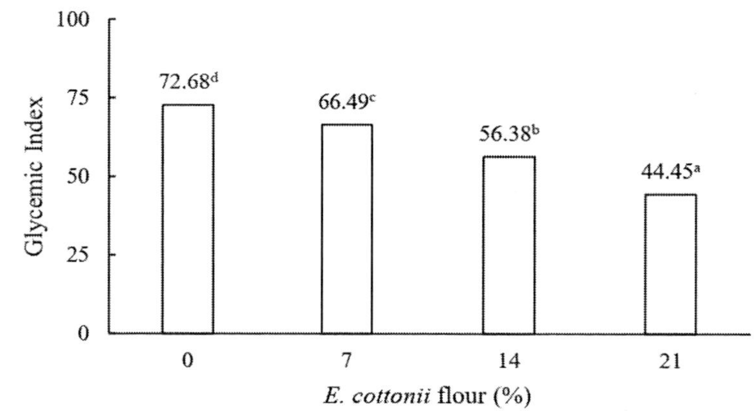

Figure 1. The glycemic index of pasta on several of *E. cottonii* flour percentage

Figure 1 showed that the glycemic index of pasta is affected by the fortification of *E. cottonii* flour. The previous study [20, 21] stated that *E cottonii* contain dietary fiber and iodine. These substances have been known affect the blood glucose. The food product that fortified with *E. cottonii* flour will decrease their glycemic index value. The soluble fiber able prevents increasing of blood glucose level caused this fiber hinder the polysaccharide hydrolysis by α-glucosidase and then inhibit the glucose absorption on the gastrointestinal [5]. Iodine of this flour also prevents increasing of blood glucose. It is caused this mineral able to induce insulin secretion [22]. Insulin is the hormone that works to initiate of glucose uptake by body cells.

IOP Conf. Series: Earth and Environmental Science **89** (2017) 012011 doi:10.1088/1755-1315/89/1/012011

4. Conclusion

The fortification of *E. cottonii* flour improve the nutrition value, iodine content, and glycemic index of pasta

References

[1] Foschia A. M., Peressini D., Sensidoni A., Brennan M. A., Brennan C. S. 2015. How combinations of dietary fibers can affect physicochemical characteristics of pasta. LWT - Food Science and Technology. 61 : 41-46

[2] Goñi I., Valentín-Gamazo C. 2003. Chickpea flour ingredient slows glycemic response to pasta in healthy volunteers. Food Chemistry. 81 : 511-515.

[3] Augustin L.S.A., Kendall C.W.C., Jenkins D.J.A., Willett W.C., Astrup A., Barclay A.W., *et al.* 2015. Glycemic index, glycemic load and glycemic response: An International Scientific Consensus Summit from the International Carbohydrate Quality Consortium (ICQC). Nutrition, Metabolism & Cardiovascular Diseases. 25 : 795-815

[4] Carrilloa P. O., Fuentes E. G., Alcántarac C. G., Queroc M. S., Martínezc M. M., Fernández T. U., Fernández P. S. 2015. Assessment of iodine nutritional status in the general population in the province of Jaén. Endocrinology Nutrition. 62 : 373-379

[5] Goff H. D, Repin N, Fabek H, El Khoury D, Gidley M J. 2017. Dietary fibre for glycaemia control: Towards a mechanistic understanding. Bioactive Carbohydrates and Dietary Fibre, In press, corrected proof, Available online 12 July 2017

[6] Ribeiro A. R., Gonçalves A., Bandarra N., Nunes M. L., Dinis M. T., Dias J., Rema P. 2016. Natural fortification of trout with dietary macroalgae and selenised-yeast increases the nutritional contribution in iodine and selenium. Food Research International. in press.

[7] Gupta S., Abu-Ghannam N. 2011. Recent developments in the application of seaweeds or seaweed extracts as a means for enhancing the safety and quality attributes of foods. Innovative Food Science and Emerging Technologies. 12 : 600–609

[8] Rasmussen L. B., Jørgensen T., Perrild H., Nils K., Krejbjerg A., Laurberg P., Pedersen I. B., Bjergved L., Ovesen L. 2014. Mandatory iodine fortification of bread and salt increases iodine excretion in adults in Denmark e A 11-year follow-up study. Clinical Nutrition. 33 : 1033-1040

[9] Rosell C. M. 2016. Fortification of Grain-Based Foods. Encyclopedia of Food Grains (Second Edition). 2 : 43-49.

[10] Nugraha G.R.H. 2015. Indeks glikemik dan kualitas pasta yang dihasilkan dari penambahan proporsi tepung *E. cottonii* yang berbeda. Skripsi. Teknologi Hasil Perikanan. Fakultas Perikanan dan Ilmu Kelautan. Universitas Brawijaya. Malang

[11] Sudarmadji S. dan Suhardi H.B. 2010. Analisa Bahan Pangan dan Pertanian. Liberty. Yogjakarta.

[12] Febrianti S., Hermin S., Atikah. 2013. Penentuan kadar iodida secara spektrofotometri berdasarkan pembentukan komplek amilum-iodium menggunakan oksidator iodat. Jurnal Kimia Student. 1 : 50-56.

[13] McCleary B.V., Draga A., Sloane N. 2013. Dietary fibre analysis in foods. Fibre-Rich and Wholegrain Foods. 25-60.

[14] Scazzina F., Dall'Asta M., Casiraghi M.C., Sieri S., Del Rio D., Pellegrini N., Brighenti F. 2016. Glycemic index and glycemic load of commercial Italian foods. Nutrition, Metabolism & Cardiovascular Diseases. 26 : 419-429

[15] Santoso J., Lestari O. A., Anugrahati N. A. 2006. Peningkatan kandungan serat makanan dan iodium pada mi kering melalui substitusi tepung terigu dengan tepung rumput laut. Jurnal Ilmu Teknologi Pangan. 4 : 131-145.

[16] Wirjatmadi B., Adriani M., Purwanti S. 2002. Pemanfaatan Eucheuma cottonii dala, meningkatkan nilai kandungan serat dan iodium dalam pembuatan mie basah. Jurnal Penelitian Medika Eksakta. 3 : 89-104.

[17] Giménez M.A., Drago S.R., De Greef D., Gonzalez R.J., Lobo M.O., Samman N.C. 2012. Rheological, functional and nutritional properties of wheat/broad bean (*Vicia faba*) flour blends for pasta formulation. Food Chemistry. 134 : 200–206

[18] Matanjun P., Mohamed S,, Mustapha N. M., Muhammad K.. 2009. Nutrient content of tropical edible seaweeds, *Eucheuma cottonii*, *Caulerpa lentillifera* and *Sargassum polycystum*. Journal of Applied Phycology. 21 : 75-80.

[19] Jannah, Sukatiningsih R., Diniyah N. 2014. Formulasi tepung komposit dari terigu, kecambah jagung, dan rumput laut pada mie kering. Jurnal Teknologi Pertanian. 15 : 15-24.

[20] Murniyani, Subaryono, Irma H. 2010. Pengolahan mie yang difortifikasi dengan ikan dan rumput laut sebagai sumber protein, serat kasar dan iodium. Jurnal Pascapanen dan Bioteknologi Kelautan dan Perikanan. 5 : 65-75.

[21] Astawan M., Wresdiyati T., Hartanta A. B. 2005. The Utilization of Seaweed as a Source of Dietary Fiber to Decrease the Serum Cholesterol in Rats. Hayati.12 : 23-27.

[22] Vosátkováa M., Janícková D. Z., Zamrazila V., Salátováa M., Hilla M., Vondra K. 2017. Ioduria and type 1 diabetes mellitus – Relationships to selected clinical markers of diabetes in adults. Journal of Applied Biomedicine. 15 : 146–150

Seaweed Fortification on Crispy Enbal as Local Food of Kei Islands

Ismael Marasabessy[1], Fien Sudirjo[1]

[1]Fisheries Processing Technology Study Program, Tual State Fisheries Polytechnic, Indonesia
Langgur-Sathean Road Km.6, Southeast Maluku, 97611, Maluku, Indonesia

Email: Ismael_mabes@yahoo.com

Abstract. One of health problems phenomenon in Indonesian and the world is increasing the degenerative disease because human's bad habits of eating that having less fiber. Source of fiber which is relatively abundant in eastern Indonesia is seaweed that is very precise to fortified on local food that aims to be more nutritious and economically valuable. The purpose of this study is to got appropriate seaweed fortification technique to produce Seaweed Crispy Enbal (SCE) as typical food from Kei islands that rich in fiber and preferred by consumers. The research was done in two stages. The first stage is to analyze quality of fiber and HCN content of seaweed and enbal flour as SCE raw material, and the two-stage is fortified fiber to enbal lempeng using two types of raw materials, namely pulp seaweed and flour seaweed. The results showed that the fiber content of seaweed *Eucheuma cottonii* and flour enbal respectively 7.01% and 4%, while HCN content less than 3 mg/kg. Fortification techniques using pulp seaweed better than others. It is because pulp seaweed produces seaweed crispy enbal with high value of sensory (really like) with having fiber content is 7.48%.

1. Introduction

The world in the last decade is haunted by the phenomenon of health problems, especially the increasing degenerative diseases, as a result of external and internal factors of the body. External factors are people's eating habits that are currently inclined to eat fast food that is high in fat but low in fiber. The internal factor is the inability of the body to regenerate cells normally also triggered by an imbalance of nutrient intake, especially food fiber. Research in the last three decades showed that fiber contributes a significant role in the prevention of degenerative diseases, but awareness about the importance of food fiber in Indonesian society lacks [8], especially for urban communities surrounded by low-nutrient foods. One source of food fiber that is very potential and easy to obtain is seaweed.

Traditionally seaweed has long been used as food and medicine because it is rich in minerals, macro elements, and other micro elements. Some types of seaweed contain important minerals that are used for body metabolism such as iodine, calcium, selenium, and fiber [2].

Southeast Maluku community or better known as the Kei community has a special local food called enbal that is local food made from cassava / toxic cassava (*Manihot* sp.) that high in cyanide acid. As a special local food, enbal is consumed by all levels of society. This food is served in a variety of ways that is fry without oil called fried *enbal*, fry mixture spread into a pan called *enbal bubuhuk* and baked called *enbal bubes* (*porna*). Crispy enbal is one of the diversified results that recently glow produced as typical of local souvenirs and quite interested in local communities and from outside Kei. Factors that affect the quality of crispy enbal are material formulas, drying techniques and baked techniques.

The Seaweed production in Kei Islands in the last 5 years is quite high and continues to increase [3]. This condition causes the supply of fiber derived from seaweed is large enough. Fiber is very important in the process of digesting food for the body. Lack of fiber can cause constipation, apenaistis, alverculity, hemorrhoids, diabetes mellitus, coronary heart disease and kidney stones [10].

For the utilization of seaweed can be more maximal as a source of fiber, and on the other hand, there is improvement of local food quality it is necessary to product diversify by fortifying the seaweed into enbal and make it more acceptable to the public communities as a unique snack by Kei Islands. The purpose of this study is to obtain the proper seaweed fortification technique to produce a rich in fiber Seaweed Crispy Enbal (SCE) and preferably.

2. Methodology

2.1. Material

The main ingredients for Seaweed Crispy Enbal (SCE) are *Eucheuma cottonii* cultivated (60 days age) that has been dried and high HCN cassava (*Manihot* sp.) 11 month age and additional ingredients include chicken eggs, butter, sugar, vanilla and milk powder.

Seaweed cultivation materials include nylon ropes of 5 and 3mm diameters, fiber ball float as the main float and plastic bottle float as float ropes (where seaweed binders). Seaweed seeds used are a type of *Eucheuma cottonii* 1 month age with weight 150 grams/seedlings.

2.2. Research Methods

This research consists of two stages. The first stage is done seaweed cultivation with longline method for 60 days, then analyzed fiber quality. Besides that, analyzed fiber content and HCN content of enbal flour (cassava flour) which will be used as raw material of Seaweed Crispy Enbal. The second stage is fortification of seaweed on the *enbal lempeng* using two types of raw materials namely seaweed pulp and seaweed flour, then analyze organoleptic and chemical qualities.

2.3. Seaweed Cultivation

Seaweed seeds used are good quality green *thallus* parts with length 10-15 cm, initial biomass 150 g, using longline method. There are 3 long lines (as replicates) are used to tie the seeds to the length of one long line 10m. The harvest is done when the seaweed is 60 days. The distance between the long line is 120 cm and the spacing between seedling is 30cm. At each corner and the center of the rope is put in the main float and the ballast. Seeds of *Eucheuma cottonii* are always kept under water. Every time the harvest, seaweed is analyzed fiber quality.

2.4. Making of Seaweed Pulp and Seaweed Flour

Seaweed pulp is made by dry seaweed soaked for 2 days until the fish is lost, then blend until it becomes mush. As for the seaweed flour is made by cutting dried seaweed size 0.5 - 1 cm and ground until smooth. Further sieved using 80 mesh sieve to produce smooth and clean seaweed flour.

2.5. Making Enbal Flour

Enbal (high cassava HCN content) peeled and washed with clean water, then grated and washed again. Further squeezed until the water content decreases. The enbal is milled and dried, then sieved using 80 mesh sieve to obtain white and clean enbal flour.

2.6. Fortification of Pulp Seaweed Into Enbal (Cassava)

Fortification of seaweed pulp is done by as much as 15% of seaweed pulp is mixed into the dough until homogeneous then spread to the plate enbal. It is then stored a night in the refrigerator and baked in the oven until cooked. For fortification with seaweed flour, which is as much as 15% seaweed flour mixed with enbal flour in the mold then baked and dried. The next process is the same as the previous treatment of the spread dough, let stand in the refrigerator, and baked. Organoleptic qualities, fiber content and HCN Seaweed Crispy Enbal used to determine the best treatment.

2.7. Research Design and Data Analysis
This study is an experimental study using Completely Randomized Design (CRD) with three replications. The variant analysis was performed on proximate and fiber data at 95% confidence level. Advanced test using Duncan's Multiple Range Test (DMRT) if there is a difference. Organoleptic data were analyzed descriptively.

3. Results and Discussion

3.1. Quality of Seaweed Fiber and Enbal Flour
The results showed the fiber content of seaweed is harvested 60 days and enbal flour respectively 7:01% and 4%, while the very low levels of HCN which is below 3mg / kg (Table 1) *Eucheuma cottonii* seaweed are cultivated in Dunwahan village Southeast Maluku has a high enough fiber content at 60 days of harvest time. This condition is possible due to the establishment of marine polysaccharides still occur continually. *Eucheuma cottonii* seaweed is containing polysaccharide in large quantities and is the structure of cell walls of seaweed that could potentially generate high fiber. According to [9, 5], the fiber is composed of three main factions namely structural polysaccharides (composed of cellulose, hemicellulose and pectin), nonstructural polysaccharides (lignin), and non-structural polysaccharides (gum, carrageenan and seaweed), if compared with other researches such as [1] where seaweed is harvested 45 days age has fiber content 0.9% and research [12] where seaweed is harvested 50 days age fiber content reaches 3%, then the quality of *Eucheuma cottonii* seaweed fiber produced high enough.

For HCN content of cassava enbal flour very small at less than 3 mg/kg and still below standard cassava flour qualities (ISO 1992), which is a maximum of 40 mg/kg and mocaf flour a maximum of 10 mg/kg so that no toxic effects on crispy enbal. The low levels of HCN due to repeated water washing treatments are then followed by extortion and curing causing more HCN to dissolve. [10], stated that HCN in food could be reduced in several ways, including soaking in water because HCN is a water-soluble compound. This is research reinforced by the results of [7] which showed the treatment cassava immersion in water for 24 hours can reduce levels of HCN to 90.10%. The value of seaweed fiber content of Eucheuma *cottonii* and HCN of Enbal (cassava) as shown in Table 1.

Table 1. Result of seaweed fiber content analysis of *Eucheuma cottonii* and HCN level of Enbal (cassava)

Raw material	Fiber (%)	HCN (mg/kg)
Seaweed	7.01	undetection
Enbal flour (cassava flour)	4	< 3

3.2. Quality of Crispy Seaweed

3.2.1. Organoleptic Quality of SCE. Level of panelist acceptance of seaweed fortified enbal with two types of raw materials so that the analyzed parameters are organoleptic quality and fiber and HCN SCE level.

The result of the organoleptic test (Figure 1) shows that the average value of organoleptic SCE fortified by seaweed pulp is preferred (value 4.08) compared to fortification treatment using seaweed flour (value 3.78). This is alleged because the seaweed pulp with wet conditions more absorbed into the enbal at the time of stand-in for a night.

All values of organoleptic parameters are above 4 (very like), except the value of fortification treatment appearance using seaweed flour which got value 2.60 (value under 3 as panelist acceptance limit). The low value of the treatment appearance is due to the resulting SCE is slightly darker color with slightly rougher surface than the fortification treatment using seaweed pulp.

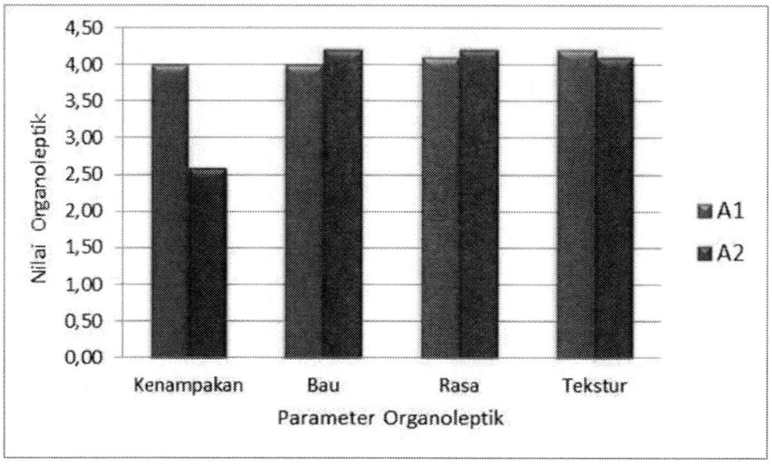

Figure 1. Graph of organoleptic value of SCE fortified by seaweed
Description: A1 = Fortification using seaweed grass
A2 = Fortification using seaweed flour

Fortification using seaweed flour mixed with enbal flour produces dry *"enbal lempeng"*. The dough used to coat the fortification of the fortified plate contains low moisture content when heated the heat penetration to the center of the product becomes slower so that the SCE surface becomes more rapidly darker (Fig. 2). Baking time is longer about 20 minutes.

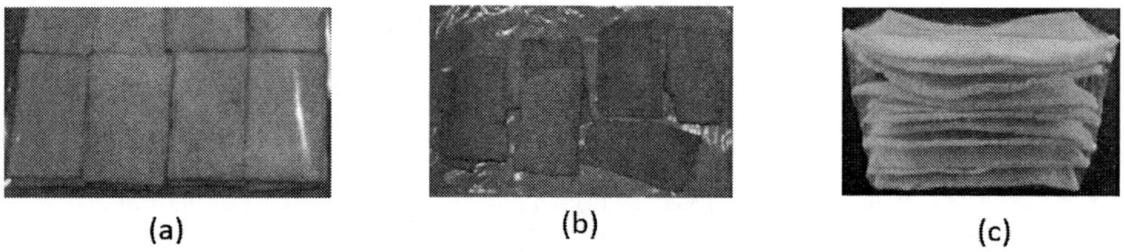

Figure 2. SCE that is fortified
Description: (a) seaweed pulp, (b) seaweed flour, and (c) that are not fortified

The condition is different from fortification using seaweed pulp. The process of wet seaweed blender added water causes the seaweed dough to be slightly liquid with a rather high water content. When heating (baking) in the oven, the SCE surface color change rate is hampered by surface water content. The slightly liquid dough makes the wet plate envelope, which causes heat penetration rather quickly to enter the center of the product so that the maturation rate is faster (about 7-10 minutes) and more evenly distributed.

SCE water content before baking is very important. According to [4], a mechanism that occurs when hot air with low moisture content is exhaled in wet food causes water to exist on the surface of the evaporated material. First of all the vapor is water on the surface of the material, then followed by water on the inside of the material. If the water on the surface of the product is low, while the texture is somewhat dense, the heat propagation to the center of the product becomes longer, while the surface is slightly dry causing the surface to rapidly turn darker.

3.2.2. Quality of Fiber and HCN SCE. Levels of fiber SCE fortified with seaweed flour slightly higher than that fortified seaweed pulp, that is each for fortification using seaweed grains 7.48% while using seaweed powder 7.77% fiber content (Table 2), but statistically, both values Is not different (sig> 0,05). When compared to the fiber content of both the above fortification treatment was much higher than the unfortified plate enbal. This condition indicates that the goal of fortification of seaweed to increase fiber in SCE is achieved. The results of this study are in line with the research of [6] by fortifying the seaweed powder in cookies to produce higher fiber values. While the nutritional and proximate value others tend to be the same and not statistically different.

The HCN concentration contained in SCE on the two treatments is very small that is <3 mg/kg still below [11] which is maximum 40mg / kg, so it does not give toxic effect on SCE. The low levels of HCN due to repeated water washing treatments cause most of the HCN to dissolve.

Table 2. Fiber level and HCN SCE fortified by seaweed

Fortification Treatment	Fiber level (%)	HCN Level (mg/kg)
Seaweed pulp	7.48	< 3
Seaweed flour	7.77	< 3

4. Conclusion

Seaweed fiber content is harvested 60 days and enbal flour respectively 7.01% and 4%, while the HCN content is very low that is below 3mg / kg. Organoleptically, the SCE fortified technique using seaweed pulp is preferred because it results in higher texture and visibility values. Cadillary SCE fortified pulp and seaweed flour are not statistically significant.

References

[1] Adhistiana R., Rahayu M.P., Ambarwati R., Herdiana E., and Vivaldy, 2008. Utilization of seaweed in the manufacture of dodol seaweed. Http://www.ipb.ac.id/seaweed making.html. [Accessed October 17, 2015].

[2] Burtin P., 2006. Nutritional Value of Seaweed. Electronic *J. Environ. Agric. Food J. Chem.* 5 (3).

[3] DKP Malra, 2010. Annual Report on Marine and Fishery Affairs of Southeast Maluku Regency.

[4] Fellows P., 2000. Food Processing Technology. Principles and Practice. Second Edition. Department of Catering Management, Oxford Polytechnic, England. (P) 313-315.

[5] Masluha D., 2013. Formulation of jelly drinks based on *Eucheuma cottonii* dan *Spirulina platensis* seaweed. Essay. Faculty of Fisheries and Marine Science IPB, Bogor.

[6] Murniyati, Subaryono and Irma H., 2010. Processing of Differentiated Noodles with Fish and Seaweed as Source of Protein, Rough and Iodine Fiber. *Journal of Postharvest and Biotechnology of Marine and Fisheries* Vol. 5 No. 1

[7] Nebiyu, A., and Getachew, E 2011 Soaking and drying of cassava roots reduced cyanogenic potential of three cassava varieties at Jimma, Southwest Ethiopia. *African Journal of Biotechnology*, 10(62), 13465-13469.

[8] Santoso A., 2011. Dietary Fiber (Dietery Fiber) and Its Benefits for Health. Journal of Magistra No. 75 Yrs. XXIII. ISSN 0215 – 9511

[9] Schneman, BO., 1986. Dietary Fiber, Physical and Chemical Properties Methods of Analysis and Physiological Effects. *Food Technology*. February (p) 104-110.

[10] Sitepu JM., 2009. Effect of Immersion Time on Decreasing Cyanide Acid Level (HCN) on Bitter cassava (Manihot esculentaGrant). [January 17, 2012].

[11] SNI [1992]. Quality Requirement of Cassava Flour. Indonesian National Standard Number 01-2997-1992. National Standardization Agency. Republic of Indonesia.

[12] Widyastuti S., 2010. The physical and chemical properties of carrageenan are extracted from seaweed *Eucheuma cottonii* and *E. Spinosumpada* of different harvest age. Agroteksos Vol. No. 20, April 2010.

IOP Conf. Series: Earth and Environmental Science **89** (2017) 012013 doi:10.1088/1755-1315/89/1/012013

Handling Technique Development of Live Carp, *Cyprinus carpio*, In Cold Dry Styrofoam Box

I Ketut Suwetja[1], Netty Salindeho[1], I Gede Prabawa Suwetja[2]

[1]Departement of Fisheries Technology, Faculty of Fisheries and Marine Science, Sam Ratulangi University, Manado 95115, Indonesia
[2]Departement of Accounting, Faculty of Economic and Business, Sam Ratulangi University, Manado, 95115, Indonesia

Email: suwetja.i.k@gmail.com

Abstract. The study focused on several following aspects: temperature and time optimation for fainting, holding media optimation, temperature and time optimation for recovery, and their correlation with mortality rate of carp, *Cyprinus carpio*. Fainting occurred at the optimum time of 11 minutes and 03 seconds, temperature of 8^0C, and holding time of 6 hours. Holding medium was rice husk. The fastest consciousness of the fish was found in 6 volt-aerated water medium. The fish consciousness after 6 hours of storing in the rice husk at the fainting temperature of 8^0C was found faster ($p < 0.05$), 11 minutes and 15 seconds, than that added with 0.02% of clove oil, 25 minutes and 16 seconds. The fish mortality rate after 6 hours of storage in the rice husk at fainting temperature of 8^0C was lower ($p < 0.05$), 46%, than that with addition of 0.02% of clove oil, 75%.

1. Introduction

Live fish transportation in Indonesia is conducted using a water-containing plastic bag filled with either oxygen addition or not, depending upon its necessity. Such a packing and transporting technique has some weaknesses, such as less efficient in weight, volume, and transport cost, decreasing water oxygen content, increased water temperature, fish poisoning from their feces, and relatively long transportation.

Nowadays, live fish transportation has been developed using a dry medium through low-temperature fainting method. It is done by cooling the water down to the fainting temperature of the fish. Low temperature is maintained in the Styrofoam box by placing some ice cube. Cool media for fish storage use either rice husk or sawdust.

One of the main goals in live fish transportation is to yield low mortality rate, and therefore, the fish must faint. Immobilization is necessarily done that the fish metabolism activity is in basal condition. At very low respiration and metabolism rate, the fish can be transported in long duration at low mortality rate. There are several immotilization techniques, such as the use of low temperature or the use of natural or artificial antimetabolite [22].

Before storing, the fish is previously wrapped with paper to avoid the fish mouth filled with the media material [17]. The wrapped fish is then placed and set in the styrofoam containing cool sawdust of 10-15°C. The fish placement is done in layers, mediafish-media etc., and the top layer is media. The media thickness is more or less 2-3 cm. To maintain the temperature stability, the Styrofoam is filled some ice cube [18].

Re-awaking carp after fainted and stored can be done by returning the fish into a sufficiently aerated tank at their habitat temperature, 25-27°C. The survival of live fish packed in unconscious condition is affected by initial condition of the fish, fainting temperature, fish origin, and shocks from overlapping at the lower level that causes fish mortality during the transportation [9, 18]

[15, 13, 11 1, 3, 6] also supported that live fish transportation was aimed to maintain the fish survival during the transportation reaching the target location. Close distance transportation does not need specific treatment, but long-distance transportation requires particular treatments to maintain the fish survivorship. Therefore, an appropriate live fish transportation technology following the commodity and condition requirements is needed. There are two live fish transportation methods, the use of water as media called wet system and the use of dry system (media without water).

The principle of live fish handling is maintaining the fish survivorship as maximum as possible up to accepted by the consumer. For this, there are several handling phases, pre-transport, transport, and post-transport handlings. Closed transportation will yield CO_2 accumulation and pH decline [7, 12, 4].

2. Methodology

Fainting media used low temperature of 8°C, and low temperature of 8°C added with 0.02% of clove oil. Storing media employed rice husk. Sawdust was not used as storing media because it has hazardous material for fish. The ice cube placed in a storing box was 1.8 kg. The fish were held for 8 hours and observed every 2 hours. Holding time treatments were 0, 2, 4, 6, and 8 hours with 2 replications.

Parameters recorded were fainting rate time [16, 21], reawaking time [2, 21], and fish mortality [10, 21]. This study used Factorial Complete Randomized Design with two factors as follows:

1. Fainting method (A) with two levels: A1 – fainting at 8°C and A2 – fainting at 8°C plus 0.02% of clove oil
2. Storage duration (B) with five levels: B1: 0 hours, B2- 2 hours, B3- 4 hours, B4- 6 hours, and B5- 8 hours.

3. Results and Discussion

3.1. Fainting Speed Time of C. carpio

Mean value of fainting speed time ranged from 1.2 to 14.16 minutes (Fig. 1). ANOVA showed that fainting method treatment, storing time, and their interaction did not significantly influence the fainting speed time ($P > 0.05$). Faster fainting time occurred in the method using clove oil since addition of clove oil into water media caused the dissolved oxygen in the water be reducing. Clove oil could function as insulator or barrier for oxygen diffusion process into the water. According to [5], dissolved oxygen decline is directly proportional to the increase in clove oil concentration, since oxygen diffusion process into the water is inhibited. The higher the clove oil concentration is, the lower the oxygen diffusion into the water will be. Eugenol as water-soluble anesthetic will reduce respiration rate of the fish. The decline in the respiration rate results in loss of all feelings on the fish body due to reduced nerve function so that nerve impulse action and delivery are inhibited [15]. According to Smith and Breet *in* [23], the ionic balance disturbance in fish brain causes the gill unable to normally function, and the osmoregulation of dissolved oxygen in the water into blood cells and gill be disturbed. Statistical analysis found that the treatment of holding duration did not give significant effect on the fainting speed time of the carp. It could result from that holding duration treatment was done after the fainting process ended so that it did not affect the fainting time

IOP Conf. Series: Earth and Environmental Science **89** (2017) 012013 doi:10.1088/1755-1315/89/1/012013

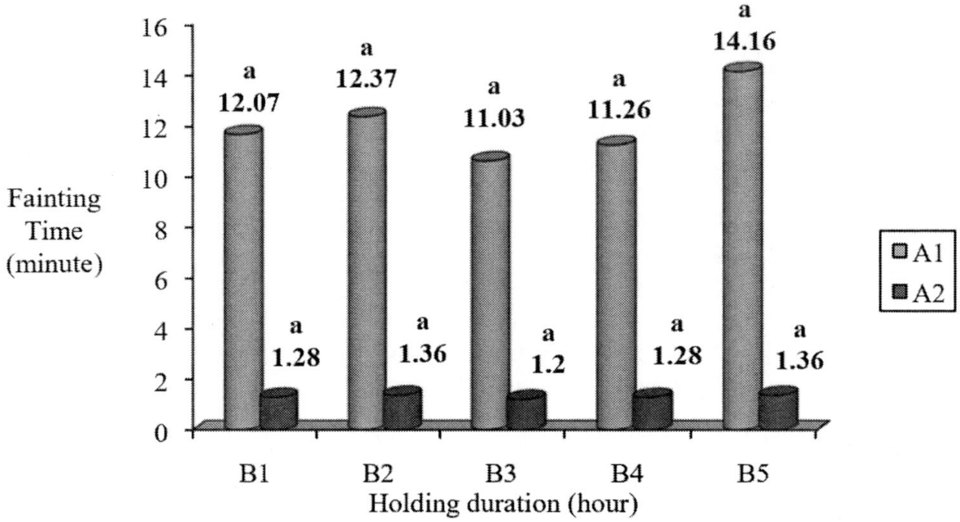

Figure 1. Fainting method and storage duration with fainting speed of *C. carpio*. A1: temperature of 8°C, A2: temperature of 8°C + 0.02% of clove oil. B1 : 0 hours, B2: 2 hours, B3: 4 hours, B4: 6 hours, B5: 8 hours.

3.2. Re-awaking Time of C. carpio

The mean value of re-awaking time of *C. carpio* ranged from 0.3 – 25.16 minutes. ANOVA indicated that at the confidence level of 95% fainting method, holding duration, and the treatment interaction gave significant effect on the re-awaking time of *C. carpio* ($P \leq 0.05$).

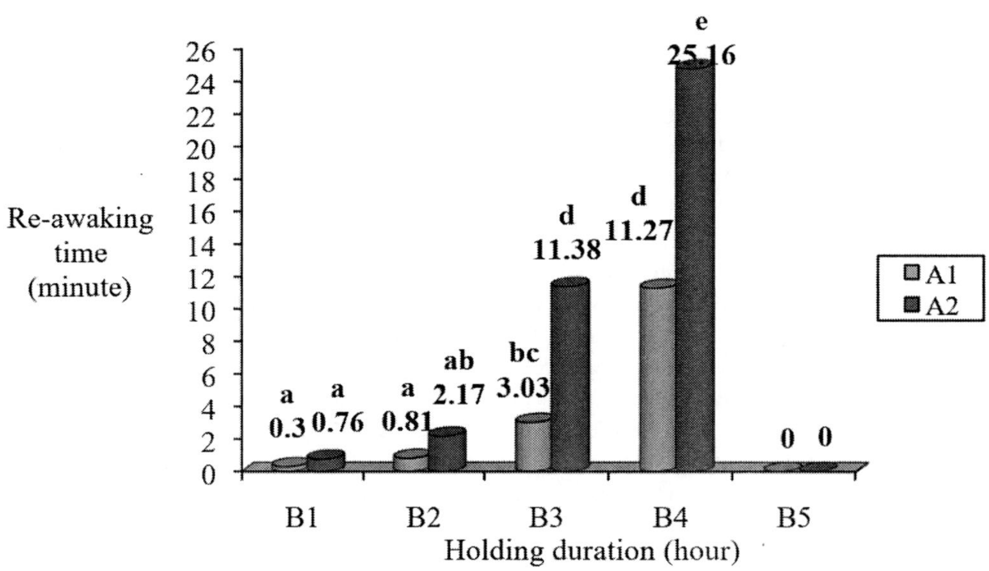

Figure 2. Fainting method and holding duration relationship with re-awaking time of *C. carpio*. A1: temperature of 8°C, A2: temperature of 8°C added with clove oil of 0.02%; B1: 0 hours, B2: 2 hours, B3: 4 hours, B4: 6 hours, B5: 8 hours.

LSD test on comparison between treatments against the e-awaking time is presented in Fig. 2. The use of fainting method at 8°C and holding duration of 0 hours (A1B1) showed the fastest recovery time, 0.3 minute and the longest recovery time was recorded in fainting method treatment of 8°C added with clove oil of 0.02% for storing duration of 6 hours (A2B4), 25.16 minutes.

This finding reflects that the use of fainting method added with clove oil of 0.02% for 6 hours of storage results in longer re-awaking time than that a without clove oil addition. It could result from that active compounds in the circulatory system of the fish body at a certain amount make the fish need longer time to return to normal condition. Eugenol as active compound in the clove oil is an antiseptic material that can weaken the nerve and disturb the nerve system [8].

Based on re-awaking time data, it is apparent that fainting method of 8°C is an effective method to faint the fish with storage duration up to 6 hours compared with the fainting method using temperature of 8°C with clove oil addition of 0.02%.

3.3. Mortality of C. carpio

The mortality rate of *C. carpio* is presented in Fig. 3. Mean mortality rate ranged from 0 to 100%.

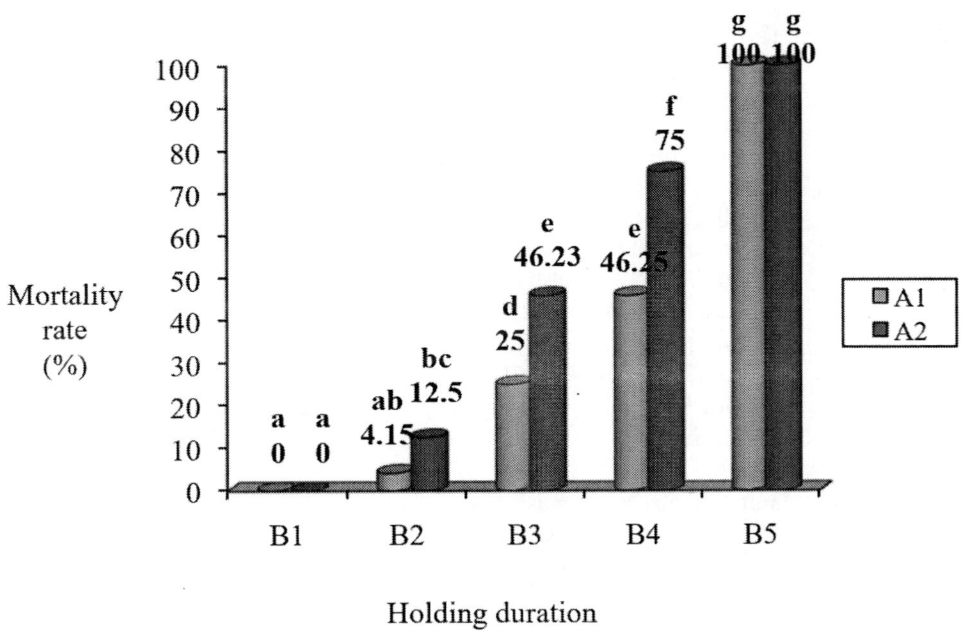

Figure 3. Fainting method and holding time relationship with the mortality rate of *C. carpio*. A1: temperature of 8°C, A2: temperature of 8°C + 0.02% of clove oil; B1: 0 hours, B2: 2 hours, B3: 4 hours, B4: 6 hours, B5: 8 hours;

ANOVA indicated that fainting method, storing duration and treatment interaction gave significant effect on the mortality rate of *C. carpio* ($P \leq 0.05$). LSD test reflected that the mortality rate of *C. carpio* for 6-hour storage using fainting method of 8°C was found lower than that in fainting method of 8°C added with clove oil of 0.02%, 46.25%, and 75%, respectively. It means that the use of fainting method of 8°C is better than that of 8°C added with clove oil of 0.02%.

This study also found that the longer the storing duration was, the higher the fish mortality could be. The fish, particularly *C. carpio*, were only capable of surviving up to 6 hours of storage. All *C. carpio* died in 8 hours of storage.

4. Conclusion

Optimum fainting time was 11.26 minutes and temperature of 8°C through treatment combination with 6-hour storage. Good holding media was rice husk, because the sawdust, despite implementation of initial washing and re-drying, the effect of wood latex on the mortality is still high. Optimum mortality rate, 46%, was obtained at the holding media of rice husk and 6 hours of storage using fainting method of 8°C without clove oil addition. Optimum re-awaking time was found at 11.27 minutes through treatment combination of 6hour holding time and optimum fainting temperature of 8°C. Re-awaking was conducted using 1.5 volts and 3.0-volt aerator, but the fish awake faster at the use of 3.0-volt aerator, since more dissolved oxygen was supplied at the 3.0-volt aerator application than that at 1.5 volts.

References

[1] Amend, D. F., Crov, T.R. & Goven, B. A., Transportation of fish in closed system; methods to control. 1982 ammonia, carbon dioxide, pH and bacterial growth. Trans. Am. Fish .Soc., 111(5):603-ll.

[2] Anggraini, D., Taqwa, F. H, and Yulisman, 2014. Mortality of carp (*Cyprinus carpio*) at the bagasse thickness substrate and different transport duration. J. Fish. And Mar. (8) : 78-89.

[3] Berka, 1986. *The Transport of live fish.* A Review. (EIFAC Technical Paper 48). Food and Agriclture Organization Of The United Nations.http://www.fao.org (29 Mei 2010).

[4] Farrell, A., Tang, S., Nomura, M., & Brauner C., 2010. *Toward Improved Public Confidence in Farmed Fish: A Canadian Perspective on Fish Welfare during Marine Transpor*t. Journal Of The World Aquaculture Society. 41, 225-239.

[5] Ferdiansah, 2000. Toxicity and anesthetic of clove oil (Eugenol Aromaticum) on catfish (*Pengasius hypophthalmus*) FPIK IPB.

[6] Gomes, L.C.; Araujo, L. C. A. R. M., & Roubach, R.., 2003. Effect of fish density during transportation on stress and mortality of juvenile tambaqui Colossoma macropomum. J. World Aquac. Soc., v.34, p.76-84, 2003a.

[7] Grottum, J., Staurnes, M., & Sigholt, T., 1997. Effect of oxygenation, aeration and pH control on water quality and survival of turbot, Scophthalmus maximus (L), kept at high densities during transport. Aquaculture Research 28, 159-164.

[8] Hart, H. 1990. Organic Chemistry. Translation. Suminar Erlangga, Jakarta.

[9] Hjeltnes, B., Waagbo, R., Finstad, B., Rosseland, B.O., Rosten, T.R., & Stefansson, S., 2008. *Transportation of fish within a closed system*. In, p. 63. Norwegian Scientific Committee for Food Safety.

[10] Jailani. 2000. Study on the effect of banana stalk as fill material on the survival rate of carp (*Cyprinus carpio*) [research paper]. Bogor : Fac. Fish. And Mar. Sci., Institut Pertanian Bogor.

[11] Junianto. 2003. Fish handling techniques. Jakarta: Penebar Swadaya.

[12] King, H., 2009. Fish transport in the aquaculture sector: An overview of the road transport of Atlantic salmon in Tasmania. J. Veterinary Behavior -Clinical Application and Research 4, 163-168.

[13] Miranti. 2011. Study on transportation of carp (Cyprinus carpio) using dry system with Styrofoam media. [PKM-AI]. Bogor: Depart. of Aquaculture. Fac.Fish and Mar. Sci. Institut Pertanian Bogor.

[14] Rinto, 2012. *Transportasi ikan hidup*. http://teknologipascapanen. blogspot.co.id/2012/02/ transportasi-ikan-hidup.html. Diakses 22 Mei 2017.

[15] Saskia, Y. Haspeni, E. and Tutik, K, 2012. Toxicity and Anesthetic Ability of clove oil (Eugenol Aromaticum) on red rainbowfish (Glassolepis nisicus). J. Fisheries Sci.. and Aquatic Resources.

[16] Septiarusli, I. E., Haetami, K., Mulyani, Y., and Dono, D. 2012. Secondary metabolite potential of futu seed (*Baringgtonia asiaica*) extract in anesthesis of Tiger grouper (*Ephinephelus fuscoguttatus*). J. Perikanan dan Ilmu Kelautan Vol. 3(3): 295-299.

[17] Suwetja, I K., A. Agustin, M.Mengga. 1993. Live carp packing experiment without water media. Paper. FPIK UNSRAT.
[18] Suwetja, I K., dan J. Pongoh, 2000. Study on rice husk utilization for live carp transportation in the dry insulated wooden box. Report. UNSRAT.
[19] Suwetja, I K., S. Rogi, and J. Pongoh. 2012. Study on sawdust utilization for live fish transportation in dry media container. Report. UNSRAT.
[20] Suwetja I.K., J. Pongoh & Prabawa I. G., 2015. The use of sawdust for live carp transportation in dry Styrofoam. [Final Repot]. Lembaga Penelitian dan Pengabdian kepada Masyarakat, Universitas Sam Ratulangi Manado.
[21] Suwetja I. K., Salindeho. N., & Suwetja, I. G., 2016. *Technical development of live carp in the dry Styrofoam for inter-regency, city, and islandmarketing in order to raise the added economic value. Year-end report on national priority research of Acceleration and Extension Masterplan of Indonesia Economic Development* - 2011-2025(Perprinas MP3EI 2011-2025. Universitas Sam Ratulangi.
[22] Wijayanti, I., Tapotabun, E. J., Salim, A., Nuer'aenaj, N., Litaay, C., Putri, R. M. S., Kaya, A. O. W., and Suwandi, R., 2011. The effect of temperature on the anesthetic condition of freshwater pemfret, *Colossoma macropomum, and freshwater lobster, Cherax quadricarinatus. Proceeding on Small Islands Development,* 2011 - ISBN: 978-60298439-2-7 : 67-76.
[23] Wilford, W.A, 1970. Effect of MS 222 on electrolyte and water content in the brain in rainbow trout. US Burean of Sport Fisheries and Wildlife Investigation in Fish Control.

IOP Conf. Series: Earth and Environmental Science **89** (2017) 012014 doi:10.1088/1755-1315/89/1/012014

Analysis Total Plate Counte (TPC) On Fresh Steak Tuna Applications Edible Coating *Caulerpa* sp During Stored at Chilling Temperature

Meigy Nelce Mailoa[1], Alfonsina Marthina Tapotubun[1], Theodora E.A.A. Matrutty[1]

Fishery Technology Study Program, Faculty of Fisheries and Marine Sciences, Pattimura University, Unpatti Campus – Poka, Jln. Mr. Chr. Soplanit Poka 97233 Ambon Maluku

Email: meigy_mailoa@yahoo.com

Abstract. A study has been conducted to determine the use of *Caulerpa* sp. Edible coatings on fresh steaks tuna against the presence of microbes during storage at chilling temperatures. In this research, two applied method of edible coating is used, that is dipping and immersion method, with chilling temperature (5°C), storage time (0, 3, 6, 9 days). Each treatment is compared to control (without an edible coating). The results showed that the application of edible coating *Caulerpa* sp in fresh steaks tuna with soaking and dipping method showed total bacteria increase during storage, but still fulfill the microbiological quality of fresh fish that is maximum total plate 5.0 x 10^5 cfu/g. Total microbes in fresh steaks tuna were soaked with immersion methods and stored up to 9 days: 2.3 x 10^5 cfu/g while total microbial on fresh steaks tuna were dipped by dipping method and stored up to 9 days ie 3.4 x 10^5 cfu/g. So it can be concluded that application method of edible coating *Caulerpa* sp on fresh steaks tuna soaked better than the method of dipping.

1. Introduction

Along with the increasing population of food needs to be increased, especially fish in the form of fresh. If fresh fish is not handled properly, then there will be a very rapid decline in quality. Biological degradation can occur due to microorganism activity in fish such as bacteria and fungi, while enzymatic damage occurs due to chemical reaction with environment causing rancidity. Changes in freshness quality can take place enzymatically, chemically and bacteriologically with organoleptic decrease [1]. For that fish preservation is necessary to maintain the quality of fish to remain good and can be consumed. Fish that will be consumed or processed must be a fish that is still in fresh condition. If fish quality has decreased, it can cause poisoning in people who consume it [2].

Currently, a new concept is being developed where preservatives as antimicrobial compounds can be made in layers or films on the food surface to maintain longer food durability during storage [3]. Edible coatings or films have been investigated capable of inhibiting moisture, oxygen, aroma, and transport of solutes [4]). Also, edible coating or film is one of the most effective methods of maintaining food quality, as a binder of color, flavor, source of nutrients, and antioxidants. Due to environmental problems, coatings are made from edible biopolymers such as proteins, polysaccharides, and lipids that are usually used as antimicrobials. [4] Edible coating is one layer that is proven to inhibit oxidation. So with edible coating can prevent the occurrence of quality degradation

IOP Conf. Series: Earth and Environmental Science **89** (2017) 012014 doi:10.1088/1755-1315/89/1/012014

and can extend shelf life [5]. Edible coating components are grouped into three, namely hydrocolloids, lipids and composites. Hydrocolloids used in the manufacture of edible coatings are proteins or polysaccharides. Polysaccharides are cellulose and derivatives, starch and derivatives, pectin, seaweed extracts (alginate, carrageenan, agar), gum (arab and gum karaya), xanthan, chitosan and others [6].

The use of edible coating on some food products has been shown to reduce or prevent quality changes and extend the shelf life of the product [3;7;5]. Edible coatings are defined as thin and continuous layers made of edible material, formed overlying food components serving as barriers to mass transfer (moisture, oxygen, light, lipids and solutes) or as an additive food, and improving the handling of a material food. Thus, food products coated with an appropriate edible coating can be protected from damage so that the shelf life can be extended.

The edible layer can be made from polysaccharides, proteins, and fat derivatives that can be used as an efficient shield against vapor and oxygen displacement. One of the natural preservatives of aquatic products that can be used in the manufacture of edible coating is the *Caulerpa* sp marine wine, a type of seaweed from the Chlorophyta group that thrives as an endemic species in the Maluku Kei Islands throughout the year and is known as "lat."

The use of Coating (coating) can be used as packaging such as wax, resin, and plastic materials for protective material[8]. The edible coating is an alternative to replacing plastic packaging because it is biodegradable as well as a barrier to control the transfer of water vapor, oxygen uptake, and lipid transfer. Edible coatings can also be used to coat products that serve as a protector of mechanically and safely damaged [9].

Fresh tuna meat steak very quickly experiences quality degradation if not handled quickly because tuna fish has high protein and fat content so that the activity of bacteria and enzymes in meat fish. One handling of tuna steak is by using coating (coating) from seagrass, so it can inhibit the activity of bacteria that can degrade the quality of the fresh tuna steaks.

Based on the above problems the authors are interested to research the existence of total bacteria in fresh tuna steaks application of edible coating *Caulerpa* sp in fresh steaks tuna with soaking and dipping method showed total bacteria increase during storage. The aim of this research is to know the effect of using "lat" on fresh tuna steaks stored at low temperature (5°C) with storage time up to 9 days. The benefits of this study provide information on the total presence of bacteria that can degrade the quality of freshly fresh tuna steaks given "lat" as edible coatings at low-temperature storage as a natural preservative.

2. Methodology

2.1. Materials and tools

The material used in the manufacture of fresh tuna steaks is fresh tuna fish obtained from the market in Arumbai Ambon City. While the material used for edible coating is from "sea" grapes obtained from Tual waters, Southeast Maluku. Glycerol and wax. The chemicals used are aquades, 0.9% NaCl, PCA. The tools used are knives, trays, duster and scales. While in the microbiological analysis is the scales, measuring glass, erlenmeyer, Petri dish, test tube, autoclave, incubator, counter colonies.
This research uses experimental method, that is by direct experiment by doing addition of edible coating from lat as inhibiting regression of fresh tuna fresh steaks kept low temperature with storage time 0, 3, 6, and 9 days.

2.2. Implementation of Research

A total of 100 grams of *Caulerpa* sp that has been washed with water to flow clean, blended until smooth. Next add glycerol as much as 1 ml and wax, then stir until homogeneous, then cook the dough on the stove to boil. After a bit cold, apply on the tuna meat ready to be processed and has been cleaned. The application of *Caulerpa* sp. The edible coating on tuna meat is done by immersion and immersion. Tuna meat that has been coating incorporated PE plastic and stored in refrigerator.

2.3. TPC Test (Total Plate Count)

The working principle of TPC analysis is the calculation of the number of bacterial colonies present in the sample with dilution as needed and done duplo. All work is done aseptically to prevent undesirable contamination and duplicate observation can improve accuracy. The number of bacterial colonies that can be calculated is a petri dish that has a bacterial colony between 30-300 colonies. Petri dishes, test tubes and pipettes before use are sterilized in oven at 180°C for 2 hours. The media was sterilized in an autoclave at a temperature of 121°C for 15 minutes at a pressure of 1 atm. After sterilization, to keep the media from freezing the media temperature is maintained at 45-55°C in a water bath. The diluent solution was prepared by dissolving 8.5 grams of NaCl in 1 liter of aquadest which was then sterilized in an autoclave at 121 ° C for 15 minutes. The sample of 10 grams was mashed first, then dissolved into a sterile diluent solution that has been contained with a volume reaching 100 ml to obtain dilution 10^{-1}. The solution is then pelleted 1 ml, then fed into a test tube containing 9 ml of a sterile diluent solution to obtain 10^{-2} dilution. And so on until obtained dilution 10^{-5}. From each tube the dilution reaction is taken by using a pipette of 1 ml then put into a sterilized petri dish. Each dilution is done in duplicate. Then each cup is moved in a circle on the table so that the PCA media evenly. After PCA freezes, the petri dish is incubated in the incubator for 48 hours at 30°C; the petri dish is placed upside down in the incubator.

2.4. Data analysis

Data analysis is done descriptively, data is presented in table form (histogram) and picture (portrait).

3. Result and Discussion

3.1. Making edible coating lat applied to fresh tuna steak

A total of 100 grams of *Caulerpa* sp that has been washed with water to flow clean, blended until smooth. Next add glycerol as much as 1 ml and wax, then stir until homogeneous, then cook the dough on the stove to boil. After a bit cold, apply to the tuna meat ready to be processed. The application of *Caulerpa* sp. The edible coating on tuna meat is done by immersion and immersion. Tuna meat that has been coating put into plastic box and PE plastic and stored in refrigerator.

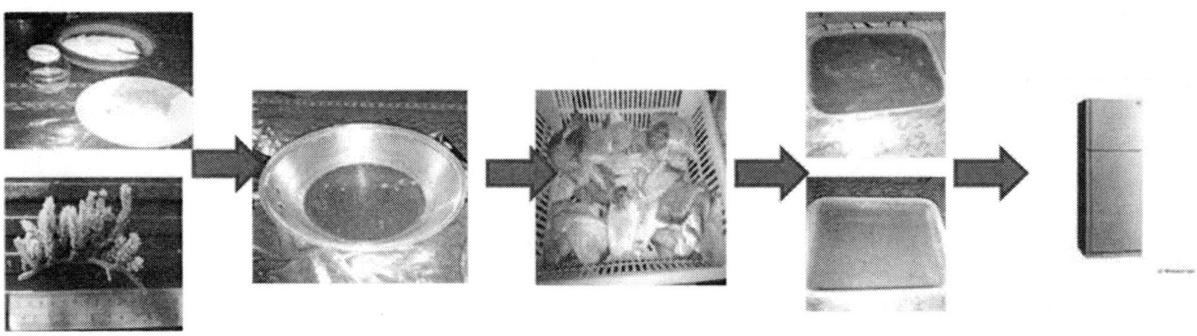

Figure 1. Edible Coating *Caulerpa* Applications at Fresh Steak Tuna

2. Total Plate Count At Fresh Steaks Tuna

Foods that contain proteins such as meat and fish are destroyed by bacteria. Food products are rarely sterile and are polluted by some microorganisms because microorganisms are widespread in the natural environment. The growth of microorganisms in or on food can result in unwanted physical and chemical changes, so the food is not feasible for consumption anymore [10]. The most prominent causes of vomiting are microorganisms and various enzymatic and nonenzymatic changes occurring after Pen, slaughter or processing [10].The calculation of colonies, based on data from each sample is only calculated dilution with the number of colonies between 30-300.This aims to minimize the probability of error in the calculation. Because the experiment was done twice (Duplo) should then use data from both repetitions by taking the average of the two data, calculated and compared with Indonesian National Standard for fresh tuna quality Results of analysis of total plate count on fresh tuna steaks applied with edible coating lat as shown In table 1.

Table 1. Total bacteria on fresh tuna steaks applied edible coating lat

Storage Duration (days)	Total Bacteria on fresh tuna steaks without edible coating lat (CFU/g)	Total Bacteria on fresh tuna steaks soaked with edible coating lat (CFU/g)	Total Bacteria on fresh tuna steaks dyed with edible coating lat (CFU/g)
0	2.2 x 10^4	**1.8 x 10^4**	**2.5 x 10^4**
3	**4.6 x 10^5**	3.3 x 10^4	3.4 x 10^4
6	3.5 x 10^6	3.1 x 10^4	4.1 x 10^4
9	2.9 x 10^8	**2.3 x 10^5**	**3.4 x 10^5**

Based on the results of TPC analysis in Table 1 above, it can be seen that fresh tuna steaks without edible coating lat have a total value ranging from 2.2 x10^4 to 2.9 x 10^8 cfu /g this indicates that the sample of fresh tuna steaks without edible coating lat is still acceptable Microbiological up to 3 days storage. While fresh tuna steaks applied by edible coating lat with soaking method indicate that as long as the total storage of bacteria has increased, and up to the storage time of 9 days fresh tuna steaks still meet the requirements of microbiology because it still meets the maximum limit of total bacteria in fish maximum 5.0x 10^5 cfu /g. The same is also seen in fresh tuna steaks treated with edible coating lat applications with dyeing methods. Total bacteria increased during storage from 2.4 x 10 ^{4} to 3.4 x10^5 cfu /g. The growth of bacterial colonies on fresh tuna steaks as shown in Figure 2.

Figure 2. Growth of Bacterial Colonies at Fresh Steaks of Tuna

Bacteria in fresh fish are concentrated on the skin, gills and stomach contents. According [11], at the beginning of storage of total bacteria found in fish is relatively no different, the bacteria contained in fish will increase with the length of storage. This is because the environment can affect the growth of bacteria to the maximum. The treatment of edible coating application on fresh tuna steak with the method of immersion to be the best because it can inhibit the decay activity better than the treatment of edible coating application on fresh tuna steak with the method of dyeing with the old storage for 9 days. Where the value of 2.3 x 10^5 cfu /g which according to SNI 01-2729.1-2006 limit TPC value is 5.0x105 colony/g.

According [12], tried the application of crushed seaweed *Caulerpa* sp. In fresh fish, can suppress the growth of bacteria and is suitable for consumption according to SNI up to 48 hours of storage at room temperature with a log TPC value of 2.7 cfu /g (24 hours) and log 3.7 cfu /g (48 hours). This means *Caulerpa* has the potential to suppress higher bacterial activity. Furthermore, [13], reported that application of 10% to 30% *C. lentilliera* sea wine showed the ability to inhibit the growth and development of bacterial groups of *Coliform* and *Escherichia coli* and according to safety standard of SNI of fresh fish. Fresh fish applied by *C. lentillifera* is feasible to be consumed until the second-day storage at room temperature and third day at cold temperature. *C. lentillifera* sea wine can inhibit the presence of pathogenic bacteria in fresh fish, and a 10% effective enough concentration is applied to inhibit Escherichia coli activity and prevent contamination of *Vibrio cholerae* and *Salmonella* sp. During storage of 2 x 24 hours at room temperature and 3 x 24 hours at the rated temperature. Also, several studies of the utilization of natural materials to maintain the quality of fresh fish as reported [14] using ethanol extract of seaweed *Padina* sp. In fresh bloated fish shows the value of TPC proper consumption according to SNI only until the storage of the 6th hour. While [15] shows the value of TPC of fresh tuna with the addition of quercetin is in the range of log 5 cfu/g up to log 8 cfu/g. The addition of 1% quercetin compound from the plant can maintain the safety of fish meat (based on histamine levels) during storage of room temperature up to 20 hours. Furthermore, [10] states that temperature is one of the most important environmental factors that affect life and growth because temperature affects the speed of metabolism and growth.

The mechanism of action of antibacterial compounds, in general, is to damage the structure of the main structure of microbial cells. This shows that the provision of edible coating lat on fresh tuna steak with soaking methods more effectively inhibits microbial growth. Microbial growth in food products is influenced by oxygen displacement during storage. The existence of edible coating then, the transfer of oxygen from the environment to food products can be inhibited so that the growth of aerobic bacteria is also hampered. Edible coating lat on fresh tuna steak can act as an inhibitor of oxygen transfer so that aerobic microbes cannot grow and develop.

4. Conclusion

Edible coating lat applied to fresh tuna steaks can give inhibition effect of fresh tuna steak stored at the low temperature (5°C) effect on total bacteria during storage at cold temperature.

References

[1] Sanger Grace 2010 Oksidasi lemak ikan tongkol (*Auxis thazard*) asap yang direndam dalam larutan ekstrak daun sirih. *Pacific Journal* 2(5):870-873

[2] Milo, S.M., L.M. E. Purwijatiningsih., S. Pranata 2013 Mutu Ikan Tongkol (*Euthynnus affinis C*) di Kabupaten Gunung Kidul dan Sleman Daerah Istimewa Jogjakarta. Skripsi. Fakultas Teknobiolgi. Universitas Atmajaya. Yogyakarta

[3] Guilbert, J.M. dan Park, C.F. Jr., 1986 The Geology of Ore Deposits. W.H. Freeman and Company: New York.

[4] Quattara B, Simard R.E, Piette G. Begin A. Holley R. A 2000 Diffusion of aceptic and propionic acidc from chitosan based antimocrobial packaging films. *J Food Sci* 65 :768-773

[5] Krochta J.M 1992 Control of mass transfer in food in edible coating and film. In : singh. R.P and M.A Wirakarrtakusumah (Eds) : anvances in food engeering. CRC Press: Boca Raton, F.L.pp. 517-538.

[6] Karnila R. 2013. Kemasan *Edible* [Bahan Ajar Pengemasan]. Fakultas Perikanan dan Ilmu Kelautan. Universitas Riau. Pekanbaru.

[7] Kester , J.J., dan Fennema, O.R 1986 Edible film and Coatings: a Review.Food Technology

[8] Karnila, R dan B. Dewita 2006 Buku Ajar Teknik Pengemasan Perikanan dan Ilmu Kelautan Universitas Riau. Pekanbaru

[9] Darni. Yuli H. Utami dan. Asriah SN 2009 Peningkatan hidrofobisitas dan sifat fisik plastik biodegradable pati tapioka dengan penambahan selulosa residu rumput laut (*Euchema spinossum*). *Jurnal Fakultas Teknik,* Universitas Lampung.

[10] Buckle KA, Edwards RA, Fleet GA, Wootton M 1987 Ilmu pangan. Penerjemah H.Purnomo dan Adiono. Jakarta: UI Press

[11] Amin dan Leksono 2001 Efektivitas Bakteri Asam Laktat dalam Menghambat Bakteri Airlangga. Jogyakarta

[12] Sabono I 2015 Studi pendahuluan mutu ikan layang (*Decapterus macrosoma*) segar yang diaplikasi hancuran anggur laut (*Caulerpa* spp.) selama penyimpanan dingin [Skripsi]. Ambon. Fakultas Perikanan dan Ilmu Kelautan Universitas Pattimura.

[13] Tapotubun AM, Savitri IKE, Matrutty TEAA 2016 Penghambatan bakteri patogen pada ikan segar yang diaplikasi *Caulerpa lentillifera*. *Jurnal Pengolahan Hasil Perikanan Indonesia* 19(3): 299-308.

[14] Husni, A., Utadi dan A. Hakim 2014 Penggunaan Ekstrak Rumput Laut *Padina* sp. untuk Peningkatan Daya Simpan Filet Nila Merah yang Disimpan pada Suhu Dingin. *Agritech*. 34 (3).

[15] Prasetiawan NR, Tri WA, Widodo FM 2013 Penghambatan pembentukan histamin pada daging ikan tongkol (Euthynnus affinis) oleh quercetin selama penyimpanan. *Jurnal Pengolahan Hasil Perikanan Indonesia* 16(2):149-157

IOP Conf. Series: Earth and Environmental Science **89** (2017) 012015 doi:10.1088/1755-1315/89/1/012015

Maturity Gonad Sea Cucumber *Holothuria scabra* Under The Month Cycle

Santi Penina Tua Rahantoknam

Management of Marine Mariculture Engineering Study Program
Tual State Fisheries Polytechnic, Jl. Raya Langgur - Sathean, Km 6. Southeast Maluku, Indonesia 97611

Email: santirahantoknam@yahoo.com

Abstract. Gonad maturity level of the sea cucumber *Holothuria scabra* is important to note for selection of parent ready spawn. Sea cucumbers are giving a reaction to the treatment of excitatory spawn mature individuals only. For the determination of the level of maturity of gonads of sea cucumbers, the necessary observation of the gonads are microscopic, macroscopic and gonad maturity gonado somatic indeks (GSI). GSI value is important to know the changes that occur in the gonads quantitatively, so that time can be presumed spawning (Effendie, 1997). Reproductive cycle can be determined by observing the evolution of GSI. The study of sea cucumbers *Holothuria scabra* gonad maturity conducted in Langgur, Southeast Maluku. Observations were made at every cycle of the moon is the full moon phase (BP) and new moon (BB) in the period January 29, 2017 until July 23, 2017. Observations *H. scabra* gonad maturity level is done with surgery, observation and calculation GSI gonad histology. GSI highest value obtained in May that full moon cycle at 90% of individuals that are in the spawning stage (phase 5), then 70% of the individuals that are in the spawning stage (phase 5) in March that the full moon cycle. The results obtained show that the peak spawning *H. scabra* period January 2017 to July 2017 occurred on the full moon cycle in May.

1. Introduction

Sea cucumbers are marine products that have significant value and are used as foodstuffs and used as medicines. Sea cucumbers are a fisheries products that have long been known and consumed by coastal communities in Indonesia and is also very well known in the countries of Continental Europe, Japan and the United States. Prospects are good for sea cucumbers commodities led to the arrest of these commodities continue to increase and be done on a large scale without regard to sustainability.

Improved retrieval of sea cucumbers from their natural habitat resulted in a decrease in the population of sea cucumber stocks are characterized by the increasing difficulty of obtaining cucumbers in nature. Resource sea cucumber has received serious attention from CITIES about extinction are likely to increase the types of resources, especially of species of sea cucumber that has commercial value [1]. Sea cucumbers have the nature of a relatively slow movement, usually live solitary and attendance is dependent on the availability of food and habitat conditions. The occurrence of overfishing due to excessive fishing will result in a lower density of population or stock in the habitat. This will affect the success of fertilization between male and female sea cucumbers so that the rate of reproduction decreases. The discrepancy between the level of resource utilization of sea cucumbers at a rate of regeneration causing low recruitment or addition of sea cucumber populations in habitats that would accelerate the extinction of sea cucumber populations.

Therefore, the availability of sea cucumber populations in the wild needs to be considered. Efforts to maintain the availability of the sea cucumber is the management of sea cucumber capture settings. Reproductive factors are among the factors that can be used as a base or an indication for the management of water resources and in particular for the survival of offspring an organism or population. This approach is employed to provide information when sea cucumber resources can be exploited in line with the sustainability of these resources to the front [10]. Many environmental factors are thought to affect the reproductive aspects such as the temperature of sea cucumbers, blooming phytoplankton and the moon cycle [7]. Reproductive aspects can also be influenced by environmental factors, among others the moon cycle [4]. [9] also supported the argument that the cycles of the moon are one of the environmental factors that can affect the reproductive cycle of sea cucumbers.

This study aims to identify and analyze the level of maturity of the gonad and gonado somatic index (GSI) *H. scabra* based on cycles of the moon, this research target is terdokumentasinya data and information about the reproductive cycle of *H. scabra* which can be used as an input determination system resource management in the waters of sea cucumbers Small Kei, Southeast Maluku.

2. Methodology

2.1. Time and place

The study was conducted during the period January 2017 - July 2017, based on the full moon cycle (BP) and new moon (BB). The research location is in the waters Langgur Small Kei, Southeast Maluku and Hatchery laboratory of Tuat State Fisheries Polytechnic.

2.2. Research design

Observations of sea cucumber gonads maturity level are based on cycles of the moon is the full moon (BP) and new moon (BB) from January 2017 to July 2017. Samples were taken every cycle of sea cucumbers in as many as 10 individuals for observation of gonad maturity.

2.3. Variable and Measurement Methods

Observations gonad maturity level (TKG) and gonado somatic index (GSI) is based on cycles of the moon. Determination of TKG and GSI done by dissected gonado sea cucumbers and observed in histology and GSI value calculation.

2.4. Data analysis

Stage of gonadal development was observed in all phases of development according to [5]. As for knowing the value of using GSI_2 GSI [2] which is the percentage of gonad weight to body weight after surgery using the equation:

$$GSI_2 = \frac{wg}{W} \times 100$$

Where:
GSI : Gonado somatic index (%)
wg : gonad weight
W : body weight

3. Results and Discussion

3.1. The maturity level of gonads

H. scabra gonad developmental stages were observed during the period January 2017 to July 2017 on each cycle of the moon is new moon phase (BB) and full moon (BP) shown in Figure 1 below.

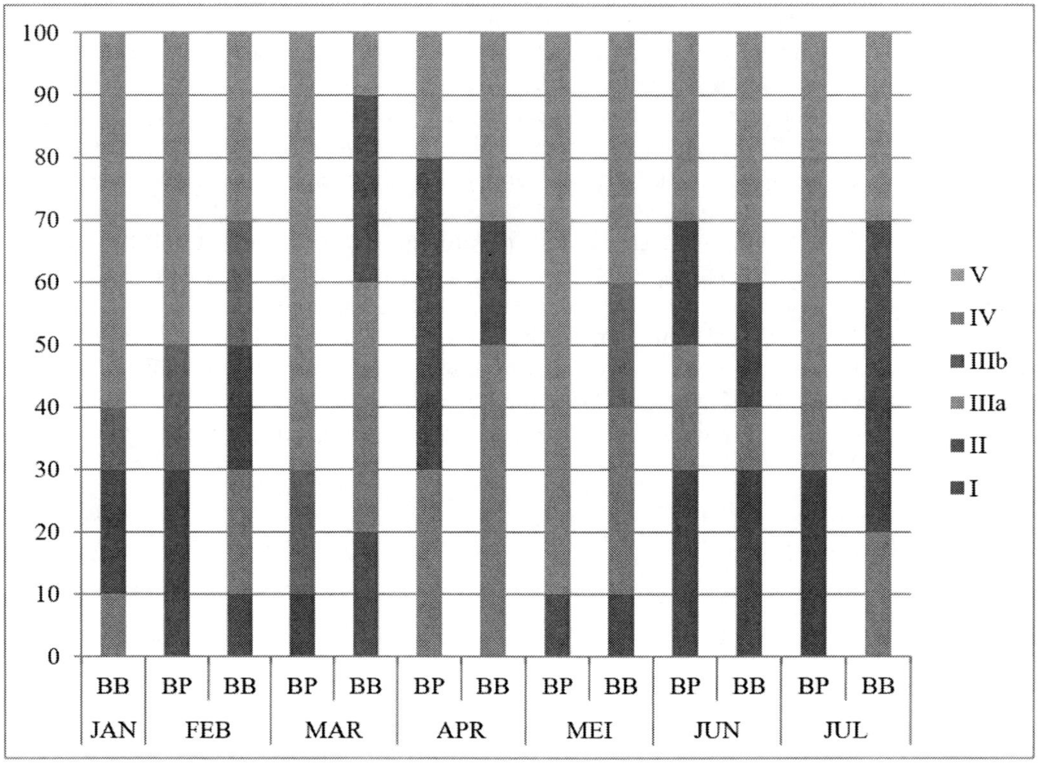

Figure 1. Percentage of gonadal development stage *H. scabra*

Where: BB: New Moon
BP: Full Moon

Figure 1 shows that the highest maturity level of gonads (spawning phase or phase V) occurs in May that the BP cycle of (90%), and in March, BP cycle spawning phase (V) equal to (70%). Cycle months in others observed that, the stage of gonad development vary where all phases of the development was observed both in the phase of emptying (I), the phase of development (II), the phase of activation of I (IIIa), phase activation II (IIIb), the maturation phase (IV) and spawning phase (V).

3.2 Gonado somatic index

The reproductive cycle of *H. scabra* can be determined by observing the evolution of GSI. Cycle *H. scabra* observed GSI shown in Figure 2 below. Sometimes the cycle GSI locally need to know because GSI sand sea cucumbers vary from one place to another [2, 11]. The highest peak of the cycle GSI observed period January 2017 - July 2017 occur at full moon cycle (BP) in May amounted to 16.8%. Average GSI is based on cycles of the moon is observed that the highest GSI occurs in cycles BP, except in January because there was no observation GSI on BP cycle. The average value of GSI *H. scabra* at the time of the full moon is higher than the new month in which the highest value occurs in May. This shows that during the reproduction process *H. scabra* research culminated in full, especially in May.

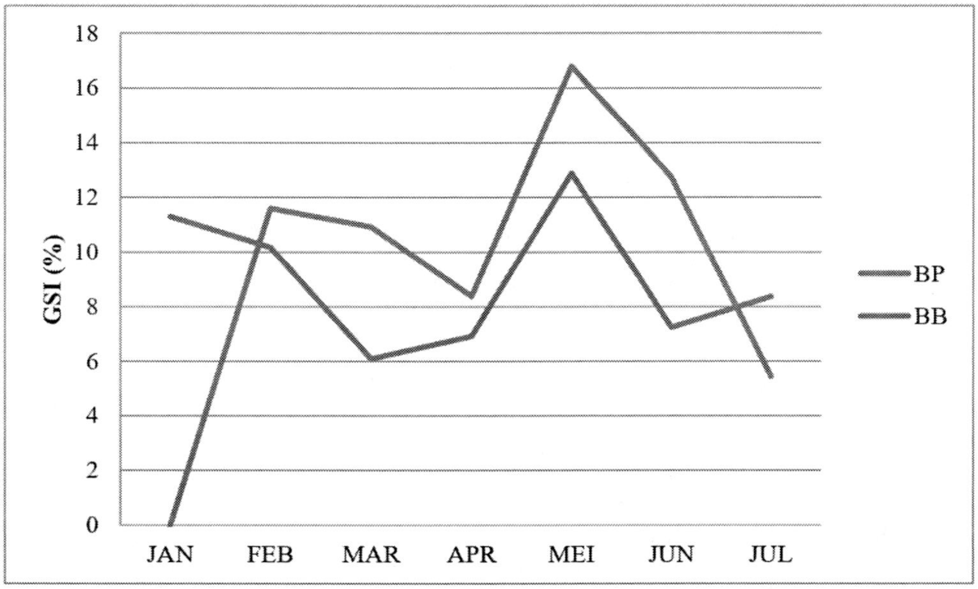

Figure 2. Mean of GSI *H. scabra* on BP and BB

The weight ratio of the gonads and GSI can be seen in Figure 3. The mean weight of the gonads was highest in June amounted to 10.9 grams BP cycle and gonad weight average lows in July of 2.7 grams on BP cycle. While the highest average on May GSI BP cycle is equal to 16.8%, then 12.88% also in May cycles BB and 12.74% in June that BP cycle. Lowest GSI during the observation period occurred in July on BP cycle is equal to 5:45%. Cycle BP in May finds that on average the highest GSI 16.8% and the highest maturity level of gonads (spawning phase or phase V) equal to (90%). [3], states that the value of the gonado somatic index (GSI) in line with the development of the gonads, the index will increase in size and the value will reach the maximum range limit during the spawning will occur. [10] also agree that the GSI will reach the maximum limit before the spawning, then decline after spawning individuals.

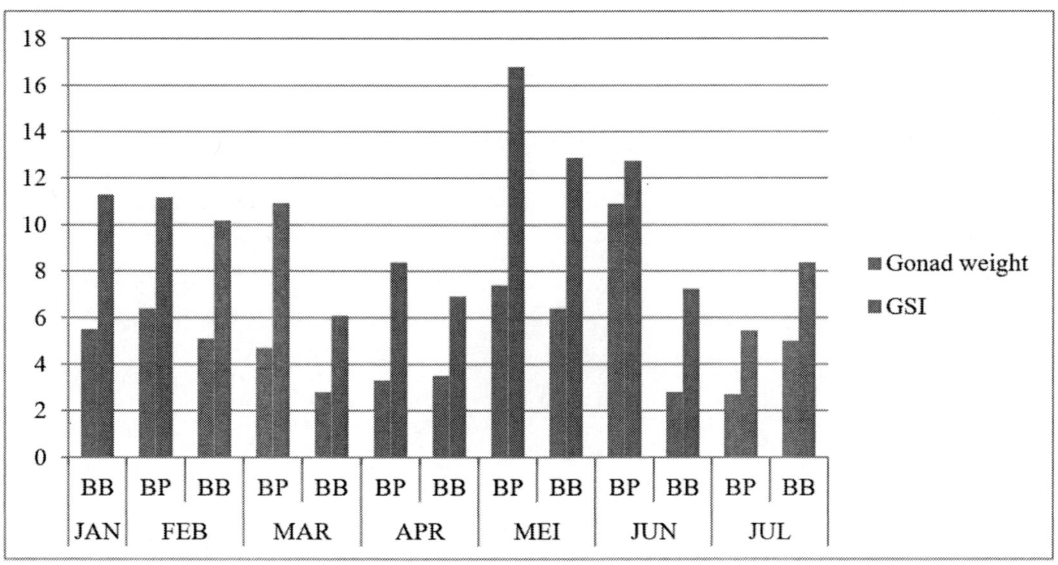

Figure 3. The mean weight of gonads and value GSI *H. scabra*

The usefulness of knowing the value of GSI was to determine the changes that occur in the gonads quantitatively, so that time can be presumed spawning [3]. [7], stated that the external factors affecting the spawning Holothuria are cycles of the moon. [6] also pointed out that can control the lunar cycle in the sea cucumber spawning by stimulating endogenous hormones. This month period or affect spawning cycle by stimulating endogenous factors. According to [9], cycles of the moon appear to collect all the stimuli that can cause sea cucumber spawning. *H. scabra* also shows lunar spawning rhythm during mating season, which spawn at dawn at the approach of a new moon or a full moon or when the highest tide [8].

4. Conclusion

Based on this research can be concluded that the average maturity of the gonad index (GSI), the highest of 16.8% occurred in the BP cycle in May and the maturity level of gonads (TKG) by 90% in the spawning phase or phases in the cycle V BP May.

References

[1] Bruckner, AW, KA, Johnson, and JD Field. 2003. Conservation strategies for sea cucumber: Can a CITIES Appendix II Listing Promote Sustainable International Trade?. SPC Beche-de-mer Information bulletin, May 2003.
[2] Conand, C. 1981. Sexual cycle of three commercially important species holothurians (Echinodermata) from the lagoon of the New Caledonia. Bull. Mar. Sci 31 (3): 523-543.
[3] Effendie, MI 1979. Methods of Fisheries Biology. Yayasan Dewi Sri. Bogor. P. 36
[4] Hartati, R. and Heri Yanti. 2004. Study of gonads Teripang latex (Holothuri vagabunda) at the time of the full moon and new moon in the waters Bandengan Jepara. A Scientific Journal on Marine Related Sciences, inhttp://ik-ijm.com/,
[5] Jiaxin, C. 1990. Brief Introduction to Mariculture of Five Selected Species in China In: Seafarming Regional Development and Demonstration Project. Bangkok. Thailand. P. 10-13
[6] Kubota, T. and M. Tomari. 1998. Reprooduction in the apodid sea cucumber Polycheira rufenses: semi lunar spawning rhythm and sex change. J. Mar. Biol. Ass. UK 78249-267.
[7] Mackey, A. 2001. Factors That Influence the Reproduction of Sea Cucumber.http://www.sci.sdsu.edu/biology/bio515/hentshel/PDFs/Mackey%283001%29.pdf, March 17, 2001.
[8] Morgan, AD 2000. Induction of spawning in the sea cucumbers Holothuria scabra (Echinodermata: Holothuroidea). J. World. Aqua. Soc 31; 186-194.
[9] Ramofafia, C., Byne, M., and Battaglene, CS 2003. Reproduction of the commercial sea cucumber Holothuria scabra (Echinodermata: Holothuroidea) in the Solomon Islands. Mar. Biol. Bull 142: 281-288.
[10] Tuwo, A and Nessa, MN 1992. Some Aspects of Biology Cucumber Important Economical. Jur. Fisheries, Fak. Livestock, Hasanuddin University, Makassar .. Page. 1-18.
[11] Tuwo, A., J. Tresnati, Syafiuddin and R. Bohari. 1996. Reproductive Biology, Parent Care and Ovulation Artificial Sand Sea Cucumber Holothuria scabra. Seminar Research Results. Manado. 13 things

Infection *Vibrio* sp. Bacteria on Kappaphycus Seaweed Varieties Brown and Green

Yuni Irmawati[1] Fien Sudirjo[2]

[1]Biotechnology Study Program, Tual State Fisheries Polytechnic, Indonesia
[2]Technology Study Program of Fishery Products, Tual State Fisheries Polytechnic, Indonesia

Email: yuni_polikant@yahoo.com

Abstract. Disease in seaweed or ice-ice, until today is still a major problem in the cultivation of seaweed. Changes in extreme environmental conditions is a trigger factor of ice-ice, which can result in seaweed susceptible to infection with pathogenic microorganisms, such as bacteria *Vibrio* sp. This research aims to determine the bacteria *Vibrio* sp. infection in seaweed *Kappaphycus* varieties of brown and green. *Vibrio* sp. bacteria isolated in the infected seaweed thallus ice-ice, grown on TCBS media, purification, gram staining and biochemical tests. *Vibrio* sp. infected to seaweed *Kappaphycus* brown and green varieties in containers controlled by different density, 10^5 CFU/ml, 10^6 CFU/ml and 10^7CFU/ml. Observations were made to change clinical effect in thallus seaweed for 14 days of observation. The results obtained show that the levels of infection bacteria *Vibrio* sp. higher in seaweed *Kappaphycus* green varieties both in density 10^5 CFU/ml, 10^6 CFU/ml and 10^7CFU/ml, when compared with varieties brown.

1. Introduction

Diseases of seaweed or ice-ice, until now still a major problem in seaweed cultivation, this attack it can cause damage to the seaweed thallus, so often farmers are in cultivation. Changes in extreme environmental conditions are the trigger factors of ice - ice disease, which can store seaweed vulnerable to infection of pathogenic microorganisms, is bacteria. The results of several studies indicate the presence of diseases in the cultivation in the sea, both fish and shrimp as well as on seaweed are the bacteria *Vibrio* sp. *Vibrio* sp. in fish and shrimp both cultivated and living in nature, has been widely known. However, infectious information directly on seaweed in controlled containers has not been widely known and published, especially in *Kappaphycus* brown and green varieties of seaweed which are the main commodities in seaweed cultivation, so this research can serve as preliminary data for further research to reduce the level of pathogenicity of *Vibrio* sp.

2. Methodology

This research was conducted in June-August 2016, and are analyzed in the microbiology laboratory, Tual State Fisheries Polytechnic. The equipment that is used consist of tube, petri dish, beaker glass, objek glass, cover glass, hotplate stirrer, needles Ose, autoclave, incubator, microscope, battery aerator, and aquarium.

Materials; seaweed, media Thiosulfate Citrate Bile Salt Sucrose (TCBS), gram staining materials (violet crystal, lugol iodine, safranin) alcohol 95%, paper oxidase, Media OF (Oxidase-Fermentation), H_2O_2 3%, aluminum foil, sandpaper, aqua, cotton, tissue, and tape.

The cultivation of seaweed begins with the filling of seawater in aquarium, installation of rope for binding of seaweed and aeration installation. Before the experiment, seaweed acclimatized for 5 days. *Kappaphycus* seaweed, brown and green varieties are placed on different aquariums, which consist of 3 treats with 2 replications and as controls. The treatments were infected with different bacterial densities of 10^5 CFU/ml, 10^6 CFU/ml and 10^7 CFU/ml by submersion method.

The samples bacteria taken from thallus of seaweed that is exposed ice-ice. By scratching the ose needle on the ice-ice section, recast on the TCBS medium and incubated in the incubator at 28^0C - 30^0C for 24-48 hours. The reconstituted bacteria, subsequently grown into Nutrient Agar medium, were scrawled on Nutrient Agar Slant, tested by gram staining, and identified by biochemical and Bergey's Manual of Determinative Bacteriology [4].

3. Results and Discussions

Gram staining on bacterial isolates showed that the colonies obtained were gram-negative bacteria because of the binding of safranin colors on the cell walls of these bacteria. [8] States that gram-negative pathogenic bacteria, its outer walls play an important role in infecting and damaging the hosted host. The results of identification of bacteria are presented in Table 1.

Table 1. Results of *Vibrio* sp. biochemical testing

Characteristic	Result
Gram staining	Gram-negative
Oxidase	+
Catalase	+
Oksidase/Fermentation	**F**
Motility	+

The results of biochemical testing of colonies obtained according to the characteristics of the bacteria *Vibrio* sp. According [5] for the characteristics of a biochemical test of bacteria *Vibrio* sp. is a positive oxidase and catalase test, fermentative oxidative/fermentative test, motile motility test, and negative gram staining. The results of observation of changes in the effect of bacterial infection *Vibrio* sp. on *Kappaphycus* brown seaweed varieties are presented in Figures 1 and Table 2, while for green varieties in Figures 2 and Table 3.

Figure 1. Changes of thallus on *Kappaphycus* brown varieties

Table 2. Observation of clinical effect of bacterial infection *Vibrio* sp. on *Kappaphycus* seaweed varieties of brown

Days	*Vibrio* sp. Density			
	10^5	10^6	10^7	Control
1	Has not seen any changes	Has not seen any changes	Has not seen any changes	Has not seen any changes
2	Has not seen any changes	Has not seen any changes	Has not seen any changes	Has not seen any changes
3	Has not seen any changes	Has not seen any changes	Has not seen any changes	Has not seen any changes
4	Has not seen any changes	Has not seen any changes	Has not seen any changes	Has not seen any changes
5	The color of the thallus begins to change	Has not seen any changes	Has not seen any changes	Has not seen any changes
6	The color of the thallus becomes slightly younger than the original color	Has not seen any changes	Has not seen any changes	Has not seen any changes
7	Thallus began to release mucus	The color of the thallus begins to change	Has not seen any changes	Has not seen any changes

IOP Conf. Series: Earth and Environmental Science **89** (2017) 012016 doi:10.1088/1755-1315/89/1/012016

8	Color differences in some branches began to be seen clearly	The color of the thallus becomes slightly younger than the original color	The color of the thallus begins to change	Has not seen any changes
9	The former part of the fault on the thallus color begins to change	Thallus began to release mucus	The color of the thallus begins to change	Has not seen any changes
10	Part of the color at the end of the thallus color begins to change	Color differences in some branches began to be seen clearly	The color of the thallus becomes slightly younger than the original color	The color of the thallus begins to change
11	At the tip of the thallus, there are white spots	The former part of the fault on the thallus color begins to change	Thallus began to release mucus	The color of the thallus becomes slightly younger than the original color
12	White spots on the end of the thallus occur almost in all parts of the branch	Part of the color at the end of the thallus color begins to change	Color differences in some branches began to be seen clearly	Thallus began to release mucus
13	White spots begin to widen	At the tip of the thallus, there are white spots	The former part of the fault on the thallus color begins to change	Color differences in some branches began to be seen clearly
14	The tip of the thallus changes color to white	White spots on the end of the thallus occur almost in all parts of the branch	The color of the thallus becomes slightly younger than the original color	The surface of the thallus begins to shrink

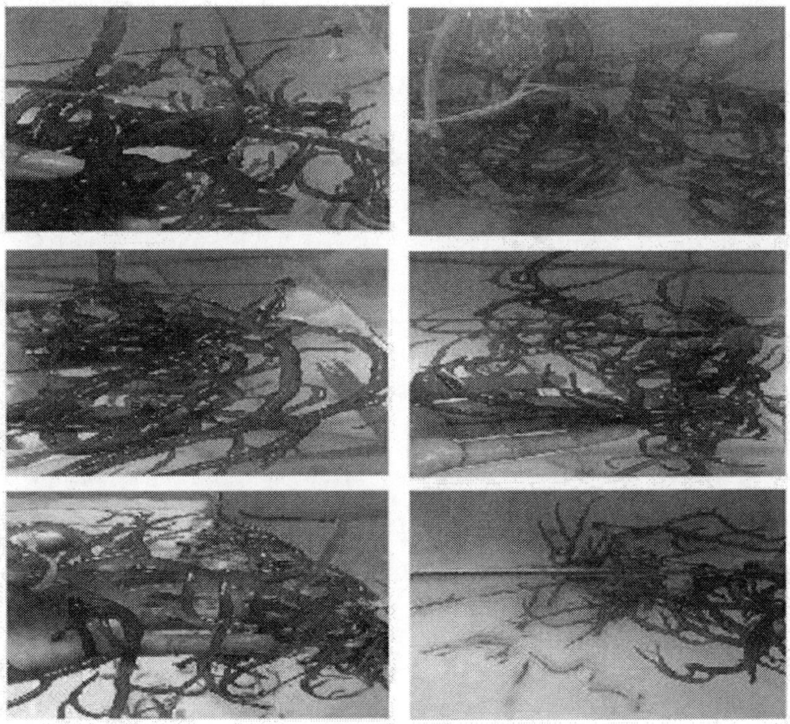

Figure 2. Changes of thallus on *Kappaphycus* green varieties

Table 3. Observation of clinical effect of bacterial infection *Vibrio* sp. on *Kappaphycus* seaweed varieties of green

Days	*Vibrio* sp. Density			
	10^5	10^6	10^7	Control
1	Has not seen any changes	Has not seen any changes	Has not seen any changes	Has not seen any changes
2	The color of the thallus begins to change	Has not seen any changes	Has not seen any changes	Has not seen any changes
3	The color of the thallus becomes slightly younger than the original color	The color of the thallus begins to change	Has not seen any changes	Has not seen any changes
4	Thallus began to release mucus	The color of the thallus becomes slightly younger than the original color	The color of the thallus begins to change	Has not seen any changes
5	Color differences in some branches began to be seen clearly	Thallus began to release mucus	The color of the thallus becomes slightly younger than the original color	Has not seen any changes
6	The former part	Color differences in	Thallus began to	Has not seen any

	of the fault on the thallus color begins to change	some branches began to be seen clearly	release mucus	changes
7	At the tip of the thallus, there are white spots	The former part of the fault on the thallus color begins to change	Color differences in some branches began to be seen clearly	Has not seen any changes
8	White spots on the end of the thallus occur almost in all parts of the branch	At the tip of the thallus, there are white spots	The former part of the fault on the thallus color begins to change	Has not seen any changes
9	White spots begin to widen	White spots on the end of the thallus occur almost in all parts of the branch	At the tip of the thallus, there are white spots	The color of the thallus begins to change
10	The tip of the thallus changes color to white	White spots begin to widen	White spots on the end of the thallus occur almost in all parts of the branch	The color of the thallus becomes slightly younger than the original color
11	The white color begins to turn pale and yellowish	The tip of the thallus changes color to white	White spots begin to widen	Thallus began to release mucus
12	Thallus had a break	The white color begins to turn pale and yellowish	The tip of the thallus changes color to white	Color differences in some branches began to be seen clearly
13	Thallus becomes easily broken when touched	Thallus had a break	The white color begins to turn pale and yellowish	The surface of the thallus begins to shrink
14	Thallus broke all over the branch	Thallus becomes easily broken when touched	Thallus had a break	The color of the thallus becomes pale

In Figures 1 and 2, it shows that the alteration of the *Kappaphycus* seaweed varieties of green varieties due to bacterial infection of *Vibrio* sp. Very different when compared with brown varieties, where almost all parts of the thallus experience bleaching, fracturing and easily broken when touched.

The dominant ice-ice disease strikes seaweed *Kappaphycus alvarezii* cultivated with the early clinical effects such as increased mucus production, rough thallus surface, wilted thallus, white spots, and bleaching of the tip of the thallus. A more severe attack of ice-ice may cause the thallus to become porous, and eventually, the infected thallus becomes fractured [2].

[11] suggested that after observation, the effect of an attack of ice-ice disease on seaweed started from the middle or tip of the thallus where the color change from yellowish brown faded to a pale white, if touched it felt slimy and two to three days later part is covered with a kind of white powder. When this part is touched or subjected to a strong enough water flow, the powder is released and the skin (epidermis) is peeled off, even visible in the tissue (cortex) and soft-looking which eventually breaks easily.

The mechanism of bacterial pathogenicity to seaweed thallus can be seen in the morphological changes of the thallus, especially the normal color changes (green and brown) to white (chlorosis). Color change (depigmentation) occurs due to the infection of pathogenic bacteria cause ice-ice disease in seaweed thallus. The change is increasing along with the increase of bacterial activity time in secreting its virulence factors to the host (seaweed thallus) [1].

Clinical effect observation results (Table 2 and Table 3) showed that in the infected with bacterial density Vibrio sp. 10^7 CFU/ml could have an impact on seaweed thallus even though it is very small and takes a long time when compared with *Vibrio* sp infection. At a density of 106 CFU/ml and 10^5 CFU/ml. [9] the minimum threshold of the presence of *Vibrio* sp. In the water is 10^4 CFU/ml, while the minimum limit of common bacteria is 10^6 CFU/ml.

Infection of *Vibrio* sp. at a density of 10^5 CFU/ml gives a very significant effect on seaweed *Kappaphycus* green varieties, which on the first day after infection, have shown clinical symptoms, i.e. changes in color in the thallus. The result of observation (day 14) also showed a very high degree of pathogenicity, in which there was a fault in all branches of the thallus. These results were very different in the brown *Kappaphycus* seaweed varieties where on the 6th day new infections showed clinical symptoms and outcomes after observation (day 14), only showed the effect of discoloration to white.

These results show that *Kappaphycus* brown seaweed varieties are more resistant to bacterial infections of *Vibrio* sp. when compared to green varieties. The endurance ability of Kappaphycus brown varieties to *Vibrio* sp. infection is thought to be due to brown varieties having higher antibacterial compounds than green varieties. [6] States that brown seaweed is one group of seaweeds that have the highest antioxidant activity when compared with red and green seaweed. Brown seaweed contains three types of hydrocolloids, namely: agar, alginate, and carrageenan [3]. Phenolic compounds are molecules that are known to act as brown algae defenses [7]. Algae activity can be used as an antiviral, antibacterial and antifungal effect on some pathogens [11].

4. Conclusions

Level of infection (pathogenicity) of bacteria *Vibrio* sp. higher in the Kappaphycus varieties of green varieties when compared with the brown varieties. Further research is needed to determine the content of brown seaweed, which is more resistant to *Vibrio* sp. and testing with other types of bacteria.

References

[1] Aris, M. 2011. Identification, Bacterial Pathogens and Utilization of Genes 16S-rRNA for the detection of Ice-Ice disease in Seaweed Cultivation. Dissertation. Bogor Agricultural Institute.

[2] Aris, M., Sukena., E. Harris., M.F. Sukadi and M. Yuhana. 2013. Molecular Identification of Pathogenic Bacteria and PCR Primer Design. *Journal of Aquaculture*. Vol 1, No 3: 43-50.

[3] Bixler, H.J., and Porse, H. 2010. A Decade of Change in The Seaweed Hydrocolloids Industry. *J. Appl. Phycol*, DOI 10.1007/s10811-010-9529-3.

[4] Brenner D J, N R. Krieg and J T. Staley. 2005. Bergey's Manual of Systematic Bacteriology. 2nd Ed. Vol. 2. The Proteobacteria. Part B. Gammaproteobacteria Springer. Michigan State University. USA.

[5] Jawetz, Melnick and Adelbergs. 2007. Medical Microbiology Issue 23. EGC Medical Book.

[6] Kelman, D., E. K. Posner, K. J. McDermid, N. K. Tabandera, P. R. Wright and A. D. Wright. 2012. Antioxidant Activity of Hawaiian Marine Algae. *Marine Drugs*, 10: 403-416.

[7] Nagai, T. and T Yukimoto. 2003. Preparation and Functional Properties of Bevarages Made from Sea Algae. *Food chem*. 81: 327-332.

[8] Qian, R-H., Z-H. Xiao, C-W. Zhang, W-Y. Chu, L-S. Wang, H-H. Zhou, Y-W. Wei and L. Yu. 2008. A Conserved Outer Membrane Protein as an Effective Vaccine Candidate from *Vibrio alginolitycus*. *Journal Aquaculture* 378: 5 – 9.

[9] Taslihan A., Ani W., Retna H., S.M. Astuti. 2004. Disease Control on Brackishwater Aquaculture, Directorate General of Fishery Center of Brackishwater Cultivation Center of Jepara.

[10] Vitor J.M, Carvalho A.F.F.U, Freitas S.M, Melo V.M.M. 2002. Antibacterial Activity of Extracts of Six Macroalgae From The Northeastern Brazilian Coast. Brazilian *Journal of Microbiology*. 33:311-313.

[11] Yulianto K and Mira S. 2009. Macro Culture Alga *Kappapycus alvarezii* (Doty) and the Symptoms of Ice-Ice Disease in Pari Island Waters. Oceanography and Limnology in Indonesia 35 (3): 325-334.

Resource-Based Capability on Development Knowledge Management Capabilities of Coastal Community

Roberto M. K. Teniwut[1], Cawalinya L. Hasyim[1], Wellem A. Teniwut[1]

[1]Fisheries Agribusiness Study Program, Tual State Fisheries Polytechnic
Langgur-Sathean Ringroad Km.6, Langgur, Southeast Maluku, Indonesia 97611

Email: teniwut_rio@yahoo.com

Abstract. Building sustainable knowledge management capabilities in the coastal area might face a whole new challenge since there are many intangible factors involved from openness on new knowledge, access and ability to use the latest technology to the various local wisdom that still in place. The aimed of this study was to identify and analyze the resource-based condition of coastal community in this area to have an empirical condition of tangible and intangible infrastructure on developing knowledge management capability coastal community in Southeast Maluku, Indonesia. We used qualitative and quantitative analysis by depth interview and questionnaire for collecting the data with multiple linear regression as our analysis method. The result provided the information on current state of resource-based capability of a coastal community in this Southeast Maluku to build a sustainability model of knowledge management capabilities especially on utilization marine and fisheries resources. The implication of this study can provide an empirical information for government, NGO and research institution to dictate on how they conducted their policy and program on developing coastal community region.

1. Introduction

Knowledge is one of the major resources to faced great challenge that currently occurred [12] while converting knowledge to become an intellectual asset which consists of human capital structural capital and relational capital are the key for competitive advantage [13]. Therefore, the successful on managing knowledge system is important for individual or community-based especially in business world [11].

Knowledge management becomes widely applied on various of the field, which has not solely exclusive to IT and IS field but also in social science, KM has evolved and drawn support from different research background [2]. Knowledge management is a strategic approach to increase added value on the effectiveness of organizational performance on facing rapidly changing social and environment [1]. Regarding the limitation of IT and IS based view, there is an increasing attention on examining the role of social community on accelerating knowledge transfer and knowledge retention [1]. Although the big challenge on social community-based knowledge management is the knowledge based on tacit dimension, while having said that, it's important to verbalize and documented knowledge that creates, available and ability transferring through a social community.

In the development of fisheries and marine development sector, maintain the sustainability and economic value at the same time is the most challenging factor. In fisheries management, there are raw data and information that obtain and collect from various sources, which is commonly practiced to make the right decision on developing a sustainable fisheries management. To face the challenge, decentralization (bottom-up) approach were adopted by many countries based on the notion that fishers can be useful sources for information for better fisheries management, where it brings us to another challenge to convince policymaker on knowledge based on fishermen and farmers instead of scientist

[4]. Thus, it is necessary to form institutionalize community-based on creating and sharing knowledge on fisheries and marine utilization and management to achieve sustainable and economic objectives.

Community's experience-based knowledge (EBK) is one of the important approaches on the implementation of knowledge base [5]. Coastal community much like in Southeast Maluku, where they have local wisdom that been institutionalized which called "sea sasi", which is based on experience that been process through trial and error for many years related to managing fisheries and marine resources in the area. Based on this evidence, the application of knowledge management based on fishermen and farmer experience should be easier in the region to convince local government to apply the KM on accelerating of sharing knowledge on managing and utilizing fisheries and marine resources in the area. Purposed of this study was to identify and analyze the empirical condition of resources-based capabilities and its impact on knowledge capabilities management. This study focused on Nonaka and Takeuchi categories on KM especially in routine knowledge assets, such as organizational structure, Technology support, and culture. A study conducted in Southeast Maluku, Indonesia, where information gathered from fishermen and farmer in the region.

2. Methodology

2.1. *Research framework*

[14] Proposed resource-based capability (RBC) approach used in this study which based on studies by Gold *et al.* (2001) where factors that influence knowledge capability management were technology, organizational structure, and organizational culture. Technology on this study refers to information technology inside and outside organization that increases the effectiveness of sharing information, where according to [10] technology can have an impact on KM when sufficient technology is in place and at the same time can bypass complex organizational structure which can increase the effectiveness of KM. The organizational structure of this study refers to all factors related to operational and command structure which include regulation, work design, rules and practice [8], where the best practice of KM has significant relation with less hierarchy on organizational structure. For organizational culture on this study is related to the situation in work which is related to knowledge sharing, coordinating, cooperating and knowledge acquiring [8-9].

We proposed some adjustment on the model of KCM and RBC that used to be applied in private business especially on the role, where fishermen, farmer and micro enterprises of marine ecotourism in the area play role as an employee and Southeast Maluku District in the organization while the leader of this region as manager. We measured RBC factors based on the perspective of the coastal community. For data collection, we used questionnaire design based on studies proposed by [6-8]. For data sampling, we used purposive sampling to select respondent based on criteria where fishermen and farmers who actively conducted fisheries and marine utilization activities from fishing, fish processing, marine ecotourism, and mariculture. We conducted our research on Southeast Maluku District, Indonesia, to be exact on 12 villages (Dudunwahan, Sitiniohoi, Ohoidertutu, Ngilngof, Matwaer, Evu, Letvuan, Sungai, Ngafan, Sathean and Selayar) as it showed in figure 1. This research conducted from February to June 2017.

2.2. *Data analysis*

We used hierarchical regression to analysis the data with SPSS 24, where the model can be seen as follow:

$$KCM = a_0 + \beta_1 T + \beta_2 OS + \beta_3 OC + \varepsilon$$

Where:

KCM	= Knowledge Capability Management
T	= Technology
OS	= Organizational Structure
OC	= Organizational Culture
ε	= Error

IOP Conf. Series: Earth and Environmental Science **89** (2017) 012017 doi:10.1088/1755-1315/89/1/012017

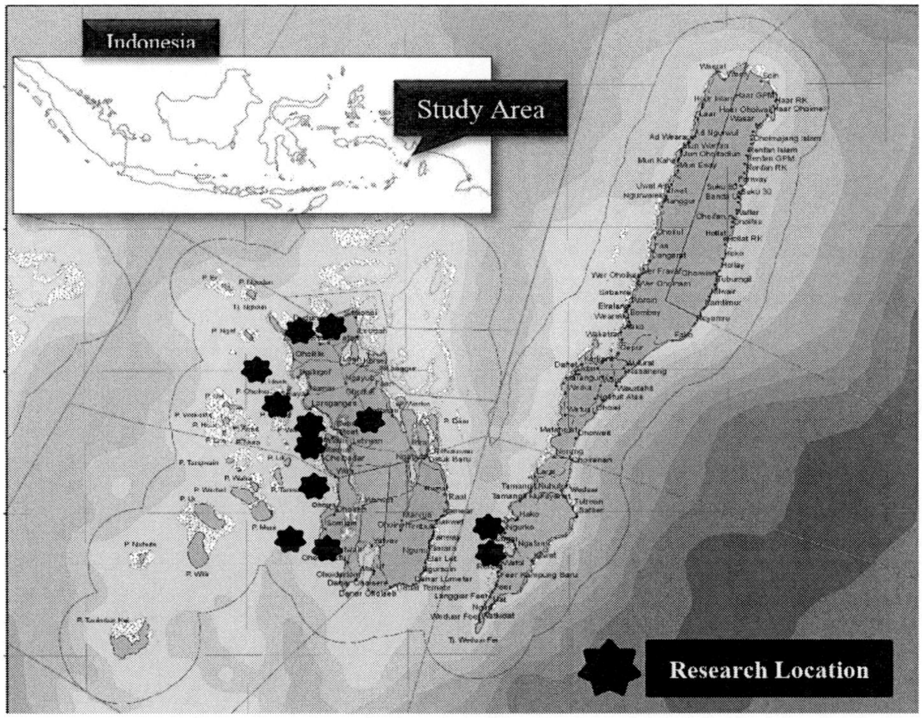

Figure 1. Research Locations

2.3. Validity and reliability test
Based on processing data, the result showed on validity and reliability test showed that all test passed, where all variables are valid based on the significant (p value < 0.05) of Pearson correlation coefficient and every questions on each variable are reliable based on cronbach alpha > 0.6.

2.4. Classical assumption test
Before we conducted discussion based on hierarchical regression result, we ought to run more test on classical assumption tests from multicollinearity, heteroscedasticity, autocorrelation, and normality. On Table 2, showed that all test passed, means that we can continue to analyze the result based on hierarchical regression result.

Table 1. Validity and Reliability Test

Reliability Test		
Variable	Cronbach Alpha	Decision
Knowledge Capability Management	0.810	Reliable
Culture	0.914	Reliable
Technology	0.908	Reliable
Structure	0.899	Reliable

Validity Test			
Variables	Coefficient	p-value	Decision
Knowledge Capability			
Management	0.789	0.000**	
KCM1	0.770	0.000**	
KCM2	0.721	0.000**	Valid
KCM3	0.774	0.000**	
KCM4	0.725	0.000**	
KCM5			

IOP Conf. Series: Earth and Environmental Science **89** (2017) 012017 doi:10.1088/1755-1315/89/1/012017

Culture			
OC1	0.865	0.000**	
OC2	0.919	0.000**	Valid
OC3	0.909	0.000**	
OC4	0.871	0.000**	
Technology			
T1	0.836	0.000**	
T2	0.872	0.000**	
T3	0.882	0.000**	Valid
T4	0.868	0.000**	
T5	0.817	0.000**	
Structure			
OS1	0.868	0.000**	
OS2	0.903	0.000**	Valid
OS3	0.906	0.000**	
OS4	0.829	0.000**	

Table 2. Classical assumption test

Classic Assumption Test			
Multicollinearity	DW = 2.010	4-dU= 1.8306; dU = 2.1694	Passed
Heteroscedasticity		Scatter Plot showed it passed the test	
Autocorrelation		All variables have < 4 VIF	
Normality		Normal P-P Plot Shows it passed	

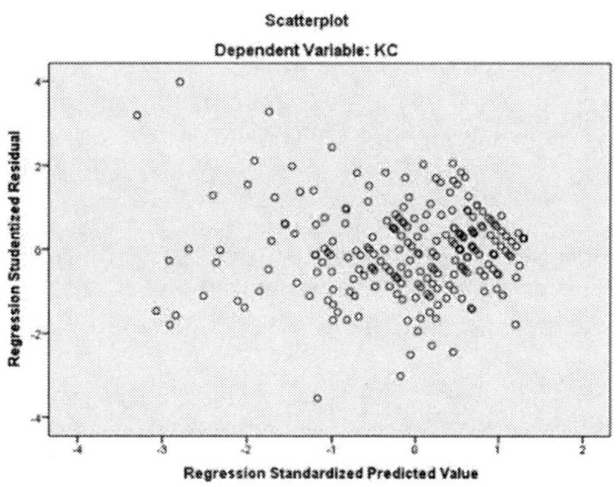

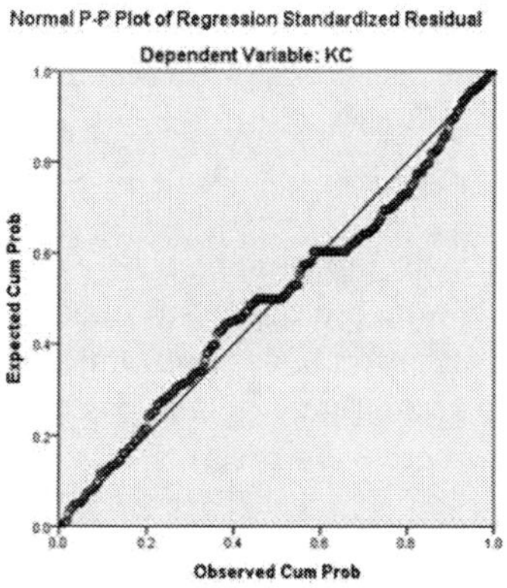

Figure 2. Scatter plot and normality plot

3. Result and Discussion

3.1. Empirical condition of knowledge need and support from local government

Our study found that most of coastal community in the region need and willing to receive any support to increase their knowledge on utilization and managing fisheries and marine resources showed in figure (a) while on the other hand, the impact felt by coastal community on the role of local government in increasing their knowledge capability was meager. Most of the coastal community stated that local government had not done any significant approach on helping the coastal community in utilize and manage fisheries and marine resources in the region (b). This result implied that there was significant need for developing a model on increasing knowledge capability management on utilizing and managing fisheries and marine resources in the region.

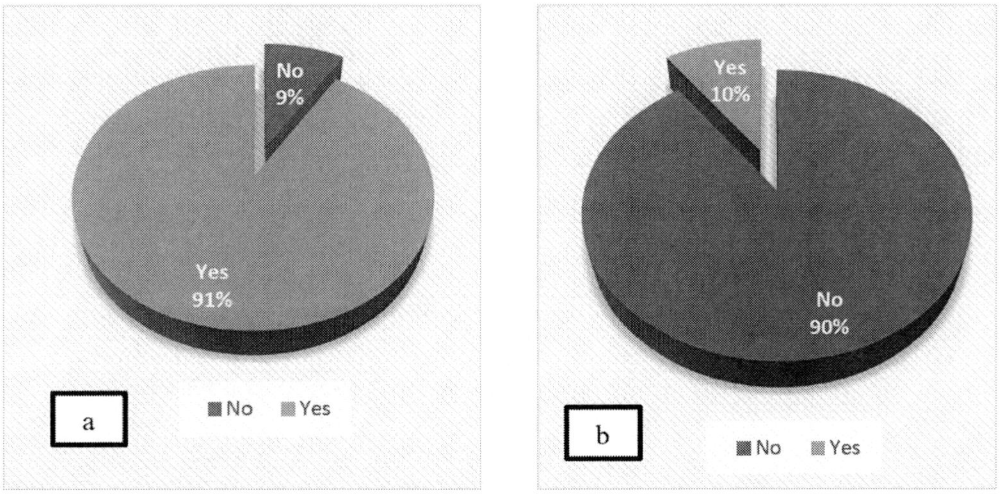

Figure 3. Comparison between need for increase the knowledge (a) and the local government role on increase the knowledge of coastal community in the region (b)

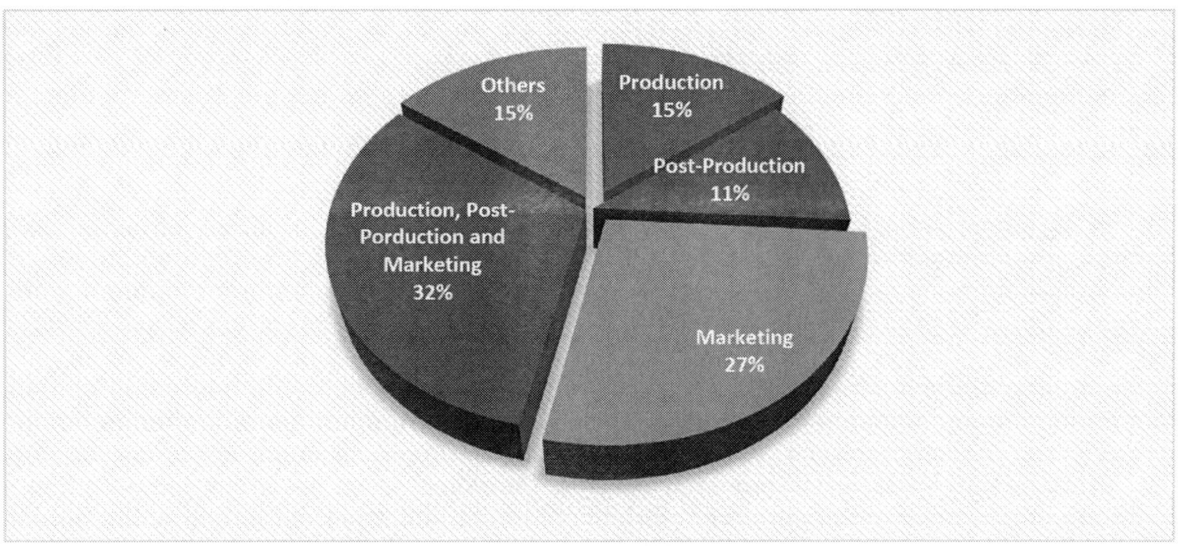

Figure 4. Problem that needs to be address

On figure 4, showed that combination of production, post-production, and marketing was the most needed to be addressed immediately, this result indicated there was a need for comprehensive training for the coastal community especially to face market competitiveness and increase the stability of their welfare in the long run. The most coastal community still conducted their business in an antiquated way, for instance, they do not have an exact schedule on fishing which would impact their production. Also they would sell their catching immediately without considering best price available in the market. For mariculture such as seaweed, the main problem was the marketing on the product and post-production. The lack of the knowledge on the post-production had in great impact on the decreasing on the quality of seaweed, in addition to harder to find better market, the profitability of the farmer become very unstable.

3.2. The relationship between Resource-based Capability and Knowledge Capability Management
On table 3, can be seen based on correlation matrix between all RBC factors (Technology, Organizational Structure, and Organizational Culture) and knowledge capability management, it appeared that all three factors of RBC have a significant and substantial relationship with knowledge capability management. As [14] pointed out in their study that in resources based capability three factors affecting knowledge capability were technology, structure, and culture. All these three factors have the great impact on the effectiveness the application of knowledge capability as it also supported by studies from [8-10].

Table 3. Relationship between RBC and KCM

Variables	Coefficient	p-value	Decision
Culture	0.679	0.000**	Significant relationship
Technology	0.790	0.000**	Significant relationship
Structure	0.710	0.000**	Significant relationship

**significant at 5%

3.3. The impact of resource-based capability on knowledge capability management
Based on the result in table 4, for three model that we have proposed, the R-square value showed for every model appears that the highest number of R-squared showed on model number 3. It means that the best model in our study where there are all three predicted of RBC on KCM. The result also showed the highest value of F test on model number 3. So it concluded that we used model number three for discussion. This result also cemented the concept proposed by [14], three factors of RBC involve on knowledge capability.

Table 4. R-square and F Test

Model	R Square	Adjusted R Square	F	Significant
Predictors: (Constant), CK	0.560	0.558	319.424	0.000**
Predictors: (Constant), CK, T	0.704	0.702	297.332	0.000**
Predictors: (Constant), CK, T, OS	0.813	0.810	360.247	0.000**

**significant at 5%

4. Conclusion
A coastal community in the region needs for support to increase their knowledge on utilization and managing fisheries and marine resources, while at the same time, the impact of local government on increasing their knowledge was insignificant [15]. Therefore there's a need to build a better model on accelerating and increasing knowledge capability of a coastal community in the region. Combination of production, post-production, and marketing was the biggest problem that needs to be addressed immediately to increase coastal community welfare. Based in the study showed that to build knowledge capability model, three factors involved on RBC had a significant relationship, and result also showed that three factors (technology, organizational culture, and organizational structure) represent the best model for KCM and all three had simultaneously effect on KCM.

References
[1] P.J. Sher, V.C. Lee 2004 Information technology as a facilitator for enhancing dynamic capabilities through knowledge management, *Information & Management* 41 (8), pp. 933–945.
[2] Wiig, K. M 2000 *Knowledge management: An emerging discipline rooted in a long history* (pp. 3-26). Boston: Butterworth-Heinemann.
[3] Brown JS, Duguid P 2001 Knowledge, and organization: a social practice perspective. *Organization Science*, 198–213.

[4] Hoefnagel, E., Burnett, A., and Wilson, D. C 2006 The knowledge base of co-management. *Developments in Aquaculture and Fisheries Science*, 36, 85-108.

[5] Wilson, D. C., del Valle, I., and Jessen, R 2006 *Introduction: the Knowledge Base as Process*. In Knowledge Base for Fisheries Management (pp. 1-25). Pergamon Press.

[6] Marquarde, M.J 1996 *Building the Learning Organization*, McGraw-Hill, New York, NY.

[7] Gold, A.H., Malhotra, A. and Segars, A.H. 2001 Knowledge management: an organizational capabilities perspective. *Journal of Management Information Systems*, Vol. 18 No. 1, pp. 185-214.

[8] Yang, C. and Chen, L.C 2007 Can organizational knowledge capabilities affect knowledge sharing behavior?. *Journal of Information Science*, Vol. 33 No. 1, pp. 95-109.

[9] Chuang, S.H 2004 A resource-based perspective on knowledge management capability and competitive advantage: an empirical investigation. *Expert Systems with Applications*, Vol. 27 No. 3, pp. 459-65.k

[10] Orlikowski, W.J 2000 Using technology and constituting structures: a practice lens for studying technology in organizations. *Organization Science*, Vol. 11 No. 4, pp. 404-28.

[11] Jimenez-Jimenez, D., and S. Valle, R 2013 Studying the effect of HRM practices on the knowledge management process. *Personnel Review*, 42(1), 28–49.

[12] Wang, S, and R. A. Noe 2010 Knowledge sharing: A review and directions for future research. *Human Resource Management Review* 20.2: 115-131.

[13] Rudež, H. N., and Mihalič, T 2007 Intellectual capital in the hotel industry: A case study from Slovenia. *International Journal of Hospitality Management*, 26(1), 188-199.

[14] Aujirapongpan, S., Vadhanasindhu, P., Chandrachai, A., and Cooparat, P 2010 Indicators of knowledge management capability for KM effectiveness. *Vine: The journal of information and knowledge management systems*, 40(2), 183-203.

[15] Teniwut W. A 2016 For sustainable revenue of fisheries sector in small islands: evidence of Maluku, Indonesia. *AACL Bioflux* 9(3):722-732.

IOP Conf. Series: Earth and Environmental Science **89** (2017) 012018 doi:10.1088/1755-1315/89/1/012018

Analysis of seaweed marketing in warbal village, Southeast Maluku Regency, Indonesia

Bruri B. Tumiwa[1], Meiskyana R. Renjaan[1], Glen B.A Somnaikubun[1] Kamilius D. Betauubun[1], Marselus Hungan[1]

[1]Fisheries Agribusiness Study Program, Tual State Fisheries Polytechnic, Indonesia Langgur-Sathean Road Km.6, Southeast Maluku, 97611, Maluku, Indonesia

Email: bruritumiwa@gmail.com

Abstract. Seaweed in Warbal Village, West Kei Kecil Subdistrict, Southeast Maluku Regency has prospects and business opportunities are adequate to give hope to farmers in improving welfare. The fact that seaweed farming has not yet provided better and maximum results as desired by the farmers. This study aims to evaluation the marketing channels, marketing margins and profit share of marketing agencies. The research is located in Warbal Village, West Kei Kecil Subdistrict, Southeast Maluku Regency which is determined purposively. The number of sample is 30 farmers taken by simple random sampling, 2 wholesaler traders and 2 collector traders taken by using snowball method. The data collection methods is interview and questionnaire directly to farmers and marketing agencies, literary method or data collector from institutions related to the research's aims. The research results show that there is two marketing channel, as follows: Channel I: farmers, wholesaler traders, collector traders, PAP; Channel II: farmers, collector traders, PAP. The magnitude of marketing margins is different between the marketing channels, and so it is with profit share of a marketing agency. On channel I, magnitude margin is IDR 3,250 and profit share is 71.11% on farmers, 17.76% on wholesaler traders and 11.09% on collector traders. On channel II, the magnitude of marketing margin is IDR 1,250 and profit share is 88.88% on farmers and 11.09% to collector traders.

1. Introduction

Indonesia has two-thirds of the region comprised of the ocean, with a rich coast of biological variety is one of the potentials that can support success in the fishery sector. Currently, efforts to manage natural resources and the environment continue to be done. This effort has shown significant progress in improving the welfare of mankind as well as achieving a harmonious and balanced environmental order. Indonesia's waters area is quite extensive, with the length of the beach more or less 81,000 km, is a fertile coastal area and can be used for fishery purposes.

Efforts to increase the production of fisheries can be pursued through cultivation, both on land and at sea. Seaweed cultivation is one type of aquaculture in the field of fisheries that have the opportunity to be developed in the territorial waters.

Seaweed is one of the potential export commodities to developing. Currently, Indonesia is still one of the important exporting countries in Asia because seaweed grows and spread almost in all Indonesian waters. Seaweed is still widely exported in the form of raw materials in the form of dried seaweed. According to the [1] as much as 70% of the production of dried seaweed raw materials are exported to China, the European Union and the Philippines. The domestic market still absorbs 30% of dried seaweed raw material.

Seaweeds are very diverse living around 900 species. For more details, the type of seaweed is 100 green algae, 200 brown algae and 600 red algae. Of the several types of seaweed that are widely cultivated in Southeast Maluku Regency is the type of seaweed *Eucheuma* scattered in some coastal areas of Warbal Island.

Seaweed cultivation business is one of the right alternatives to be developed in this region, considering Southeast Maluku Regency is a very low area of rainfall. Therefore, a new breakthrough by exploiting coastal area to develop this seaweed cultivation business. One of the areas that develop a lot of seaweed cultivation is in the Warbal Village Subdistrict West Kei Kecil.

Seaweed cultivation has long been carried out by the community, but the marketing of seaweed products is still faced with various problems that include weak bargaining position of farmers so that the price of seaweed is determined more by traders without much-involving farmers as producers.

Problems often faced by seaweed farmers in Warbal Village Subdistrict West Kei Kecil in the development of the economy associated with the marketing of the results. Efficient marketing of seaweed products is certainly a major factor determining whether or not productivity increases. Information on seaweed marketing conducted by field extension officers is felt not optimal because until now the farmers still have not get clear marketing information about everything related to seaweed product marketing. As a result, farmers can not take part in seaweed price determination. Farmers only as price recipients are not price determinants. Thus the bargaining position of the farmers (Bargaining position) is weak. This is often a force for traders, on the other hand, farmers depend on seaweed livelihoods. For that, we need to research marketing analysis of seaweed in Warbal Village Subdistrict West Kei Kecil Southeast Maluku Regency. Because the success of farming activities depends not only on the efforts and work of the farmers of other things themselves but also supported by many of them is a clear marketing institution so that people know more where the results of their farming brought to market.

[6] states that the market as a place to sell goods and services so that consumers can meet their needs and desires. As for marketing institutions, the market is a place to conduct business activities by implementing certain marketing functions so that marketing institutions can gain profit. According to [5], Marketing is a system of business activities aimed at planning, pricing, promoting and distributing goods and services that satisfy the needs of existing buyers and potential buyers, nor does it end at the time of sale or transaction. All decisions taken in the field of marketing must be demonstrated to determine the product, market, price, promotion, and distribution system. Agricultural marketing is a process of the flow of a commodity in which the transfer of property and creation takes place, to place and use the forms undertaken by marketing institutions by executing one or more marketing functions [6]. Marketing is any business activity related to the current delivery of goods or services from producers to consumers. Marketing encompasses all activities of the company in adapting to the environment creatively and profitably [3].

Marketing is an important management function needed to create demand for products sold. According to [2], marketing functions are grouped into 3 (three) functions, among others:

1. Exchange function.
   Is an activity related to the transfer of property rights of goods and services marketed. This function consists of storage functions, processing functions, and sales functions.
2. Physical function.
   It is all actions related to the goods that give rise to the use of place, shape and time. Physical functions include storage functions, processing functions, and transport functions.
3. The function of the facility.
   It is all action related to the exchange that occurs between producers and consumers. The function of this facility consists of standardization functions, payment functions, and market information functions.

In marketing a farm product definitely, requires a marketing agency. A marketing agency is a business entity or an individual conducting marketing, distributing services and commodities from producer to final consumer and having a relationship with a business entity or other. Marketing institutions arise because of the desire of consumers to obtain commodities by the time, place and shape that consumers want. The task of the marketing agency is to carry out marketing functions in the form of marketing margins. Agricultural production marketing institutions are well known by [4] are sellers traders, wholesalers traders, collectors traders, wholesalers, exporters/importers, brokers, auctions, and retailers.

On the other hand, one of the dominant factors affecting a marketing system regarding efficiency is by looking at the conditions of channels, marketing institutions, and market structures. Some of the advantages gained by knowing the marketing channel of a commodity include knowing which paths more than one channel will be made and will also make it easier to find the margins received by every marketing agency involved in the marketing system [3]

High-level marketing channels can be found, when the marketing runs smoothly, especially in countries or developed regions, from products to Obtaining information on end customers and controlling production can be said to increase if the number of levels of marketing channels increases.

According to [2], marketing channels are the bodies that organize the activities or functions of marketing by which goods move from the producer to the consumer. The purpose of this research is to analyze the marketing of seaweed, especially the marketing chain, studying the efficiency level of seaweed marketing in Southeast Maluku Regency.

2. Methodology

This research was conducted in the Warbal Village Subdistrict West Kei Kecil Southeast Maluku Regency, which is determined purposively with the consideration that Warbal Village is one of the villages producing seaweed in Southeast Maluku Regency. The study period starts from January to March 2016.

2.1 Method

The population in this study consists of seaweed farmers and traders involved in marketing seaweed. In this research, sampling method using snowball sampling method, that is first level sampling, in this case, is simple random sampling farmer (simple random sampling) that is selected 30 peasants from seaweed farmer population number as much 130 people are considered to represent farmers who seize seaweed. While the next sample, in this case, is the trader is determined by the farmer where the selected farmer is identified (Excavated data), then these respondents are asked to identify another responder (traders) who are part of the target population. This process will cease when the sample size is considered adequate. Based on information from farmers: wholesaler trader amounted to 2 people and collector traders amounted to 2 people. For more details can be seen in Table (1)below:

Table 1 The population of the study

Kind of marketing agency	Population	Sample
Farmers	130	30
Wholesaler trader	2	2
Collector trader	2	2

Observation and data collection in this research consist of:

1. Primary data is data directly taken from farmers and marketing institutions through interviews and questionnaires covering the land area, production, the price of sales and purchases, and marketing costs.
2. Secondary data is data obtained from relevant agencies and literature-literature that is still associated with this research.

IOP Conf. Series: Earth and Environmental Science **89** (2017) 012018 doi:10.1088/1755-1315/89/1/012018

2.2 Data analysis

The data obtained in this study are analyzed and calculated based on the following formula:

a. Calculation of marketing margins

MP = KP + BP or MP = Pr - Pf

Description:

MP = Marketing margin

KP = Benefits of marketing

BP = Cost of marketing

Pr = Price at consumer level (PAP)

Pf = Price at producer farmer level

b. The calculation of the profit share of ith marketing agency

$$Ski = \frac{Ki}{Pr - Pf} \times 100\%$$

Description:

Ski = Section (Share) profit marketing agency i^{th}

Ki = Profit marketing agency i^{th}

Pr = Price at consumer level

Pf = Price at producer level

c. Calculation of share (share) price of farmers

$$Sf = \frac{Pf}{Pr} \times 100\%$$

Description:

Sf = The price of farmers

Pf = Farmer level price (manufacturer)

Pr = Prices at consumer level (inter island traders)

d. Calculation Share marketing agency costs of ith

$$Sbi = \frac{bi}{Pr - Pf} \times 100\%$$

Description:

Sbi = Share cost of i-th marketing agency (i1 = Wholesaler Trader, i2 = Collector Trader)

bi = Marketing cost of ith marketing agency

Pji = The selling price of ith marketing agency

Pbi = The purchase price of ith marketing agency

Pf = Farmer level price (producer)

Pr = Prices at end consumer level (inter-island traders)

Marketing margin is also called total marketing margin (total M) where M total = M1 + M2 + M3 + ... Mn, which is the marketing margin of each marketing agency, so the distribution of margin can be explained as follows:

$$Dmi = \frac{Mi}{M\ total} \times 100\%$$

Description:

Dmi = Distribution margin group marketing agency ke - i

Mi = Marketing margin of ith marketing agency (i1 = middleman, i2 = collector).

M total = Pr-Pf

IOP Conf. Series: Earth and Environmental Science **89** (2017) 012018 doi:10.1088/1755-1315/89/1/012018

3. Results and Discussions

The marketing agencies involved in marketing *Eucheuma* seaweed species in Warbal village are:

1. Producers are farmers who do seaweed farming business around the coast (coast). The land used to cultivate seaweed is a loose sea that is controlled by the state, so farmers only have the right to use. The limits of land used by the number of stretch straps owned by each farmer and the control of the land are not permanently owned but only controlled as long as they do cultivation activities. Seed production is partially harvested used as seeds back and partially dried for sale to traders. Seaweed drying is carried out on stilts or tarps. Drying is done for approximately four days if the weather conditions are sunny.
2. Traders of middlemen are traders who buy directly to the farmers in the village. Generally, seaweed purchased is seaweed that has been dried by producers or seaweed farmers who have been packed with sacks that contain an average of 60-80 kg of seaweed. Collector traders buy dried seaweed at farmers for Rp 8,000.
3. Collector traders are traders who buy seaweed from middlemen traders and also farmers who are located in the capital Regency of southeast Maluku. Collectors traders have a large capital so that they can accommodate while seaweed to wait for a suitable price or higher price.

3.1 Marketing Channel

In general, the marketing chain of dried seaweed in the research location is relatively simple through several marketing institutions such as middlemen and merchant traders. Based on the results of interviews and field observations, farmers in Warbal village sell their products to middlemen traders in the village, because they are no longer laboriously bringing the proceeds to the local traders and do not incur substantial expenses and time and effort. The process of bargaining occurs between farmers and traders, but in general, farmers are always in a difficult position because in the end traders middlemen and collectors that determine the selling price. This is not disputed by the farmers because the farmers and traders already know each other. Also, farmers tend to immediately sell the results because of the economic needs of the family.

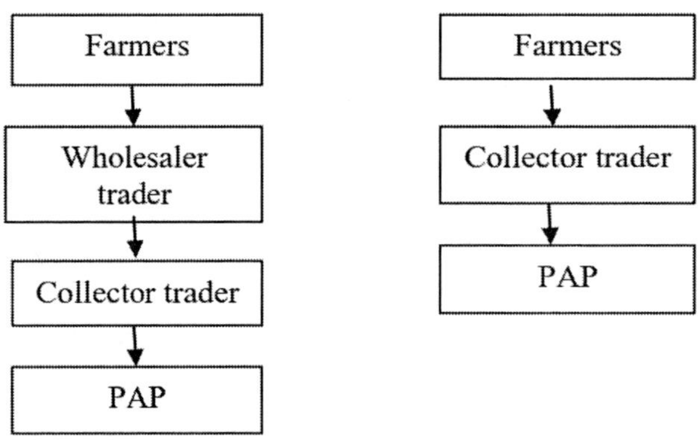

Figure 1. The marketing channel of seaweed farmers in Warbal Village.

From Figure 1 shows that the marketing of seaweed from farmers to exported through two channels:

1. The farmer sells to the wholesaler trader then through the collector trader and finally distributed to the inter-island trader (PAP).
2. Farmers sell to collector trader and then distributed to inter-island traders (PAP).

The number of farmers as many as 30 people, who sell on the model of marketing channel I as many as 17 people or 56.67% of the number of products marketed is as much as 7920 kg (7.92 t), with the distance from the farmhouse to wholesaler trader are ± 10 km. Respondent farmers choose marketing channel I because of the urgency of household needs, distance from farmer's house to inter-island trader is too far, and limited transportation cost. In the marketing channel II model, the number of farmers as many as 13 people or 43.33%, with the number of products marketed is 9775 kg (9,77 t) with the distance from the farmer's house to the collector merchant is ± 23 km. Farmers choose channel II because the selling price is bigger, have capital or transportation cost, and some have own vehicle/transportation.

3.2 Marketing Margin

Marketing margin is one of the indicators in determining marketing efficiency which is formulated as marketing cost plus profit or difference of price paid by the final consumer (PAP) and price received by producer farmer. If marketing margins are high, then marketing is said to be inefficient marketing system. In this case the intermediary trader as the cause of the high margin of marketing. Margin marketing includes two elements, namely: (1) marketing costs are costs that are issued by intermediary traders in carrying out marketing functions that include transportation costs, packaging, storage risk and other costs. (2) the benefits to the marketing agencies involved.

In Table (2), it is seen that the share of prices received by farmers is 71.11 from the price paid by consumers, or 28.88% (3250/11250 x 100%) is the farmer's marketing margin. Of the total margin of 28.88%, it turns out 19.55% (2200/11250 x 100%) is the marketing costs incurred by the two marketing agencies involved. The distribution of marketing margin of 3250 is seen from the ratio between cost and profit (π / c) unevenly. It turns out that from Rp.100 the cost incurred by farmers can earn a profit of Rp.1042, the middleman earns a profit of Rp.81, the collector trader earns a profit of Rp.13. Farmer profits are bigger because the marketing costs of the farmers are smaller than marketing costs incurred by middlemen and collector trader.

Table 2 Marketing channels of margin distribution, share price and marketing benefits of seaweed in WarbaL Village (Channel I).

No	Marketing agency	Value (Rp / kg)	Margin distribution (%)	Share price (%)	π /c
1	Farmers (Pf)				
	Selling price	8000	-	71,11	10.42
	Cost	700			
	Advantages	7300			
2	Wholesaler trader				
	Purchase price	8000	-	-	0.81
	Cost:				
	Transport	100	3.07	0.88	
	Sack	500	15.38	4.44	
	Rafia rope	500	15.38	4.44	
	Selling price	10000	-	-	
	Marketing margins	2000	-	-	
	Advantages	900	27.69	8.00	
3	Collector trader				
	Purchase price	10000	-	-	0.13
	Cost:				
	Transport	100	3.07	0.88	
	Sack	500	15.38	4.44	
	Rafia rope	500	15.38	4.44	
	Selling price	11250	-	-	
	Marketing margins	1250	-	-	
	Advantages	150	4.61	1.33	
4	PAP / Final Consumer (Pr)				
	Purchase price	11250			
Total marketing margin		3250	100.00	100.00	

Information:
Margin Distribution = cost ratio Profitability with margin marketing x 100%
Share price = Price ratio, cost, and profit with price paid by consumer x 100%
π / c = Profit ratio with cost

In Table (3), it is seen that the share of the price received by the farmer is 88.88 from the price paid by the consumer, or 11.11% (1250/11250 x 100%) is the farmer's marketing margin. From the total margin of 11.11%, it turns out that 9.77% (1100/11250 x 100%) is the marketing cost incurred by the two marketing agencies involved. The marketing margin distribution of 1250 is seen from the ratio between cost and profit (π / c) uneven. It turns out that from Rp.100 the cost incurred by the farmers can earn a profit of Rp.1328 and the collector trader earns a profit of Rp.13. Profits of farmers are greater because marketing costs incurred by farmers is smaller than the marketing costs incurred by collector trader.

Table 3 Marketing channels of margin distribution, share price and marketing benefits of seaweed in WarbaL Village (Channel II).

No	Marketing agency	Value (Rp / kg)	Margin distribution (%)	Share price (%)	π /c
1	Farmers (Pf)				
	Selling price	10000	-	88.88	13.28
	Cost	700			
	Advantages	9300			
2	Collector trader				
	Purchase price is	10000	-	-	0.13
	Cost:				
	Transport	100	8	0.88	
	Sack	500	40	4.44	
	Rafia rope	500	40	4.44	
	Selling price	11250	-	-	
	Marketing margins	1250	-	-	
	Advantages	150	12	1.33	
3	PAP / Final Consumer (Pr)				
	Purchase price	11250			
Total marketing margin		1250	100.00	100.00	

Table 4 Recapture of purchase price level, selling price, marketing cost, profit, marketing margin and share of selling price on each marketing channel.

No	Marketing agency	Marketing channel I (Rp/kg)	Marketing channel I (Rp/kg)
1	Farmers :		
	Selling price	8000	10000
	Cost	700	700
	Advantages	7300	9300
2	Wholesaler trader		
	Purchase price	8000	
	Marketing cost	1100	
	Selling price	10000	
	Advantages	900	
3	Collector trader		
	Purchase price	10000	10000

Marketing cost	1100	1100
Selling price	11250	11250
Advantages	150	150
Total marketing margin	3250	1250
Share the selling price of the farmer	71.11	88.88

4. Conclusion

Marketing of seaweed in Warbal Village, West Kei Kecil Subdistrict, Southeast Maluku Regency, through two (2) marketing channels: Channel I: farmers, wholesaler traders, collector traders, PAP; Channel II: farmers, collector traders, PAP. Based on the margin and price share analysis, it can be seen that marketing margin on a channel I is Rp. 3250 and on channel II of Rp. 1250, while the share of farmer's price on channel I is 71,11%, Wholesaler Trader 17,76%, Collector Trader 11,09% and at channel II, the share of farmer price is 88,88%, Collector Trader 11,09%. This happens because the marketing costs and selling price of each packaging agency on each marketing channel are different.

Acknowledgements
Authors would like to express their gratitude to head of fisheries agribusiness study program and informants in Warbal village for supporting this research.

References

[1] Marine and Fisheries Department 2016 Final Report of Fishery Condition in West Kei Kecil Subdistrict. Southeast Maluku Regency.
[2] Hanafiah and Saefuddin 2006 Fishery Products Commerce. University of Indonesia. Jakarta.
[3] Kottler, Philip and Gary Armstrong 2000 Marketing Basics. Prenhallindo. Jakarta.
[4] Masyrofie 1994 *Agricultural Product Commerce*. Faculty of Agriculture Brawijaya Universitas. Malang.
[5] Soekartawi 2002 *Farming Analysis*. University of Indonesia. Jakarta.
[6] Sudiyono 2001 *Agricultural Marketing*. University of Muhamadiyah. Malang.

Real Time Business Analytics for Buying or Selling Transaction on Commodity Warehouse Receipt System

Taufik Djatna[1], Wellem A. Teniwut[2], Nina Hairiyah[3], Marimin[1]
[1]Department of Agroindustrial Technology, Faculty of Agricultural Technology, Bogor Agricultural
[2]Fisheries Agribusiness Study Program, Tual State Fisheries Polytechnic, Langgur, Indonesia
[3]Agro-Industrial Technology Study Program, Tanah Laut State Polytechnic, South Kalimantan, Indonesia

Email: taufikdjatna@ipb.ac.id

Abstract. The requirement for smooth information such as buying and selling is essential for commodity warehouse receipt system such as dried seaweed and their stakeholders to transact for an operational transaction. Transactions of buying or selling a commodity warehouse receipt system are a risky process due to the fluctuations in dynamic commodity prices. An integrated system to determine the condition of the real time was needed to make a decision-making transaction by the owner or prospective buyer. The primary motivation of this study is to propose computational methods to trace market tendency for either buying or selling processes. The empirical results reveal that feature selection gain ratio and *k*-NN outperforms other forecasting models, implying that the proposed approach is a promising alternative to the stock market tendency of warehouse receipt document exploration with accurate level rate is 95.03 %.

1. Introduction

Warehouse receipt system is a financial system for storing agricultural commodity and values to cope with a fluctuation in prices. Commodities stored in the warehouse will be secured by a document warehouse receipt issued by the warehouse management and has been officially registered with the National Registration Centre. Warehouse receipt document can be traded (buying and selling) on the stock market by the holder of the warehouse receipt document. Currently, open access to the internet and mobile telecommunication enabled each party in warehouse receipt system to transact in a farm of buying or selling. Parties may consist of farmers, traders, financial institution or else. They need an integrated tool such as business intelligence system anywhere and anytime with the real time data access. A warehouse receipt is a document of proof of ownership of goods stored in a registered warehouse specifically published by the warehouse manager. Warehouse Receipt can be transferred enough with an endorsement. Receipt Warehouse with it being "Negotiable." Warehouse here means it can be various, depending on the stored commodities, ranging from dried seaweed, chocolate, coffee, rice, to crude palm oil (CPO). This warehouse receipt can be used as collateral for bank loans. Since the warehouse receipt is a proof of ownership, the warehouse receipt may be traded, traded, exchanged or used as collateral for the loan, or may be utilized for the delivery of goods in derivative transactions such as futures contracts.

But unfortunately, the use of warehouse receipts is still insufficient because most countries are not willing to accept the concept of proof of ownership of the moving goods. Usually, the evidence of ownership exists only for immovable goods.

The scope of Business Intelligent (BI) should include making the best use of information for strategic, tactical, and operational needs [1]. The purposes in building BI strategy is to help business with long-term planning, help middle management with tactical reporting, and help operations with day-to-day decision-making to run the business efficiently. BI is all about providing people with the information they need to do their jobs more efficiently. A wide range of BI services needs to be given to meet a wide variety of requirements. A scope of BI strategy should be determined by the business drivers and business goals. The scope should always account for the changing business requirements to keep the BI strategy aligned with business (Figure 1).

BI framework brings together data governance, data architecture, technical architecture, data integration, data quality, end-user information delivery, data security, etc. to empower the BI initiatives. A framework should set standards that BI participants must adhere to. The framework should connect significant components that are part of overall BI vision. Establishing BI competency center or center of excellence as part of the BI framework will help in integrating BI best practices with the ongoing BI work and the BI environment of the enterprise.

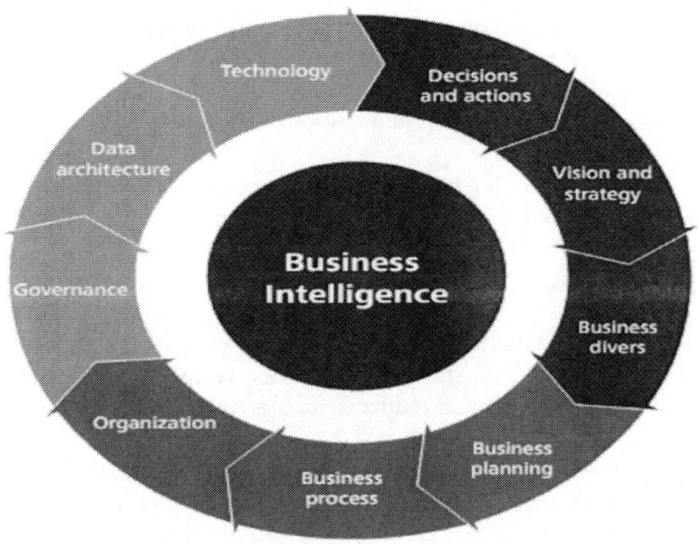

Figure 1. BI aligning with enterprise strategy to deliver value [1]

The overarching methodology that refers to the skills and technologies to explore past business performance to make better decisions is called Business Analytics. Business Analytics uses data, statistical and quantitative analysis, predictive modeling, and optimization to make businesses work better [2]. Business analytics are needed to make the best decision on the sale or purchase transaction tendencies in real time by the owner or prospective buyer. The position of business analytics on enabling intelligent business all in the enterprise is located in the upper position (Figure 2).
BI strategy should aim to support the complete breadth of decision-making ability in the enterprise. Strategic decisions deal with the long-term planning, performed by top management, focus mostly on demographic, and industry trends. These address broad issues to achieve general objectives. Strategic decisions are where BI traditionally has been implemented. Businesses today want more than just strategic insight from their BI implementations.

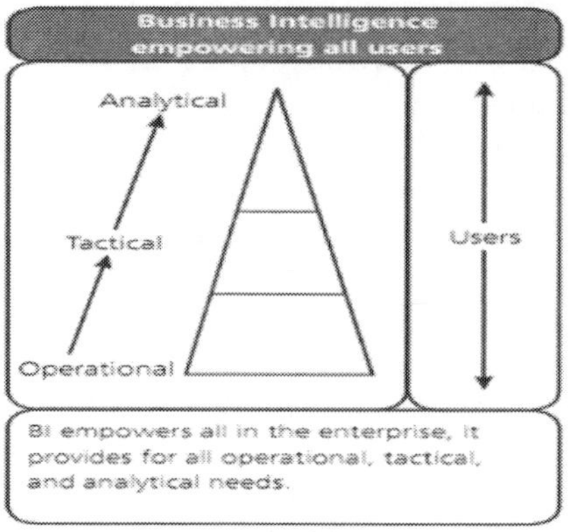

Figure 2. BI enabling all in the enterprise [1]

There is an expectation from BI to enable better execution of the numerous tactical and operational decisions that enterprise makes every day. For example, a product manager has to decide the discount schedule or pricing decision for a product every day relies on the market price on condition. Since the market has high volatility and noisy environment, tactical and operational decisions are the drivers for day-to-day management of the business at different levels. These decisions had a smaller business impact when it measured in silos as compared to strategic decisions. When the information gathered and integrated, multiple tactical and operational decisions will have higher and significantly help driving business better. BI strategy should embody the approach to enable better decision-making at all levels of the enterprise.

The stock market tendency has a big challenge task due to its high volatility price environment. There have many studies using artificial neural networks (ANNs) in this area. The early days of these studies focused on the application of ANNs to stock market prediction, such as [3-5]. Recent research tends to hybridize several artificial intelligence (AI) techniques [6]. Some researchers tend to include novel factors in the learning process. Kohara et al. [6] Incorporated prior knowledge to improve the performance of stock market prediction. Kim and Han [7] proposed a genetic algorithms approach to feature discretization and the determination of connection weights for ANN to predict the stock price index. They suggested that their approach reduced the dimensionality of the feature space and enhanced the prediction performance.

The main limitations of ANN above can be overcome, a novel intelligent learning algorithm, feature selection with gain ratio and *k*-Nearest Neighbour (*k*-NN) approach are proposed in this study. We focus on *k*-NN techniques because of their ease of interpretability to determine the trend of sales or purchases in stock market warehouse receipt documents that depend on many parameters. The primary objective of this study is to propose computational methods to trace market tendency both for buying or selling processes warehouse receipt document using feature selection gain ratio and *k*-NN technique.

2. Model Building Process

2.1. Feature Selection with Gain Ratio

It is well known that implementing feature selection improves the accuracy of a classifier. The degree of improvement will depend on many factors: the type of classifier, the effectiveness of the feature selection and the quality of the features. In the case of the simple *k*-NN classifier, the feature selection deletes noisy features and reduces the feature-space dimension. Moreover, for an ensemble of classifiers, the feature selection can promote diversity among the ensemble members and can improve their local specialization. The potential for an ensemble to be more accurate than its constituent members depends on the diversity among its members [11].

The information gain measure favors attributes with many values over those with few values [12]. Gain ratio (GR) overcomes this problem by introducing an extra term taking into account how the feature splits the data:

$$GR\ (S,A) = \frac{IG\ (S,A)}{SI\ (S,A)}$$

$$SI\ (S,A) = \sum_{i=1}^{d} \frac{s_i}{s} . \log_2 \frac{s_i}{s} \quad (1)$$

SI is d subsets of examples resulting from partitioning S by the d-valued feature A. Since the SI term can be zero in some special cases, we define: GR(S, A) = IG(S, A) if SI (S, A) = 0 for feature A. For the most part, this improvement over IG proves significant in the evaluation presented here.

2.2. k-NN Based Classifiers

Real-time business analytic for buying or selling transaction on a commodity warehouse receipt system is using a technique of classification in data mining that is *k*-NN. *K-NN* classifiers are instance-based algorithms by taking a conceptually straight forward approach to approximate real or discrete valued target functions. The learning process includes storing the presented data. All instances correspond to points in an n-dimensional space and the nearest neighbors of a given query that already defined regarding the standard Euclidean distance [8]. The probability of a query q belongs to a class c that can be calculated as follows:

$$dist(X_1, X_2) = \sqrt{\sum_{i=1}^{n} (x_{1i} - x_{2i})^2} \quad (2)$$

K is the set of nearest neighbors, kc the class of k and d(k,q) the Euclidean distance of k from q.

Despite their simplicity, k-NN classifiers suffer a serious drawback. The distance between items is calculated based on all the attributes. This implies any features that are in fact irrelevant for classification have the same impact at relevant features. This noise-sensitivity leads to miss-classification problems and degradation in the system accuracy. Such behavior is well known in the literature and is usually referred as a curse of dimensionality [11]. In experimental research section, we will show that this problem affects heavily the classification accuracy of the system. In fact, not only noisy features that affect the classifier accuracy but correlated features may also cause problems.

However, *k*-NN classifiers that used in conjunction with effective feature subset selection techniques are readily interpretable and can provide important insight into a weak theory domain. Black-box classifiers (e.g., neural nets) do not offer the same insight. To overcome the problem of the high dimensional feature space, it is possible to use different strategies. In the next section, we present a set of ensemble methods that we have applied to simplify the decision surface of the *k*-NN deals with. This simplification is obtained in the ensemble members by reducing the number of classes that used to train the *k*-NN or by reducing the feature space dimensionality.

In this study, prediction of decision classification is made to determine the best time to sell, buy or hold the warehouse receipt document by owners or prospective buyers in real time based on transaction history data.

2.3. The Model Architecture

The architecture of real-time business analytic for buying or selling transaction in this research illustrated in Figure 3. From the vast amount of information available in an input, real-time business analytic that using feature selection gain ratio and *k*-NN will prediction the best condition for sell, hold or buy transact for the owner or prospective buyer.

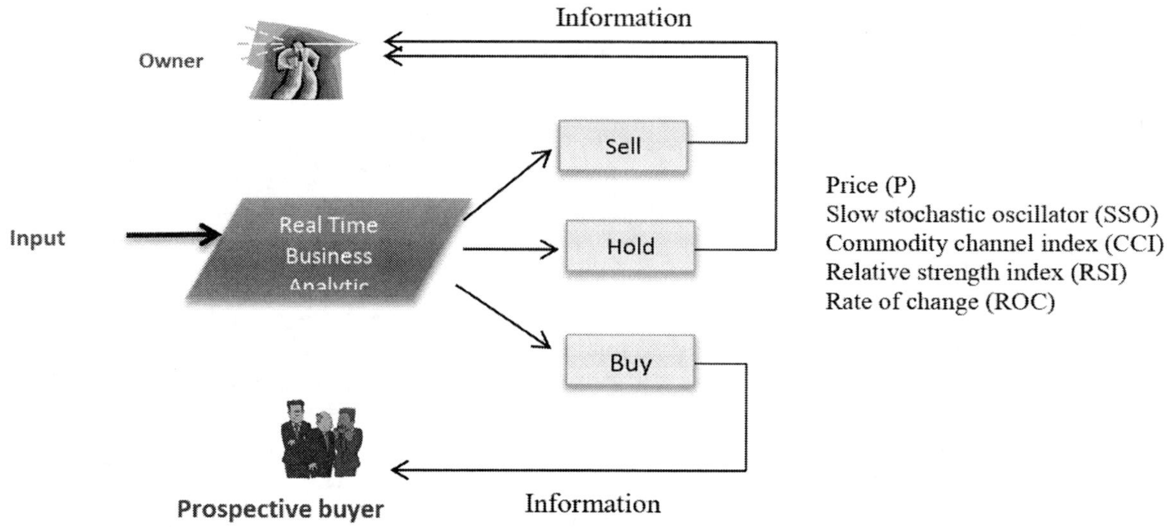

Figure 3. The architecture of real-time business analytic for buying or selling transaction warehouse receipt document

3. Empirical Study

3.1. Research Data

The research data that used in this study is generated from upper and lower data range. Since we attempt to mine the stock index movement tendency, technical indicators are used as input variables. This study selects 18 technical indicators to make up the initial attributes, as being determined by the review of domain expert and prior research [13-15]. The description of initially selected attributes is presented in Table 1.

Table 1. Initially selected feature indicators and their formulas

Feature indicators	Formula
Price (P)	x_t , (t = 1, 2, …, n)
Stochastic oscillator (SO)	$(\frac{x_t - x_i(m)}{x_A(m) - x_i(m)})$
Moving stochastic oscillator (MSO)	$\frac{1}{m}\sum_{i=l-m+1}^{t}(SO_{t-i})$
Slow stochastic oscillator (SSO)	$\frac{1}{m}\sum_{i=l-m+1}^{t}(MSO_{t-i})$
Rate of change (ROC)	$\frac{x_t}{x_{t-m}}$
Momentum (M)	$x_t - x_{t-m}$
Moving average (MA)	$\frac{1}{m}\sum_{i=l-m+1}^{t}(X_i)$

IOP Conf. Series: Earth and Environmental Science **89** (2017) 012019 doi:10.1088/1755-1315/89/1/012019

Moving variance (MV)	$\frac{1}{m}\sum_{i=l-m+1}^{t}(X_i - X_i)$
Moving variance ratio (MVR)	$\frac{MVt^2}{MV_{t-m}}$
Exponential moving average (EMA)	$a \times x_t + (1-a) \times x_{t-m}$
Moving average convergence & divergence (MACD)	$\sum_{i=l-m+1}^{t} EMA_{20}(i) - \sum_{i=l-m+1}^{t} EMA_{40}(i)$
Accumulation/ distribution oscillator (ADO)	$((x_l(m) - x_t)/(x_h(m) - x_l(m))$
Disparity 5 (D5)	$\frac{x_t}{MA_5}$
Disparity10 (D10)	$\frac{x_t}{MA_{10}}$
Price oscillator (OSCP)	$\frac{(MA_5 - MA_{10})}{MA_5}$
Commodity channel index (CCI)	$\frac{(M_t - SM_t)}{0.015D_t}$ $where\ M_t = x_h(t) + x(t) + x_l(t), SM_t = \sum_{i=l-m+1}^{t}\frac{M_i}{m}, D_t = \sum_{i=l-m+1}^{t}\lvert M_i - SM_t\rvert/m$
Relative strength index (RSI)	$100 - \frac{100}{1+RS}$ $where\ RS = \frac{\sum_{i=l-m+1}^{t}(x(i)-x(i-1))}{\sum_{i=l-m+1}^{t}(x(i)-x(i-1))}$
Linear regression line (LRL)	$\frac{m \times \sum_{i=l-m+1}^{t} i \times x(i) - \sum_{i=l-m+1}^{t} i \times \sum_{i=l-m+1}^{t} x(i)}{m \times \sum_{i=l-m+1}^{t} i^2 - (\sum_{i=l-m+1}^{t} i)^2}$

3.2. Experiment Result

Figure 4 displays the knowledge flow of computational methods to obtain a prediction market trend towards buying or selling a warehouse receipt documents. This knowledge flow and data analysis is using Orange Software.

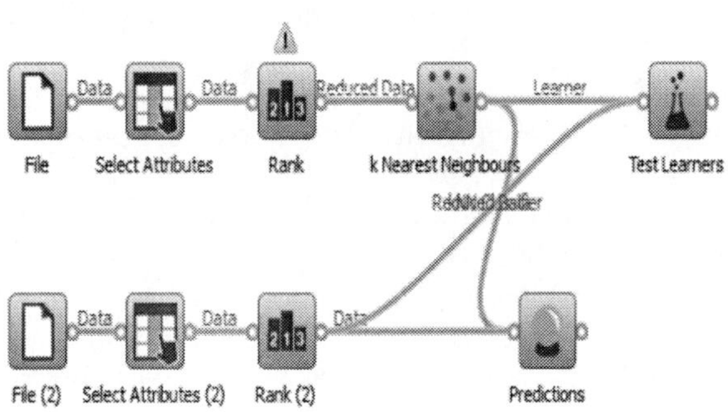

Figure 4. The knowledge flow

From 18 parameters in Table 1, after the sort by using a feature selection gain ratio method obtained five most influential parameters on the prediction of selling and buying warehouse receipt document.

The parameters we consider are price (P), a slow stochastic oscillator (SSO), commodity channel index (CCI), relative strength index (RSI) and rate of change (ROC) as in Table 2.

Table 2. Result of feature selection gain ratio

Attribute	Gain ratio
Price (P)	0.001
Slow stochastic oscillator (SSO)	0.016
Commodity channel index (CCI)	0.005
Relative strength index (RSI)	0.006
Rate of change (ROC)	0.009

Datasets are used to classify the *k*-NN uses five parameters generated in the feature selection stage with the gain ratio

Table 3. Result of prediction time to buying or selling using *k*-NN

MSE	RMSE	MAE	RRSE	R2
4.9617	2.2174	2.083	1.4284	-1.0403

Each of the models described in the last section is estimated and validated by in sample data. The model estimation selection process is then followed by an empirical evaluation based on the out-of-sample data. At this stage, the relative performance of the models is measured by MSE, RMSE, MAE, RRSE and R2. Table 3 reports the experimental results. The results show that the prediction means square error value of 4.9617 which means that the resulting level of accuracy rate is 95.03%.

4. Conclusions

This study uses a feature selection method of computing the ratio of profits and k-NN to track trends in the market either to buying or selling transaction process. Regarding empirical results, we find that the whole forecasting model can be used as an alternative solution to buy or sell a stock market trend of warehouse receipt documents with accurate rate is 95.03%.

References

[1] Pant, P 2009 Business intelligence (BI) How to build successful BI strategy. Deloitte Consulting LLP 1-17.

[2] Kamijo, K., Tanigawa, T 1990 Stock Price Pattern Recognition: A Recurrent Neural Network Approach. In: *Proceedings of the International Joint Conference on Neural Networks*, San Diego, CA 215-221.

[3] Yoon, Y., Swales, G 1991 Predicting Stock Price Performance: A Neural Network Approach. In: *Proceedings of the 24thAnnual Hawaii International Conference on System Sciences*. Hawaii 156-162.

[4] Trippi, R.R., DeSieno, D 1992 Trading Equity Index Futures with a Neural Network. *Journal of Portfolio Management* 19 309-317.

[5] Witten, I.H., Frank, E 2000 Data Mining: Practical Machine Learning Tools and Techniques with Java Implementations. Morgan Kaufmann Publishers, San Francisco, CA.

[6] Kohara, K., Ishikawa, T., Fukuhara, Y., Nakamura 1997 Stock Price Prediction Using Prior Knowledge and Neural Networks. *International Journal of Intelligent Systems in Accounting, Finance and Management* 6 11-22.

[7] Tsaih, R., Hsu, Y., Lai, C.C 1998 Forecasting S&P 500 Index Futures with a Hybrid AI system. *Decision Support Systems* 23 161-174.

[8] Quah, T.S., Srinivasan, B 1999 Improving Returns on Stock Investment through Neural Network Selection. *Expert Systems with Applications* 17 295-301.

[9] Kim, K., Han, I 2000 Genetic Algorithm Approach to Feature Discretization in Artificial Neural Networks for the Prediction of Stock Price Index. *Expert Systems with Applications* 19 (2000) 125-132.
[10] Zhang, G., Patuwo, B.E., Hu, M.Y 1998 Forecasting with Artificial Neural Networks: the State of the Art. *International Journal of Forecasting* 14: 35-62.
[11] T M. Mitchell 1997 Machine Learning, McGraw-Hill.
[12] Kupinski, A.M., Giger, M.L 1999 Feature selection with limited datasets. *Medical Physics* 26: 2176-2182.
[13] Hann, T.H., Steurer, E 1996 Much Ado about Nothing.Exchange Rates Forecasting: Neural Networks vs. Linear Models Using Monthly and Weekly Data. Neurocomputing 10: 323-339
[14] Yu, X.H.: Can Backpropagation Error Surface not have Local Minima.
[15] G. Zenobi, P. Cunningham 2001 Using Diversity in Preparing Ensemble of Classifiers Based on Different Subsets to Minimize Generalization Error, *12th European Conference on Machine Learning (ECML 2001),* L. De Readt & P. Flach (Ed.), 576-587, Springer Verlag, 2001.
[16] Orange Software. University of Ljubljana. Website : orange.biolab.si.

IOP Conf. Series: Earth and Environmental Science **89** (2017) 012020 doi:10.1088/1755-1315/89/1/012020

Postponing Labor in Fisheries, Tourism and Agriculture Sectors: Rural Eastern Indonesian University Students in Java

Christopher Foertsch
Department of Anthropology, Doctoral Student University of Victoria, British Columbia, Canada

Email: chrisfoertsch@gmail.com

Abstract This paper explores the migration of Eastern Indonesian university students who come to Java for education. Often from rural, economically disadvantaged regions such as the Kei Islands in Southeast Maluku, and Nusa Tenggara Timur (NTT), these young adults delay joining fisheries, agriculture, or tourism sectors. Instead, these relatively high-performing students travel to the "center of the country" seeking skills and experiences promised by higher education in Javanese urban centers. This qualitative, anthropological research complements other, more technical and economic approaches. Based on interview and observational data, a complicated portrait emerges of these bright young people from fishing and farming communities in Maluku and NTT. Many idealistically plan to return to their home communities, hoping to improve local fishing and farming methods or to work as teachers, civil servants, or tour guides. Others do not intend to return home, where they think jobs are scarce and traditional livelihoods unattractive. Analysis of this generation's perspective has critical implications for educators and policymakers wishing to prevent a "brain drain" of their educated native sons and daughters, whose experience and skills could contribute importantly to the various socio-economic demands present in island regions, including fisheries and agriculture, conservation, tourism, and employment.

1. Introduction

Migration has been part of the human experience since prehistoric times, the movement of individuals and groups to feed their families seek economic opportunities and adventure. In what is now the Republic of Indonesia, seafaring and inland peoples have been encountering each other through trade, war, and political affiliation for millennia [10]. In recent centuries this movement has been linked to centralized political and economic power radiating from a Javanese Center out to peripheral Outer Islands, through 17th and 18th century Dutch efforts to monopolize the spice trade in the archipelago, later through the Dutch East Indies colonial project, and since 1945 through the Java-centric government of the Republic of Indonesia. This center-periphery dynamic consolidates power and wealth in the metropole—Java—drawing human and natural resources from Outer areas for the benefit of the Center, leaving peripheral regions such as Maluku and East Nusa Tenggara (NTT) comparatively underdeveloped in aspects of infrastructure, employment, and education. It is in the context of this regional economic inequality that young people from Eastern Indonesia and elsewhere migrate to the cities of Java for higher education.

IOP Conf. Series: Earth and Environmental Science **89** (2017) 012020 doi:10.1088/1755-1315/89/1/012020

At the time of Indonesian Independence after World War II, there were few options for higher education, mostly in Java and reserved for children of the elite [11-12]. Since then, the numbers of Indonesians achieving basic, middle, and higher education has grown steadily, yet inequalities in access and quality persist [3, 6, 8].

This ethnographic account of the experience of educational migrants from islands in NTT and the Kei Islands of Southeast Maluku builds on the author's MA thesis fieldwork in Malang, East Java [5], and on his ongoing (2016—2017) Fulbright Research Fellowship in Yogyakarta and Malang on Java, the province of NTT, and the Kei Islands of Southeast Maluku. This paper seeks to create a partial portrait of this population of young working-age people, who choose to pursue higher education rather than directly enter the workforce for various motivations. It also hopes to describe some of the factors and processes that drive this migration, as well as some of the personal, economic and societal effects of this increasing and ongoing phenomenon of educational migration in Indonesia. Particular attention is paid to a comparison between the inland rural, agrarian region of Manggarai, Flores, NTT, and the Kei Islands of Southeast Maluku, where small-scale agriculture, fishing, and other marine livelihoods and industries predominate.

2. Methodology

This research is based on data collected by the author through semi-structured interviews, and unstructured observations and conversations with students at universities, polytechnics, academies and *Sekolah Tinggi* (higher schools), as well as with alumni and educators in the Javanese cities of Yogyakarta and Malang, urban educational hubs that attract students from all over Indonesia. Data also come from interviews and observations on the island of Flores, NTT, the home island of a large number of study participants. As the research is ongoing, data do not yet include field observations and interviews in the Kei Islands themselves, relying for the time being on interviews conducted in Java with Kei Islanders studying there.

To date, qualitative data include 44 semi-structured interviews with a total of 87 participants, recorded in 2015 and 2017. Of these, 22 come from the administrative Districts of Manggarai, Flores, NTT, while 8 are from the Kei Islands of Southeast Maluku. (Other participants are primarily from other islands in NTT, such as Timor, Sumba, Adonara, and others.) Analysis of the text of these interviews complements the researcher's written observations and notes from organized and informal activities. While individual accounts can and do differ depending on personal experience, through a sufficient sample size, some common themes and trends emerge upon analysis [4].Participants were recruited through the researcher's work as a guest lecturer and affiliate at Universitas Kanjuruhan Malang, as well as through what Bernard calls "chain referral sampling," also known as "snowball sampling" [2].

Formal authorization and ethical protocol adherence for working with human subjects are respectively under the auspices of RISTEK (the Indonesian Ministry of Research and Technology) and the Institutional Review Board (IRB) at the researcher's former institution, Oregon State University.

The use of qualitative methods and ethnographic methods is intended to complement the quantitative and technical methodology frequently used in other fields, giving voice to the human experience at the center of a given topic [2,4,9].

3. Results and Discussion

The majority of the participants in this study are from small rural families in villages where farming and small-scale fisheries are the predominant occupations, while a few have parents who are teachers, civil servants or small business owners. The overwhelming majority are among the first generation to attend a university, including some whose parents are yet illiterate. Previous generations, and even older brothers and sisters of the current generation, had little opportunity or means to achieve an education beyond a basic level due to barriers of cost and access.

A librarian in his 50's now working at a *Sekolah Tinggi* in Flores related his experience growing up in a remote Manggaraian village. After finishing elementary school, to attend junior and the senior high school he was obliged to live with relatives in the district capital of Ruteng, at that time a two-day journey by foot. Very few of his contemporaries were sufficiently motivated and financially supported by their families to accept such difficulties and costs to pursue education. Even fewer continued to university education in Java as he did, working in restaurants and in the library at his school in East Java to pay tuition and living expenses to supplement the meager funds his parents were able to send him. Since the time Mr. Silvestrus (all names are pseudonyms) attended university in the 1980's, more Catholic and national elementary, middle, and high schools have opened in and near his home village, and improved transportation means that a trip to the district capitol is only a few hours by bus.

A similar evolution has happened in the still remote village of Wukir, East Manggarai, where increased opportunities for elementary, junior and middle school education have increased in recent decades, and transportation is marginally better (though still arduous, especially in the rainy season). According to informants there, including teachers, students and returned alumni, now nearly all junior high (SMP—*Sekolah Menengah Pertama*) graduates go on to high school (SMA—*Sekolah Menengah Atas*), and the majority of SMA graduates aspire to attend higher education (with one or two exceptions from graduating classes of 35-40). A group of middle school teachers told me that just ten years ago they estimate that only some 50% of their students completed high school and went on to university or other higher education. The senior teachers at the Catholic SMP (built in the early 1980's) can count on their fingers the few students from the '80's and '90's who went on to higher education. The founding of additional private and public SD (*Sekolah dasar*—elementary schools), SMP, and SMA's in the area in the 2000's produced more graduates ready for higher education, and now many of those university graduates have returned to teach in those same schools, in turn preparing and priming the youngest generation to follow in their footsteps. Many returned graduates also staff the local government, health, and agriculture institutions.

This dramatic change in only a few decades is exceptional, and cannot be generalized, except as an example of a community that values basic, middle, and higher education and has incorporated this value into the local culture.

Informants from Kei reported a drastically different story, suggesting that they are among a minority of their peers who attend higher education. Still, it seems to be at a higher rate than their parents' generation, which could suggest that their community is at an earlier point on the growth curve than their Manggaraian contemporaries. Interestingly, among the small sample of Kei students in this study, two were from a family of teachers and PNS (civil servants) in the city of Tual, and three others were their cousins, having attended SMA in the same city. Though unconfirmed as of this writing, it's plausible that these lucky few can study in Java thanks to historical benefits from privileged family networks that favor education and access to power and position to particular groups in Kei [1,7].

Despite improved access to education, and an increased societal and familial support in recent years (as in the case of Wukir and other participants' Manggaraian villages), sending a son or daughter to attend university in Java or anywhere is still a significant financial sacrifice for a family. For a family from a rural farming or fishing community—such as in Manggarai and Kei—income is unpredictable and seasonal, unlike their counterparts who hold the few civil service positions (*Pegawai Negeri Sipil*—PNS) and get a monthly salary from the government. From their meager earnings, families must pay the substantial cost of transportation to Java, as well as tuition and cost of living.

This is in addition to the cost borne by families in terms of giving up a son or daughter whose absence means they will be unable to contribute to labor-intensive farming, fishing, gardening and household duties, not to mention years of paying for SD, SMP, and SMA tuition, uniforms, and fees. Costs rise even more when more than one child chooses to attend university.

Despite these costs and barriers, increasing numbers of young people (and their families) are choosing to do so. A look into the motivations and perceived benefits is important to understanding this phenomenon.

3.1 Motivations for Educational Migration

When answering the question "why did you choose to study in Java?" respondents' responses clustered around a handful of reasons, remarkably similar regardless of their geographic origin. A principal motivator is common to youth the world over, a yearning for adventure, to seek experience (*pengalaman*) in cities far from home, to see things and people from other parts of the country. They want to see first-hand the *Pusat Negeri* (the Center of the Nation) they had only known through TV, movies, and stories from older relatives who preceded them in this journey. Related to this is a coming-of-age sense of personal independence (*kemandirian*) from parents, to prove to themselves and their community that they can find grown-up success through this endeavor. In this way, educational migration is a relatively recent form of mobility that respondents call *merantau* (wander), which has primarily been a search for work and income to send home. Common destinations of the *merantau* of respondents or their relatives are to cities for work in construction or factories, to Bali for employment in the tourist industry, to Malaysia, Kalimantan, or Papua to work in palm oil plantations, mining, or service sectors. In fact, some slightly older respondents in their 30's or 40's reported that they had originally set out for this kind of economic *merantau* with no intention of pursuing higher education at the outset. Mr. Davidus, a high school teacher in Wukir, first studied for a few months at a seminary in Eastern Flores, before he got a call from a friend in Jakarta looking to hire workers for a motorcycle parts warehouse, where he worked for five years, sending remittances home and saving a bit for himself. While in Jakarta he took the opportunity to get a teaching degree, graduating in 2013, and finally returned to East Manggarai after a total of ten years away. Four of his classmates from his SMA had done nearly the same thing, also working in the national capitol and returning with degrees from the same *Sekolah Tinggi* as him.

When Eastern students decide to study in Malang or other cities in Java, they report that the cost of living there is lower than in other cities—lower for example than in Kupang, the capitol of NTT—and there is at least a chance they can find side jobs to supplement money their family can send them. Besides also having a pleasant climate, Malang has a reputation for being a welcoming place for Eastern students who want to study at one of the 55 campuses there. Yogyakarta is of course also known as an "educational city," with literally several hundred campuses, small and large. Respondents reported that they chose Javanese campuses because the study program they wanted to pursue wasn't offered nearer to home, or if it was offered, the quality of lecturers, technology and facilities was perceived to be lower than in Java. According to a lecturer and administrator at STKIP in Ruteng—the capitol of Manggarai District, Flores—her campus was the only one in the province of NTT with the highest accreditation rating of A, yet the number of study programs they offer is only six, limited to teaching, divinity and other humanities.

Underlying motivations to study in the first place vary but also cluster around a few themes. Some wish to gain skills and certification to bring back home to find work in education, government, healthcare, or in private businesses if such options exist. Some of this is a professed desire to help their family and community, which are often economically and educationally disadvantaged, especially when compared to Java. Others want to become entrepreneurs, such as Karlos, from Tual, Kei, who wants to start an import business in his hometown to sell particular products he has come to enjoy during his time in Java. Andi, a second-year veterinary student from Wukir, Manggarai wants to start a feed store and raise improved breeds of cattle when he finishes his degree. Still, others profess a desire to continue their *merantau* to other places where they hope to find satisfying jobs since they think it will be unlikely to find such employment at home. This troubling pessimism was prominent among several of the respondents from the Kei Islands.

3.2 Challenges and Realities in Java
When Eastern students reach Java and begin higher education, they face a series of obstacles beyond the economic ones already discussed above. Often unfamiliar with admissions procedures at prestigious state universities, they can arrive in Java too late to take the entrance exam, or perform poorly if they do take it, relatively unprepared compared to Javanese peers who attended prestigious SMA's with higher standards and resources than found in rural Eastern schools. As such, they must enroll in low cost, less prestige private universities with lower accreditation scores. Additionally, as ethnic and religious minorities, they are exposed to other kinds of prejudice and discrimination, particularly in housing choices, and in negative perceptions by other groups, mainly Javanese.

To cope with these challenges, Eastern students forge "new families" of mutual support during their time in Java, younger siblings (*adik*) relying on older relatives (*kakak*) from their home towns and also

IOP Conf. Series: Earth and Environmental Science **89** (2017) 012020 doi:10.1088/1755-1315/89/1/012020

from other islands [5]. Many students followed *kakak* to Malang, Yogyakarta or elsewhere expressly because they (and their parents) knew they could count on their elders' guidance and protection. Still, a substantial portion of students is unable to graduate for a handful of reasons, including mismanaging money, big city temptations, violent conflicts with other groups, alcohol or drug abuse, and pregnancy. A girl from Kei Besar estimated that 10-15% of her peers would be unable to graduate due to one of those reasons.

3.3 After University
Despite the challenges and barriers, it would appear that the number of Eastern students graduating from universities in Java is growing. For previous generations, when this number was fewer, there was a good chance they would be able to find the civil service or teaching position in their home areas. But chances are slimmer in the current job climate, which is flooded with graduates and local governments are no longer hiring like they were during previous periods, particularly during the *reformasi* era "*pemekaran*" of the 2000's when new Districts (*Kabupaten*) and Subdistricts (*Kecamatan*) required staffing. Others do return home and can teach in new private or state schools, but often with meager pay that must be supplemented with farming or other subsistence activities. Nurses and midwives often must take volunteer positions at the local Health Center (*Puskesmas*) since paid positions are unavailable.

Some graduates try to find employment in Java, but face challenges in this realm too, facing stiff competition from Javanese job candidates who attended prestigious state universities such as Brawijaya in Malang or Gajah Mada in Yogyakarta. Others continue their *menta*l to other islands to look for work-related or not to their degrees, sending remittances home as they can. In small, remote Wukir, Manggarai, study participants reported that they had relatives living all across the country and the world—an NGO employee in Surabaya, teachers and an oil rigger in Papua, a forestry officer in Kalimantan, and clergy in Germany, the Philippines, and West Africa.

A half-dozen informants are university lecturers who have lived in Java for decades and taken spouses from there and raised families. Their experience as early Eastern pioneers at Javanese universities in Java was difficult and expensive, requiring sacrifices from their entire families. In their estimation, returning home after graduation wouldn't have afforded them a chance to recoup their families' financial investment in their academic career. So they found higher paying work than they could get back home, and theirs is a story of one-way migration, with occasional trips to Flores or Adonara every few years to visit relatives.

Wukir is remarkable that many did, in fact, return to their home village, even though chances of making money are slim there. Some of it is a practical economic decision, especially if one's family holds land since it is then possible to feed the family, even if a salary is low or nonexistent.

Another important motivator to return home (at least to Wukir), despite the drawbacks, is the strong sense of community and roots, not only with living relatives and neighbors but also a strong faith in the protection and respect of ancestors (*leluhur*) who also inhabit the area and watch over the residents. This strong adherence to local *adat* is a reason given by several people in Wukir for the relatively high standard of well-being there, despite poverty and the hard life of a farmer. This well-being, according to a well-educated and prominent figure at the local health facility, includes spirituality, health, and education, values shared by the *leluhur* as well, who want the best for their people. She thought this was not the case in other places, with less emphasis on education, and where *adat* was also not as carefully followed as in Wukir. It will be interesting to explore whether there is a further correlation between adherence to *adat* and the average level of education in other areas of NTT and Maluku.

4. Conclusion
What does this mean for the workforce in rural and island towns and villages, the sending communities of these educational migrants? They choose to leave home, to *merantau*, at an age when their youth and energy could be very useful to their parents in traditional occupations, such as farming and fishing. They leave and become consumers of education, spending their families' savings, instead of undertaking a classic *merantau* in search of medium-term economic gain in foreign and domestic labor

to bring home cash to build a house, buy land or a boat, or pay for a dowry. They are hoping their investment will pay off in the longer term, with increased economic prospects over a lifetime. Clearly, the twenty-first century requires even the sons and daughters of farmers and fishermen to have at least basic education to be successful participants in society and the economy. Stereotypically, it had been true for decades that anyone who went further and attended university would be rewarded with a prestigious, well-paying position. But this is no longer guaranteed.

What does it mean that the median level of education is increasing in places like Wukir and Kei to include ever more university graduates who might or might not be able to find work that corresponds to their degrees or that compensates the time and resources their communities invested in them? In Wukir, it would seem that the local culture has changed in just a few decades so that now nearly all young people attend SMP, SMA then wants to go on to higher education. Despite poverty, people hold "school parties" (*pesta Sekolah*) to raise funds for young sons and daughters heading off to attend school. These parties are perceived as a form of *gotong royong* (mutual assistance), and elders perform *adat* ceremonies to seek the blessing of the ancestors for the success of the party and the student. Only a decade or two earlier, parental pride and familial success were measured in weddings and births. Now a young woman's parents have added additional educational criteria for prospective sons-in-law. A young man without a college degree is a less attractive candidate for marriage than a graduate.

What is the difference then between Kei, where respondents report being bored, pessimistic about finding work and meaning to move elsewhere after graduation, compared to a place like Wukir, where education is prized, and young graduates still frequently choose to return following their *merantau* on other islands? At the time of writing, the researcher had yet not carried out his fieldwork in Kei, so the answer to this question is pending input from students, families, and colleagues there. As well, an analysis of the quantitative elements (statistical data on population, income, literacy, education, etc.) of this story would help to triangulate whether these qualitative observations can be generalized. From an anthropological perspective, it's appealing to consider if there is a correlation between *adat* adherence, education, and attitudes about one's home community. Perhaps *adat* is an indicator of community health and well-being, and thus something that can attract young people back home after their educational migration, where they can apply their new knowledge and experience for the benefit of their community, as well as their ancestors.

Communities can try to promote interest in and preservation of *adat* by encouraging students to conduct their final theses (*skripsi*) on topics related to the traditions and wisdom of their elders. Besides recording and saving this knowledge in written and electronic form, it would also serve to increase young people's affiliation with their home communities, and this could make coming home more attractive to university graduates. Coming home is also more attractive if there is gainful employment to be had. Perhaps local, provincial, national scholarships for students entering particular fields identified as especially in demand by the local economy would be beneficial, as might small business loans and entrepreneurship training for young people attempting to boost their local economy. Likely, these measures have been applied already with varying measures of success, and future fieldwork will explore their effects, benefits, and drawbacks.

As mentioned earlier, this research is ongoing, and this paper and presentation are intended to stimulate discussion and highlight these human aspects of the workforce from the island and rural areas in Eastern Indonesia. The researcher welcomes feedback from interested colleagues.

References

[1] Adhuri, Dedy S.2013 *Selling the Sea, Fishing for Power*. Acton, A.C.T.: ANU E Press.

[2] Bernard, H. Russell 2011 *Research Methods in Anthropology*. 5th Ed. Lanham, Maryland: AltaMira Press.

[3] Bjork, Christopher. 2005 *Indonesian Education: Teachers, Schools, and Central Bureaucracy*. East Asia History, Politics, Sociology, Culture. New York: Routledge.

[4] Creswell, John W 2014 *Research Design*. 4th ed. Los Angeles, California: SAGE.

[5] Foertsch, Christopher R. 2016. "Educational Migration in Indonesia: An Ethnography of Eastern Indonesian Students in Malang, Java." Corvallis, Oregon: Oregon State University.

IOP Conf. Series: Earth and Environmental Science **89** (2017) 012020 doi:10.1088/1755-1315/89/1/012020

[6] Hill, Hal, and Thie Kan Wie. 2013. Indonesian Universities: Rapid Growth, Major Challenges In *Education in Indonesia*, 160–79.

[7] Hooe, Todd Ryan 2012 Little Kingdoms': Adat and Inequality in the Kei Islands, Eastern Indonesia. University of Pittsburgh. http://search.proquest.com/openview/bda1377d95edf49bb8d62aa778126954/1?pq-origsite=gscholar&cbl=18750&diss=y.

[8] Logli, Chiara 2016 Higher Education in Indonesia: Contemporary Challenges in Governance, Access, and Quality." In *Handbook of Asian Higher Education*. New York City: Palgrave Macmillan.

[9] Patton, Michael 2002 *Qualitative Research and Evaluation Methods*. 3rd ed. Thousand Oaks, California: Sage Publications.

[10] Ricklefs, M.C 2008 *A History of Modern Indonesia since C. 1200*. 4th ed. Stanford, CA: Stanford University Press.

[11] Murray, Thomas, R. 1973 *A Chronicle of Indonesian Higher Education, the First Half Century, 1920-1970*. Singapore: Chopmen.

[12] Vickers, Adrian 2013 *A History of Modern Indonesia*. 2nd ed. Cambridge, UK: Cambridge University Press.

Family vs Village-Based: Intangible View on the Sustainable of Seaweed Farming

Wellem A. Teniwut[1], Yuliana K. Teniwut[2], Roberto M. K. Teniwut[1], Cawalinya L. Hasyim[1]
[1]Fisheries Agribusiness Study Program, Tual State Fisheries Polytechnic, Langgur-Sathean Ringroad Km.6, Langgur, Southeast Maluku, Indonesia 97611
[2]Department of Agroindustrial Technology, Faculty of Agricultural Technology, Bogor Agricultural, Indonesia, 16680

Email: wateniwut@polikant.ac.id

Abstract. Compare to other fishery activities for instance fish mariculture and catching fisheries, seaweed farming is considered easier. Also, the market for seaweed is wider and will keep growing. Thus, makes seaweed farming as one of the fastest commodity to improve the welfare of a coastal community. There are technical and non-technical factors in seaweed farming management, for non-technical on this intangible factors vary between family-based and village-based management, therefore aimed of this study was to simulate farmers decision to choose between family-based and village-based on seaweed managing system trigger by intangible factors. We conducted our study in Southeast Maluku, data collecting conducted from October to December 2016 by depth interview and questionnaires on seaweed farmers. We used logistic regression to compare each intangible factors on family and village-based seaweed farming management. The result showed that for family-based management farmers were willing to transfer their knowledge among each member in the household. For village-based revealed that farmers with higher education background tend to work on village-based, also, the result also stated that in village-based management member were those who have better capability and skill, at the same time village-based management had a small probability for conflict to occur compared to family-based.

1. Introduction

Family-based farming management tends to be associated with an inadequate system that conducted their farming system scattered around the area [13] with no particular time frame for farming activity during each farming season also the owner and operational business runs by same people [1]. In some rural area in a developed country like Canada, many family-based farm system now has adjusted to becoming more industrialized by adopted recent technology [2]. Although, for an undeveloped country like in West Africa for instance, the challenge for family-based farming was entirely differed which related to government policy, market structure, resources and potential demand [4]. The gap between family-based farming in developed and undeveloped on their ability to adjust brings more attention to the important role of each government to address the problem that family-based farming faced, especially in the country like Indonesia [6]

In Indonesia, Kei Islands, in particular, are the rural coastal area which located in the east on Indonesia which makes this area is considered isolated from a big city where also the center of the industry since most of the large cities located on the west side. The geographical condition added difference challenge to family-based farming especially in the fisheries sector in this area, which is related to information, selling price projection and delivery cost. Most commonly farming activity in

Kei Islands is seaweed farming where most of remaining seaweed farmers are family-based with only a few left of village-based seaweed farming group. Village-based seaweed farming is a group that formed by local government consist of a farmer from the same village that is supported with tools and equipment to conducted seaweed farming. The local government of Southeast Maluku District has roughly over 100 village-based farming system under their management, but since 2012 number of active village-based farming group keeps decreasing [5]

Both family-based seaweed farming and village-based seaweed farming are not sustained either on their productivity and their durability. Each of farming based has their particular problem to deal with, for family-based, it's about the financial aspect where for village-based technical factors are factors come to attention. In Addition to management and professionalism problem on both family and village-based seaweed farming, conflict among members related to profit sharing and capability's gap among members also factors that have to put into consideration [3]. This study aimed of this study was to simulate farmers decision to choose between family-based and village-based on seaweed managing system trigger by intangible factors, where intangible aspects that put into consideration in this study were education background, willingness to transfer knowledge, conflict avoidance, trust among member, capability of each member, sex, and gender. The outcome of this study is to provide real insight to local government on maintaining the productivity of seaweed in the area by forming the better seaweed farming system in the area.

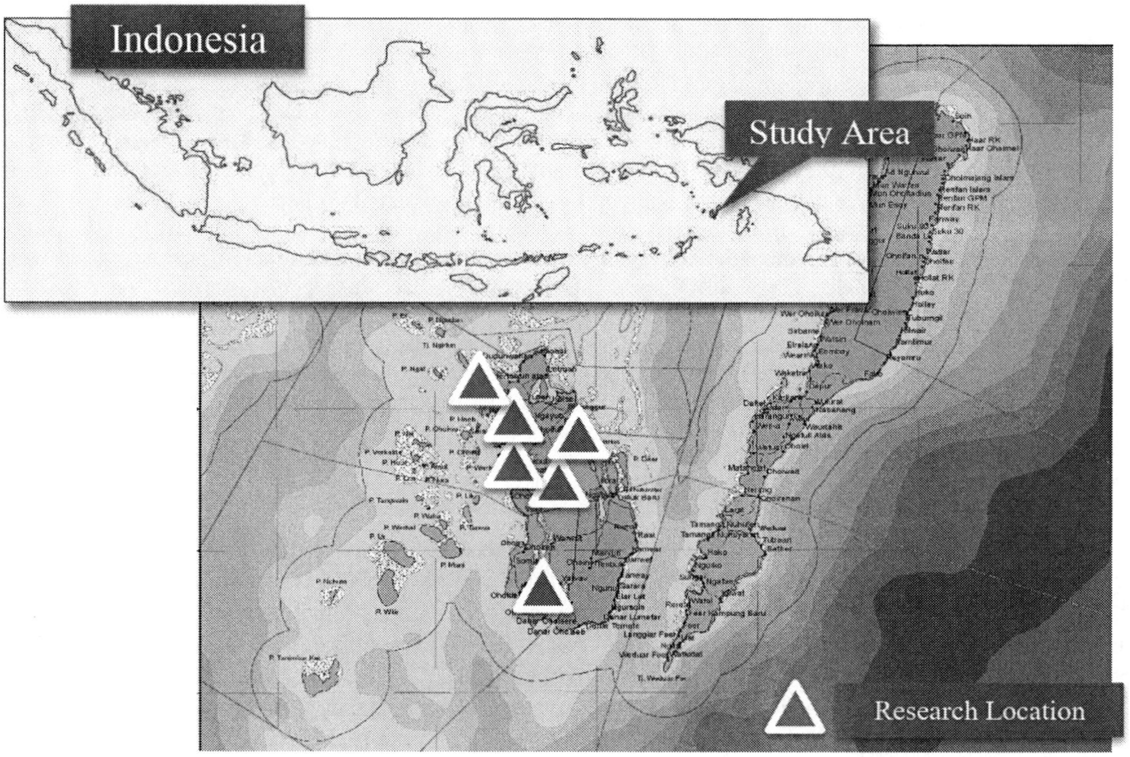

Figure 1. Study area

2. Methodology

This study measures the durability and sustainability regarding production activities based on intangible assets approach, where [8] stated that intangible assets refer to intellectual capital, where Roos proposed intellectual capital that consists of two types of capital, first human capital and second structure capital. Attitude, relationship, transfer of knowledge for the development of each member's capability, conflict management, and trust are intangible factors that have to be considered on conducted business, which also can become assets for competitive advantage.

For collecting the data we used structured questionnaire, for trust variable we adopted concept proposed by [7], that trust is one the most important sources of competitive advantage where one person belief on their partner behavior to uphold their commitment which they have agreed upon. Each group may have a different degree of capability [9] either of family-based or village-based group, therefore, capability used in this study emphasize more on operational capabilities, where [10] stated that operational capability as ability to perform tasks for better operational function of the organization or in this study groups of seaweed farming system. For conflict, we adopted study from [11] proposed to identify conflict which refers to a situation where specific identity (such as family, business owner, etc.) act inconsistently with their respected role. For transfer knowledge, we adopted studies by [12] of their model on transfer knowledge between from people inside the organization who has more knowledge and experience to people with less knowledge and experience, the type of knowledge that being transfer focus on tacit knowledge.

2.1. Data analysis

This study conducted in seaweed villages in Southeast Maluku District, Indonesia. We used purposive sampling method approach on six villages that being divided into three clusters. The first cluster is villages that have the high productivity of seaweed namely Letvuan and Ohoidertutu, for the second cluster consists of villages that have the average productivity of seaweed namely Namar and Ohoililir and the third cluster includes villages that have the low productivity of seaweed Sathean and Debut (Figure 1). Total 160 questionnaires being spread to seaweed farmers of these villages but only 136 questionnaires to meet the requirement for processing data. We used logistic regression for data analysis with STATA 12.0 software.

Econometric model in this study can be seen as follows:

$$\Pr[y_i^f = 1|X_i] = F(\beta_0 + Age_i\beta_{Age} + Sex_i\beta_{Sex} + Edu_i\beta_{Edu} + TK_i\beta_{TK} + Tr_i\beta_{Tr} + Cf_i\beta_{Cf} + Cap_i\beta_{Cap}) \quad (1)$$

$$\Pr[y_i^v = 1|X_i] = F(\beta_0 + Age_i\beta_{Age} + Sex_i\beta_{Sex} + Edu_i\beta_{Edu} + TK_i\beta_{TK} + Tr_i\beta_{Tr} + Cf_i\beta_{Cf} + Cap_i\beta_{Cap}) \quad (2)$$

Where y_i^f and y_i^v represent binary variable for family-based (*f*) and Village-based (*v*). If y_i^f farmers prefer to work on family-based seaweed farming, then it equals one and zero if otherwise, whereas, y_i^v equals one if farmers prefer to work in the village-based system and otherwise equals zero. For explanatory variables, we have seven categories age, sex, education, transfer knowledge (TK), Trust (Tr), Conflict (C) and Capability (Cap).

2.2. Respondent Characteristics

As it shooed on table 1, over half of farmers on this study were 25-44 years old with roughly 39% were 45 to 70 years old. Most of the farmers on this study were male around 82.4% and female 17.6%. Based on educational background, the result showed a fascinating result which also described the real condition of seaweed farmers in the area, where over 95% of seaweed farmers in the area did not have sufficient educational background to the least at diploma level. This may cause a problem since seaweed cultivation for productivity is to be considered easy compared to other marine culture, but it requires some basic knowledge on downstream and upstream to make this business can be sustained in rapid competition. Therefore, local government needs to conduct an intensive and structured training on seaweed cultivation and post-production process to maintain the durability of seaweed productivity in the area. Over half of farmers in the study have conducted seaweed farming between 6 to 12 years; this means respondent of our study has enough experience on seaweed cultivation although because of the lack of education background, still lack on an effort to maximize the revenue of farmers especially in post-production.

Table 1. Respondent Characteristics

	Frequency	Percentage
Age		
<25 years old	7	5.1
25-34 years old	37	27.2
35-44 years old	39	28.7
45-54 years old	30	22.1
>54 years old	23	16.9
Sex		
Male	112	82.4
Female	24	17.6
Education		
No Formal Educational Background	40	29.4
Elementary	35	25.7
Junior High	57	41.9
High School	2	1.5
College	2	1.5
Years conducting seaweed farming		
< 3 years	14	10.3
3-5 years	17	12.5
6-8 years	55	40.4
9-12 years	36	26.5
> 12 years	14	10.3

N=136

3. Result and Discussion

3.1. Family-based seaweed farming management system

In Table 2 can be seen the estimation results of equation (1) for intangible view on family-based seaweed farming system in Southeast Maluku. The value of the pseudo-R^2 is 0.206, the likelihood ratio test (LR = 13.12; Prob>chi^2 = 0.069) suggests that the explanatory variables are jointly significant at the 10 percent level. These statistics indicate that the model fits the data relatively well for a binary regression model.

Logistic regression results showed that farmers who willing to transfer their knowledge to a member of seaweed farming would prefer to work in family-based seaweed farming, where all socio-economy indicators and other intangible factors have a non-significant effect on farmers who choose family-based seaweed farming. Willingness to transfer knowledge in the family member is rational and predicted behavior. Since blood relation will provide the high emotional bond to diminish barrier from one farmer to share their knowledge with their close relatives.

Table 2. Logistic regression result on family-based seaweed farming management system

Variable	Coefficient		p-Value
Socio-Economy			
Age	0.045		0.256
Education	-0.005		0.991
Years of seaweed farming	-0.111		0.465
Intangible Factors			
Transfer Knowledge	2.437		0.014**
Trust	0.846		0.280
Capability	0.383		0.725
Conflict avoidance	0.351		0.644
Log-likelihood		-22.861	
Pseudo-R^2		0.206	
Likelihood ratio		13.12	
Prob>chi^2		0.069	
Hosmer-Lemeshow		2.93	

3.2. Village-based seaweed farming management system

In Table 3 can be seen the estimation results of equation (1) for intangible view on family-based seaweed farming system in Southeast Maluku. The value of the pseudo-R^2 is 0.117, the likelihood ratio test (LR = 19.44; Prob>chi^2 = 0.006) suggests that the explanatory variables are jointly significant at the 1 percent level. These statistics indicate that the model fits the data relatively well for a binary regression model.

Based our finding on socio-economy factors, education had a negative effect on farmers who choose to work on the village-based system. This result means that farmers in the area to against the idea of working with another member of the village when they have better education level than another farmer in the same village. This result also aligns with the logistic result on the family-based system, because the more educated farmers mean they have more knowledge than those how to have less, in addition, willingness to transfer knowledge heavily related to emotionally attached factor in the coastal community. Although farmers who prefer to join village-based managing system have high expectation to enhance their capability by learning from another member from the village, the result also indicated that those prefer to avoid conflict on conducted seaweed farming tends to choose village-based managing system.

Table 3. Logistic regression result on village-based seaweed farming management system

Variable	Coefficient		p-Value
Socio-Economy			
Age	0.002		0.891
Education	-0.562		0.017**
Years of seaweed farming	0.041		0.566
Intangible Factors			
Transfer Knowledge	0.822		0.190
Trust	-0.691		0.133
Capability	2.099		0.007**
Conflict avoidance	0.639		0.083*
Log-likelihood		-77.302	
Likelihood ratio		19.44	
Pseudo-R^2		0.117	
Prob>chi^2		0.006	
Hosmer-Lemeshow		2.49	

4. Conclusion

In coastal area much like in Kei Islands where the understanding of business in ideal form is still lacking also the influence of local wisdom and custom belief from the community around the area makes its necessary to carefully crafted better seaweed farming management system to maintain the sustainability on the profitability. This study concluded that combination of village-based and family-based seaweed farming managing system is the best form regarding intangible factors. Each managing system has its advantages for the family-based system it has to have at least one member with better capability in seaweed farming because it will sustain in the long run because our study showed that willingness to transfer knowledge on this system among other intangible factors measured. In the other side, in village-based managing system member in this system tend to have intention increase their ability from learning from another member in this scheme at the same time it has a low potential of conflict occurrence. Therefore we proposed combination system where the local government needs to focus on village-based system and make is as center on the developing of farmers on every village before they return to conduct seaweed farming with their family.

References

[1] Calus, M. and Van Huylenbroeck, G 2010 The Persistence of Family Farming: a Review of Explanatory Socio-economic and Historical Factors. *Journal of Comparative Family Studies* Volume XXXXI (5) 639-660.

[2] Smithers, J., and Johnson, P 2004 The dynamics of family farming in North Huron county, Ontario. Part I. Development trajectories. *The Canadian Geographer/Le Géographe Canadien*, 48(2), 191-208.

[3] Indian Ocean Commision (IOC) 2012 Implementation of a Regional Fisheries Strategy For The Eastern-Southern Africa and India Ocean Region; The Farming of Seaweed, *Smartfish Agrotec Report*/SF/2012/28.

[4] Toulmin, C., & Guèye, B 2005 Is there a future for family farming in West Africa?. *IDS Bulletin*, 36(2), 23-29.

[5] Teniwut W. A dan Kabalmay. J 2014 Studi Empiris: Evaluasi Usaha Budidaya Rumput Laut Di Kabupaten Maluku Tenggara. Prosiding Seminar ilmiah Tahunan ke-2. Langgur, Indonesia. 16 Desember 2014 [in Indonesian].

[6] Teniwut W. A 2016 For sustainable revenue of fisheries sector in small islands: evidence of Maluku, Indonesia. *AACL Bioflux* 9(3):722-732.

[7] Sundaramurthy, C 2008 Sustaining trust within family businesses. *Family Business Review*, 21(1), 89-102.

[8] Baxter, R., & Matear, S 2004 Measuring intangible value in business-to-business buyer–seller relationships: An intellectual capital perspective. *Industrial Marketing Management*, 33(6), 491-500.

[9] Mahmood, I. P., Zhu, H., and Zajac, E. J 2011 Where can capabilities come from? Network ties and capability acquisition in business groups. *Strategic Management Journal*, 32(8), 820-848.

[10] Nada, N., and Ali, Z 2015 Service value creation capability model to assess the service innovation capability in SMEs. *Procedia CIRP*, 30, 390-395.

[11] Shepherd, D., and Haynie, J. M 2009 Family business, identity conflict, and an expedited entrepreneurial process: A process of resolving identity conflict. *Entrepreneurship theory and practice*, 33(6), 1245-1264.

[12] Cabrera-Suárez, K., De Saá-Pérez, P., and García-Almeida, D 2001 The succession process from a resource-and knowledge-based view of the family firm. *Family Business Review*, 14(1), 37-46.

[13] Yang, H., Zhang, X., & Zehnder, A. J 2003 Water scarcity, pricing mechanism and institutional reform in northern China irrigated agriculture. Agricultural Water Management, 61(2), 143-161.

A conceptual mitigation model for asymmetric information of supply chain in seaweed cultivation

Wellem A. Teniwut[1], Kamilius D. Betaubun[1], Marimin[2], Taufik Djatna[2]
[1]Fisheries Agribusiness Study Program, Tual State Fisheries Polytechnic, Langgur, Indonesia, 97611
[2]Department of Agroindustrial Technology, Faculty of Agricultural Technology, Bogor Agricultural, Indonesia, 16680

Email: wateniwut@polikant.ac.id

Abstract. Seaweed cultivation has a better advantage over other fisheries activity in terms of easiness on conducting the production and multiplier effect on coastal community welfare. The effect of seaweed farming on the prosperity of coastal community in Southeast Maluku started to take place in 2008, although in 2012 either number of production and workforce is declining rapidly. By solving this problem, this article also provided with identifying and analyzing the supply chain of seaweed cultivation in Southeast Maluku. Based on this analysis we have found that one of the main reasons of declining seaweed production and the number seaweed farmers was asymmetric information that occurred on seaweed supply chain in Southeast Maluku. The component of asymmetric risk was the quality of the seeds, price, information and technology and the knowledge of actual market of seaweed, especially by seaweed farmers. Therefore, it is essential to make a conceptual model on mitigation of asymmetric information on the supply chain of seaweed production. We proposed a conceptual model based on four perspectives, first was goal, criteria and sub-criteria, actor and the solution to mitigate asymmetric information supply chain on seaweed cultivation.

1. Introduction

An efficiency of a supply chain is one of the keys to the success of modern business sector especially in this century, although, the higher efficiency of supply chain also tend to increase the complexity of supply chain risks [1]. To optimize supply chain, every member needs to consider supply chain risks. Studies showed by [13] that each it's necessary to put supplier failure risks into consideration, while [14] stated that distributor needs to act accordingly to face the growing of product cycle. The risks in the supply chain could be the result of a natural disaster [15] bad demand forecasting and price fluctuation [16-17] and inaccurate of supplier quantity [18].

[19] proposed supply and demand-based of supply chain risk, related to potential disruptive on supply chain from downstream to upstream, where normal information flow would play a key factor in this matter. Information disruption in the supply chain is one of the risks may occur in a supply chain, thus integration information concept proposed by experts to address asymmetric information problem in supply chain [2]. Rapid competition, instability in internal of each business, the change in the environment caused by nature or human-made and high volatile politic situation are sources of risk that every business sector needs to deal with recently [3]. The risk of information from original supplier to final user is known as information risk [4]. Therefore, it's necessary to provide a model on mitigation risk on asymmetric information in the supply chain.

The mitigation model on asymmetric information in the supply chain in this study adopted studies by [5-6] with their concept of "pre-disruption" and "post-disruption" or also known as "prevention" and "Responses." This approach also means that mitigation model in supply chain risk on the occurring of asymmetric information refers to every effort to identify risks and manage to prevent from happens and if distortion in information already happens then take a necessary action to make sure this situation handled. [33] stated that to avoid the risk in a supply chain to happen there were three steps to take which first was risks identification, risks assessment and risks mitigation following in that order accordingly.

This concept pointed of view also shared by [7] although, in addition to those three steps, their study expands "to-do list" for risk mitigation in supply chain into few more detail steps include risk estimation and risk control. The actor who also responsible and have the resources to influence the risk and distortion that already happen also take an essential part of mitigation model of asymmetries information supply chain. Studied from [20] showed that sharing information among supply chain members could reduce risk.

Seaweed cultivation in Southeast Maluku Regency, Indonesia, face a significant challenge started from 2012 where selling price of seaweed dropped from IDR 12.000 – IDR 15.000 /Kg to IDR 5.000/Kg. Several main reasons for this to happen was the incidence of ice-ice disease at seaweed farms that caused massive crop failure that also had an influence on the price of seaweed because of the declined on the quality on seaweed. The other reason was the behavior of local distributor related to an effort to influence the price of seaweed by holding the information on the quantity of demand of seaweed to the farmers so they could dictate the selling price [8]. Recently, in addition to those two factors, farmers also face the problem with a supplier of seeds, where farmers have found significant difficulties on the search for high quality on seaweed's seed, this due to the lack of basic knowledge by farmers. This situation indicated that the problem on seaweed cultivation in Southeast Maluku occurred because of there is an asymmetric information on the supply chain of seaweed in this area, from farmers to distributor and supplier and farmers. Based on the description above, this study aimed to provide a conceptual mitigation model on asymmetric information of seaweed cultivation in Southeast Maluku. Therefore the purpose of this study was first, to identify the risks in asymmetric information on supply chain especially in seaweed cultivation. Second, to determine and analyze the actor and the solution on the possible risk that been identify and third, was to build a conceptual model to mitigate asymmetric information on the supply chain in seaweed cultivation.

2. Methodology

In this study, we used an integrated approach to manage the distortion on asymmetric information in seaweed cultivation by identifying the possible risks that can occur and how to deal with the risks when it happens (Figure 1). This study took a case study of seaweed cultivation in Southeast Maluku, Indonesia.

2.1. Identification of risks in asymmetric information on supply chain

Based on the study by [9] that to identify risks in the supply chain, the commonly used of methodology was experts interview, where in this case, we have interviewed with experienced practitioners of seaweed cultivation from Southeast Maluku. We also interviewed with bureaucrats and academician who have in-depth knowledge of what happened in seaweed cultivation in this area. Also, we also conducted focus group discussion in two of the largest production of seaweed in this area. To enrich the data, we also used survey which also proposed by [5] where we surveyed villages that have seaweed cultivation in last three years to have comprehensive information regarding the problems occurred in seaweed cultivation. We also added all the empirical data with a literature review to compare sources and the risks, also review on similar cases in other countries. The summary of the methodology showed in Figure 1.

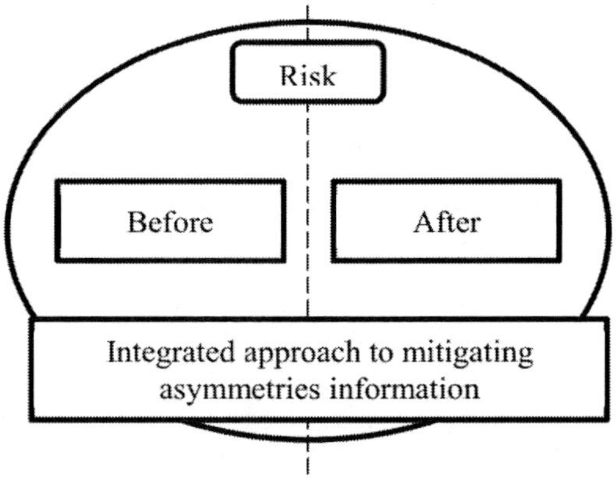

Figure 1. Methodology Approach Risks Identification

2.2. Identification solution on the possible risk in asymmetric information on supply chain

Failing on providing any related and sufficient information among member of the supply chain can increase the possibility of risks [11]. [10] Stated the strategies to mitigate risks where its approaches that every business sector can utilize to manage the possible risk and reduce the effect on supply chain were avoidance; control; cooperation and flexibility. For better formulation on mitigation model on asymmetric information in seaweed cultivation in Southeast Maluku, it is important to identify and analyze the supply chain in this area. Identify the supply chain stream, determine the condition of all members related to this supply chain and analyze the product and production of seaweed. On figure 3, the method for identifying the solution on the risks in the occurrence of asymmetric information in seaweed supply chain, started by identify supply chain of seaweed cultivation and analyze the constraints factor that keeps this supply chain to be optimum.

2.3. Conceptual mitigation model on asymmetric information of seaweed cultivation

In dealing with many factors that have to be considered in better decision making when many factors have to be weighed in against competing priorities then Multi-Criteria Decision-Making (MCDM) can be the solution. Many studies have used MCDM for risk mitigation [21-23]. One of the most extensively used approaches of MCDM is AHP (Analytical Hierarchy Process). AHP has been widely used for risk mitigation in supply chain [24-26, 31]

The AHP approach, developed by Satty. AHP integrates measures into one overall score to have a ranking of decision alternatives result usually provides a simplifying in multiple criterion problems by decomposing it into a multilevel hierarchical structure. [12] Pointed out that AHP is a useful tool for the understanding decision-making process that been provided because it can be view in hierarchical graphical form.

Therefore, to build a conceptual mitigation model on asymmetric information in seaweed cultivation in Southeast Maluku, we used AHP approach, to select the best alternative to mitigate the asymmetric information in the supply chain of seaweed farming. The risk factors and the solutions include key elements to manage the risks along with the actor that have resources to do so are combine to select the best alternative on mitigating asymmetric information on seaweed cultivation supply chain (Figure 2).

3. Results and Discussion

3.1. Risks on seaweed farming supply chain in Southeast Maluku

In supply chain of seaweed cultivation in this area, three main risks have occurred since the declining of seaweed production in 2012. First, the lack of trust between supply chain members is the reason each of them behaviors differently to maximize their profitability [28]. With more trust among members in the supply chain, the responsiveness as the result supply chain will function dynamically [29]. On the relationship between supplier and farmers, both parties want to seize the opportunity to gain the most each other. Farmers want a high-quality seed with a lower price or with a lower price they can have more quantity of the seeds. In the other hand suppliers also want farmers to pay more for seeds and with the lack of information that farmers do not have on the supply and availability of high quality of seeds. While at the same time the relations between farmers and distributors also have a lack of trust among them, farmers want their product to be pay in a competitive price and with a hope that distributor willing to buy their product immediately without any waiting time. Whereas, distributor used the lack of knowledge of farmers on demand and price that available in the market outside the area by offering noncompetitive price.

Second, on the risk of seaweed, cultivation supply chain was fluctuation on the selling price in this area [8]. Due to the geographical condition of Southeast Maluku that considered as an archipelagic region, the distance between this area and areas in Maluku Province is significant; besides, there is no end user of seaweed product in this area. In this condition, the distributor has more power on influencing the selling price of seaweed product than farmers, because they are the only buyers available in the area. Distributor in some cases forced to buy seaweed from farmers in early period before the selling price and demand of seaweed increased there are using information that they have to control the price. Farmers also take part of fluctuation on selling price. They sell their unqualified seaweed, to get fast money, they conducted an early harvest, which is not accordingly the suitable with time for the crop. Therefore, the quality of seaweed is low, and that implied on the selling price of seaweed.

The third risk on seaweed supply chain was decreasing on the profitability of seaweed. Lack of basic and applied knowledge on seaweed cultivation by farmers makes their profits are not sustainable. Aside from the knowledge the lack of ability on their part to obtain information regarding demand and selling price in seaweed, which it will cause declining on the profit [30]. Southeast Maluku has the marine resources and due to the characteristic of this area provide a perfect area for seaweed cultivation, though a significant number of farmers in this region who conducted seaweed cultivation have small access in information technology available related on seaweed farming.

In the other hand, the supplier in this region also faced the same problem on decreasing on their profitability, because the number of seaweed farmers is also decreasing since 2012, the demand for seeds also declining. For the distributor, they found some difficulties on supply seaweed to market outside the area, because the production of seaweed in this area keep dropping. Distributors also take a necessary action to dealing with this problem by providing tools and seeds to farmers with an incentive to pay seaweed in certain price, but in these circumstances, for other farmers will gain the loss because it narrowed their market since distributor provide their supply. Therefore, it's necessary to exchange sufficient information among members of the supply chain to increase profitability and manage the uncertainty to have successful supply chain [32].

IOP Conf. Series: Earth and Environmental Science **89** (2017) 012022
doi:10.1088/1755-1315/89/1/012022

3.2. A conceptual mitigation model on seaweed cultivation supply chain based on AHP approach

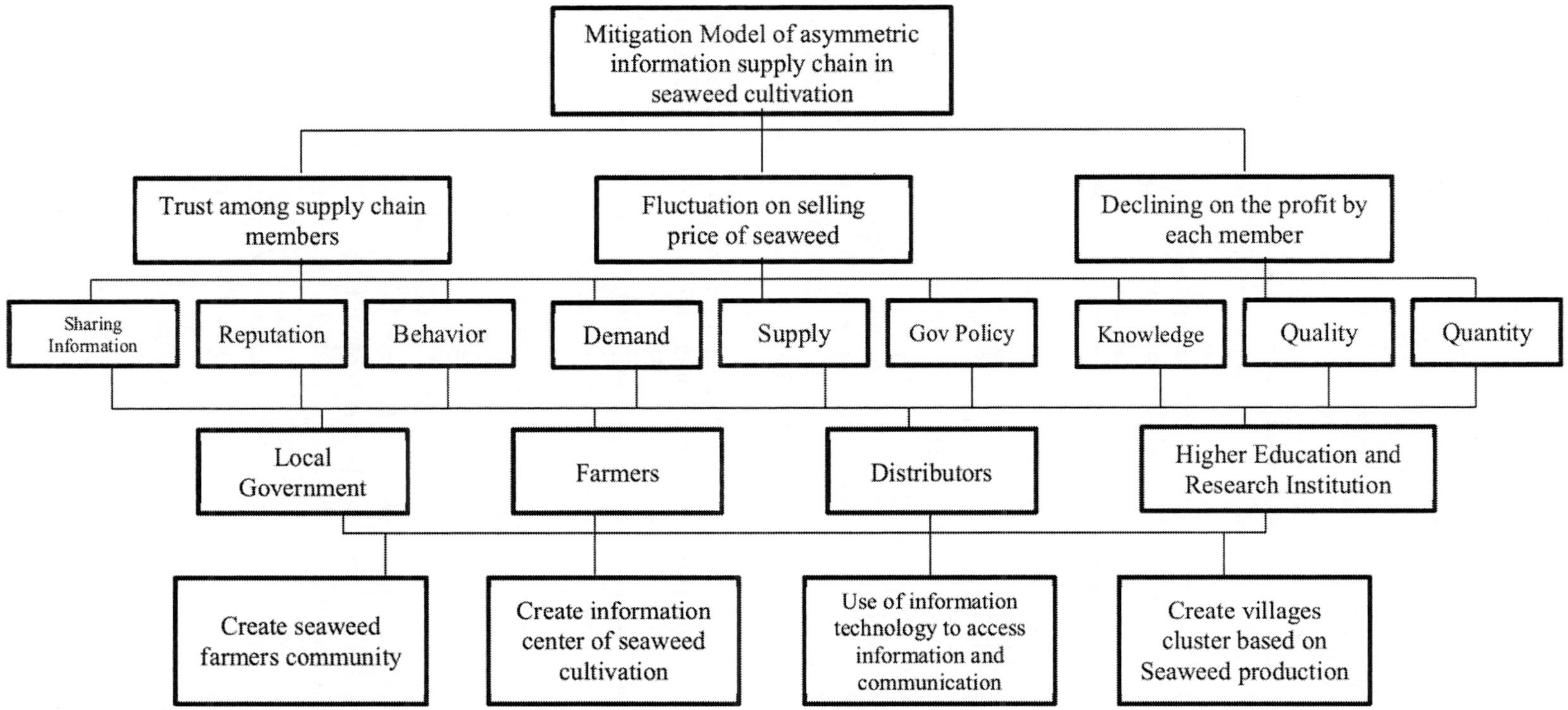

Figure 2. Hierarchy of AHP Model on mitigation asymmetries information on seaweed cultivation supply chain

4. Conclusion

Three risks occurred in the supply chain of seaweed cultivation. Trust among supply chain members, fluctuation on selling price of seaweed, declining on the profit. Key factors that can mitigate the risks to happen are information sharing, reputation, behavior for trust among supply chain members; demand, supply, government policy for fluctuation on selling price and knowledge on science and technology, quality, quantity for declining in profit. Actors that can help to mitigate the asymmetric information that occurred in the seaweed supply chain are local government, farmer distributors, higher education, and research institution. Based on all aspects considered, we proposed alternatives of mitigating asymmetric information in seaweed cultivation supply chain Southeast Maluku. First, create the community of seaweed farmers in the area. Second, create the information center related to seaweed farming. Third, use of information technology to access information and communicate with farmers and distributor outside the area and forth, create a cluster based on the productivity of villages in the area to easier the coordination among farmers, supplier, and distributors.

References

[1] Behdani, Behzad and Adhitya, Arief and Lukszo, Zofia and Srinivasan, Rajagopalan, 2012 How to Handle Disruptions in Supply Chains – An Integrated Framework and a Review of Literature Available at SSRN: https://ssrn.com/abstract=2114201 or http://dx.doi.org/10.2139/ssrn.2114201

[2] Kulp, S. C., Lee, H. L., and Ofek, E 2004 Manufacturer benefits from information integration with retail customers. Management science, 50(4), 431-444.

[3] Nyoman Pujawan, I., and Geraldin, L. H 2009 House of risk: a model for proactive supply chain risk management. *Business Process Management Journal*, *15*(6), 953-967.

[4] Juttner, U., Peck, H. and Christopher, M 2003 Supply chain risk management: outlining an agenda for future research, *International Journal of Logistics: Research and Application*, Vol. 6 No. 4, pp. 197-210.

[5] Thun, J. and Hoenig, D 2009 An Empirical Analysis of Supply Chain Risk Management in the German Automotive Industry, International Journal of Production Economics 131(1): 242-249.

[6] Dinis, N. M. M. T 2010 *Impact of Demographics on Supply Chain Risk Management Attitudes: Prevention Vs Response* (Doctoral dissertation, Massachusetts Institute of Technology, Engineering Systems Division).

[7] Adhitya, A., Srinivasan, R. and Karimi, I.A 2009 Supply chain risk identification using a HAZOP-based approach, *AIChE Journal* 55(6): 1447-1463.

[8] Teniwut W. A dan Kabalmay. J 2014. Studi Empiris: Evaluasi Usaha Budidaya Rumput Laut Di Kabupaten Maluku Tenggara. *Prosiding Seminar ilmiah Tahunan ke-2.* Langgur, Indonesia. 16 Desember 2014 [in Indonesian].

[9] Yang, Y.C 2010 Impact of the container security initiative on Taiwan's shipping industry, *Maritime Policy and Management* 37(7): 699-722.

[10] Miller, K. D 1992 A framework for integrated risk management in international business. *Journal of international business studies*, *23*(2), 311-331.

[11] Van der Vorst, J. G., and Beulens, A. J 2002 Identifying sources of uncertainty to generate supply chain redesign strategies. *International Journal of Physical Distribution & Logistics Management*, *32*(6), 409-430.

[12] Marimin. 2004 Teknik dan Aplikasi Pengambilan Keputusan Kriteria Majemuk. Gramedia Widiasarana Indonesia. Jakarta [In Indonesian].

[13] Lee, T.Y.S 2008 Supply Chain Risk Management, *Int. J. Information and Decision Sciences* 1: 98–114.
[14] Mills, J. F., and Camek, V 2004 The risks, threats and opportunities of disintermediation: A distributor's view. *International Journal of Physical Distribution & Logistics Management*, 34(9), 714-727.
[15] Martha, Joseph and Sunil Subbakrishna 2002 Targeting a Just-in-Case Supply Chain for the Inevitable Next Disaster, *Supply Chain Management Review*, Vol. 6, No. 5, pp. 18-23.
[16] Christopher, M. and Lee, H 2004 Mitigating supply chain risk through improved confidence, *International Journal of Physical Distribution & Logistics Management*, Vol. 34 No. 5, pp. 388-96.
[17] Juan Yang, Haorui Liu, Xuedou Yu, and Fenghua Xiao 2016 Emergency Coordination Model of Fresh Agricultural Products' Three-Level Supply Chain with Asymmetric Information. *Mathematical Problems in Engineering*, vol. 2016, Article ID 2780807, 9 pages, 2016.
[18] Zsidisin, G.A 2003 Managerial perceptions of supply risk *Journal of Supply Chain Management*, Vol. 39 No. 1, pp. 14-25.
[19] Jüttner, U 2005 Supply chain risk management: Understanding the business requirements from a practitioner perspective. *The International Journal of Logistics Management*, 16(1), 120-141.
[20] Frazier, G. L., Maltz, E., Antia, K. D., and Rindfleisch, A 2009 Distributor sharing of strategic information with suppliers. *Journal of Marketing*, 73(4), 31-43.
[21] Enyinda, C. I., Mbah, C. H., & Ogbuehi, A. 2010 An empirical analysis of risk mitigation in the pharmaceutical industry supply chain: A developing-country perspective. Thunderbird International Business Review, 52(1), 45-54.
[22] Levy, J. K 2005 Multiple criteria decision making and decision support systems for flood risk management. Stochastic *Environmental Research and Risk Assessment*, 19(6), 438-447.
[23] KarimiAzari, A., Mousavi, N., Mousavi, S. F., and Hosseini, S 2011 Risk assessment model selection in construction industry. *Expert Systems with Applications*, 38(8), 9105-9111.
[24] Aminbakhsh, S., Gunduz, M., and Sonmez, R 2013 Safety risk assessment using analytic hierarchy process (AHP) during planning and budgeting of construction projects. *Journal of safety research*, 46, 99-105.
[25] Gaudenzi, B., and Borghesi, A. 2006 Managing risks in the supply chain using the AHP method. *The International Journal of Logistics Management*, 17(1), 114-136.
[26] Zayed, T., Amer, M., and Pan, J 2008 Assessing risk and uncertainty inherent in Chinese highway projects using AHP. *International journal of project management*, 26(4), 408-419.
[27] Chan, F. T., and Kumar, N 2007 Global supplier development considering risk factors using fuzzy extended AHP-based approach. *Omega*, 35(4), 417-431.
[28] Kwon, I. W. G., and Suh, T 2004 Factors affecting the level of trust and commitment in supply chain relationships. *Journal of supply chain management*, 40(1), 4-14.
[29] Handfield, R. B., and Bechtel, C 2002 The role of trust and relationship structure in improving supply chain responsiveness. Industrial marketing management, 31(4), 367-382.

[30] Hallikas, J., Virolainen, V.M. and Tuominen, M 2002 Risk analysis and assessment in network environments: a dyadic case study, *International Journal of Production Economics*, Vol. 78, pp. 45-55.
[31] Septiani, W., Marimin, M., Herdiyeni, Y., and Haditjaroko, L 2016 Method and approach Mapping for Agri-food Supply Chain Risk Management: A literature review. *International Journal of Supply Chain Management*, 5(2), 51-64.
[32] La Londe, B 2002 Who Can You Trust These Days? *Supply Chain Management Review*, May-June 2002, pp. 9-10.
[33] Oehmen, J., Ziegenbein, A., Alard, R. and Schönsleben, P 2009 System-oriented supply chain risk management. *Production Planning and Control*, 20(4), pp.343-361.

Social-Ecological System in Seagrass Ecosystem Management at Kotania Bay Waters, Western Seram, Indonesia

Mintje Wawo

Marine Science and Fishery Faculty Pattimura University, Ambon, Indonesia

Email: minowawo@yahoo.com

Abstract. The concept of the Social-Ecological System (SES) of the coastal region, can be found in the seagrass ecosystem in the Kotania Bay Waters. Seagrass ecosystem as one of the productive ecosystem is part of an ecological system that can influence and influenced social system, in this case by people living around the seagrass ecosystem. This aim to estimating the socio-ecological vulnerability system of the seagrass ecosystem in the Kotania Bay Waters, the Linkage Matrix is used (de Chazal et al., 2008). This linkage matrix was created to determine the perception and understanding of the community on the ecosystem services provided by the seagrass ecosystem through the appraisal of various stakeholders. The results show that social values are rooted in the public perception of ecosystem goods and services, which are rarely considered. The ecological and economic value of natural resources is increasingly being used to determine the priority areas in the planning and management of coastal areas. The social value that exists in natural resources is highly recognized in conservation.

1. Introduction

The study of seagrass ecosystem management can be said very little. In the national and international constellation, management of mangrove and coral reef ecosystems dominate the coastal and marine ecosystem management. Internationally, at least there is now MFF (Mangrove for the Future) and CTI (Coral Triangle Initiative) which became an icon of coastal and marine ecosystem management. As part of the vegetation, the primary function of seagrass is to absorb carbon to play a role in the carbon balancing process in the biosphere [1].

The function of ecosystem management is not only physical dimension to maintain the sustainability of coastal and marine resource system, but also has a social aspect. Because the existing social system in the coastal and marine areas (example coastal communities that have interacted dynamically with coastal and marine resources) is one of the elements to be considered in the management of coastal and marine ecosystems, including the management of seagrass ecosystems. Integration of ecosystem-based management by considering the dynamics of social systems in it, known as the Social-Ecological System Approach [2].

The concept of a model combining human systems and natural systems is also revealed by [3]. Communities rely heavily on natural systems, resulting in a continuous flow of ecosystem goods and services. The same as between the seagrass ecosystem and the people living nearby. The concept of ecosystem services has been proposed as a meaningful framework for natural resource management [4]. In management, to understand the importance of ecosystems in the economy and decision-making policy is to establish the relationship between what ecosystems (goods and services) provide and how individuals value them. Figure 1 shows this relationship with a simple model: the ecosystem and its

IOP Conf. Series: Earth and Environmental Science **89** (2017) 012023 doi:10.1088/1755-1315/89/1/012023

functions and processes provide output in the form of goods and services, resulting in benefits to the population which can then be measured as an increase in human welfare.

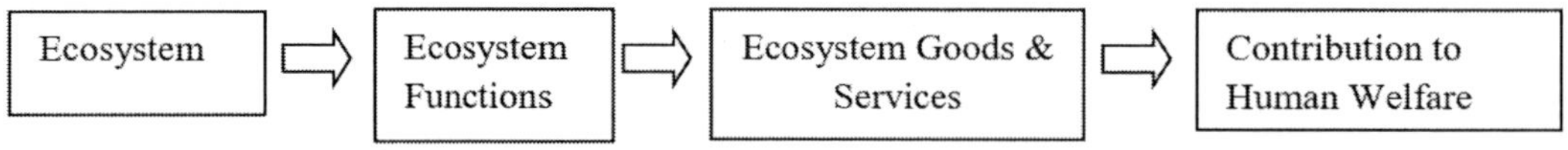

Figure 1. The relationship between ecosystem and human welfare [5].

The existence of the seagrass ecosystem in the waters of Kotania bay is currently decreasing extent [6]. With the reduced scope of the seagrass ecosystem, the services provided automatically also reduced. Whereas according to some research results on the seagrass ecosystem in the waters of Kotania bay, there are various services provided by the ecosystem (Table 1).

Table 1. Goods and services provided by the seagrass ecosystem in the Kotania Bay waters

Seagrass Ecosystem Services	Seagrass Ecosystem Goods
As a place of seaweed cultivation	35 species of seaweed, Gracilaria sp. and *Hypnea* sp. (more difficult to find).
As the habitat of some organisms	Nine species of sea cucumbers, only two species are economically important (*Holothuria scabra and Stichopus variegatus*). Whereas in 1984 there are four species of economically important.
	Six species of economically important mollusks (*Anadara antiquate, Fragum fragum. Pinna* sp., *Modiulus modiulus, Strombus* sp., and *Tridacna* sp.). Abundance is diminishing.
	99 species of fish, its existence is decreasing.
	Dugong (*Dugong dugong*) and penyu hijau (*Chelonia mydas*), the population continues to decline.

Source: [7]

Based on the above description, the sustainability status of seagrass ecosystem in Kotania bay waters is understood as a matter worthy to be studied in the framework of development and conservation of seagrass ecosystem. The management of seagrass ecosystems related to the status of seagrass ecosystems and coastal communities, whose lives depend on seagrass ecosystems and their interrelations, needs to be further investigated using the Social-Ecological System approach. So the research is aimed to estimate the socio-ecological vulnerability system of seagrass ecosystem in Kotania Bay waters, where the management of seagrass ecosystem with Matrix Method of Linkage based on ecosystem services will again influence the existence of seagrass ecosystem and also its community system.

2. Methodology

This study was conducted from May to July 2011, at Kotania Bay Waters, Western Seram, Maluku Province. Sampling sites were determined through PRA activity using community mapping approach that relates to resource uses on seagrass ecosystem by coastal communities. Participatory rural approach (PRA) is an approach has been written as "how to" guide stakeholders to employing participatory decision-making techniques for resource management in coastal zone [8]. Survey ecology and common method were used to collect data.

To know the perception and understanding of community on seagrass ecosystem, using the approach of Linkage Matrix [9]. The Linkage Matrix is an assessment method of SES vulnerability that is implicitly related to the evaluation of various stakeholders in the resolution of vulnerability estimates in the same or different places. Three steps must be taken namely:

1. Various ecosystem services are identified and assessed by different stakeholders.
2. A series of matrices are used to quantify sequentially related to social and ecology information. The matrices organize and integrate information from the initial change of the ecosystem to the determination of changes in ecosystem services.
3. The ecosystem traits underlying the delivery of ecosystem services are identified, followed by quantification and measurement of some intangible ecosystem services including the classification of the existence of ecosystem services.

To estimate the socio-ecological vulnerability system of the seagrass ecosystem, used five matrixes (ecosystem services matrix, descriptors matrix, ecosystem properties matrix, land-use attributes matrix and Utilization Attributes matrix) are transparent and flexible to classify and link social and ecological information so that they can be integrated. By combining the results of ecological surveys, land use and social surveys, matrix is filled with scores or values that describe the relationship between rows and columns. Scores or assessments show how stakeholder groups assess each ecosystem service and assessment form categorized in 3 simple values according to the Likert scale: (-) = negative, (0) = unimportant and (+) = positive. Information is based on interviews with individuals.

Explanation of scores used are as follows:

- A positive value (+) means stakeholders directly do the activity (gain/ benefit), others do activities that benefit stakeholders, or stakeholders have a positive assessment of ecosystem services.
- Zero (0) means that the stakeholder does not perform the activity and the activity does not have an impact on the assessed ecosystem services or the services of the ecosystem are considered unimportant.
- A negative value (-) means an activity or representation of a region is seen to have a negative effect on ecosystem services.

3. Results and Discussion

3.1. Ecological system

The extent of seagrass beds in the waters of the Kotania bay is 823,615 ha, with the presence of 7 species of seagrass namely *Enhalus acoroides, Thalassia hemprichii*, and *Halophila ovalis* from Hydrocharitacea, and four species of *Cymodocea rotundata, Cymodocea serrulata, Halodule pinifolia*, and *Syringodium isoetifolium* from Cymodoceaceae [10]. In Indonesia found as many as 13 species of seagrass. The spread of seagrass species at the study sites varied. Five species can be found starting from Buntal Island, Tatumbu Island, Osi Island, Burung Island, Marsegu Island to Barnusang cape. Six species are found only in Loupessy village, while seven species of seagrasses are found in Tamanjaya village.

Average cover percentage of seagrass in Kotania Bay waters can be seen in Figure 2, generally of 50.59%, with a range of 29.23% (in Loupessy village) up to 65.17% (on Marsegu Island). The percentage distribution of high seagrass cover is likely to be in open water area of Marsegu Island rather than the area that is in the bay waters that jutted into/mainland. Based on the above-mentioned seagrass status, the seagrass status conditions in the Kotania Bay Waters are in less good or less healthy condition. The conditions of seagrass beds on Marsegu Island are still in good or healthy condition, and seagrasses in Loupessy village are in worst shape.

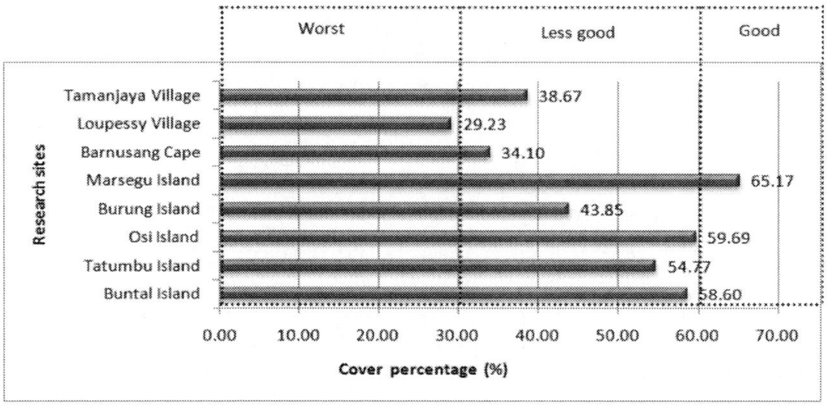

Figure 2. Average Cover Percentage of Seagrass Based on Research Sites and Status according to State Minister of Environment in Decision no. 200 in 2004

3.2. Social system

Coastal communities located in the three research sites (Osi Island village, Buntal island and Wael village) are from Buton/Southeast Sulawesi who has lived there for many years, and the majority are Muslim. They have mixed livelihoods according to season conditions for a year. The reasons for the selection of these three locations because the coastal communities have a major work that takes place on the seagrass ecosystem. The 70 respondents were selected in Wael village (27 people), Osi Island village (41 people) and Buntal Island (2 persons). Utilization of damaging coastal resources is still often done by people who are in the Kotania bay, to meet the needs of his life such as fishing activities using explosives (bombs) and ornamental fishing with potassium and coral collection activities for sale as building material (PRA results). Coral sales conducted by the community at a price of IDR 150 thousand /m^3.

The livelihoods of coastal communities (fishermen) in the Kotania Bay are distinguished using fishing gear used to utilize marine organism in the seagrass ecosystem, such as gill nets, set net, trap, handline fishing and "bameti". Based on the result of research, the dominant livelihood is handline fishing (47,24%) followed by bameti (27,15%), gill net (13,47%), set net and trap (12.14%). The catch on the seagrass ecosystem in the form of economically important fish. Various utilization activities are undertaken by fishermen located in coastal areas of the Kotania Bay, to meet their living needs. On the seagrass ecosystem operates several fishing gears that are used by local fishermen to catch the organism to fulfill their life needs, such as gillnet, set net, trap, handline fishing and "bameti". These utilization activities are activities undertaken by the community, which utilizes the services that providers of the seagrass ecosystem, evident by the average revenue per year from fishing operations (gillnet, set net and trap) received by the community of IDR. 30-44 million.

Communities inhabiting the coastal of Kotania Bay mostly have livelihoods as fishermen because they have been hereditary (94.29%), not for their desires, easy to access or no alternative (Figure 3). Fishermen who move on this seagrass ecosystem mostly come from Osi Island village (54.29%), followed by Wael Village (41.43%) and Ambon Island (4.29%). They have settled for (31-40) years at the study site (34.29%). Fishermen education on the coastal of Kotania Bay is mostly only at Primary School (92.86%).

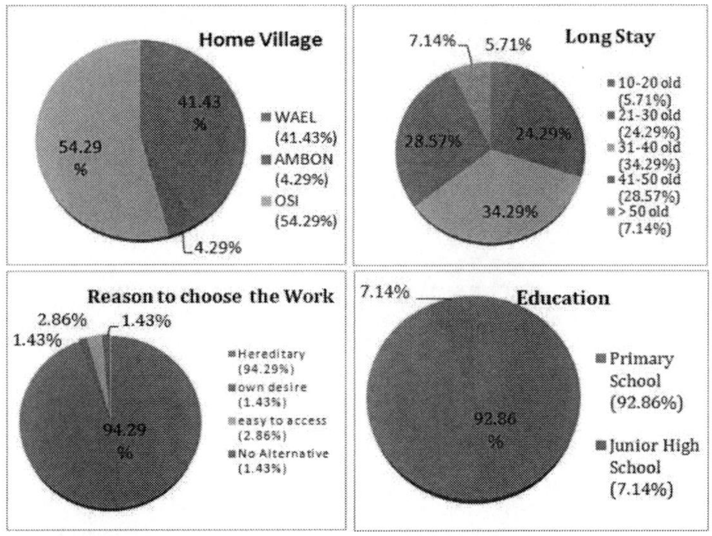

Figure 3. Characteristics of fisherman at research sites

3.3. Linkage Matrix

Social-Ecological System concept located in the coastal area can be found in the seagrass ecosystem in the waters of Kotania Bay. Seagrass ecosystem as one of the productive ecosystem is part of an ecological system that can influence and changed social system, in this case by the people living around the seagrass ecosystem. Seagrass ecosystems that have a role as a fish habitat are one of the main places for fishermen to take and exploit fish resources. Thus it forms a social system of society in the shape of fishing communities that depend on the existence of the ecosystem of seagrass (ecological system). On the other hand, the excessive exploitation done by the society, for the higher economic needs, makes the seagrass ecosystem disrupted. Utilization that is not balanced with conservation activities causes worsening of existing seagrass ecosystem. Social and ecological systems are two systems that have connectivity in the form of a relationship of interdependence. Changes in the social system will result in changes in ecological systems as well as changes in the ecological system will change the existing social system. In other words, the seagrass ecosystem is very vulnerable due to the behavior of the surrounding community. Human behavior more often increases the vulnerability of ecosystems rather than increasing resilience. Every action and human intervention has the potential to reduce ecological toughness because the goal of the human activity is generally to control or alter the processes occurring within an ecosystem [11].

The concept of vulnerability has a clear history of biophysical and social sciences. More recently, biophysical and social orientations have been simultaneously used in testing the vulnerability of the environment and human or socio-ecological systems [9]. In that aspect, the relationship between the property of ecosystem and increased human well-being forms a framework known as ecosystem services. The ecological and economic value of natural resources is increasingly being used to determine the priority areas in planning and managing coastal areas. Social value is rooted in the public perception of ecosystem goods and services, which are rarely considered. These values include the social value existing in natural resources, highly recognized in conservation that offers a comprehensive understanding of a process toward change in the socio-ecological system [12]. Based on human perception, natural ecosystems provide not only life support services but also other services that actively support human life, such as recreation and aesthetics [13]. The community is a part directly related to environmental problems because people live for long periods and who feel the first time the impact. Public perception is the key to managing environmental challenges.

The ecosystem services matrix is a matrix containing stakeholder identification and assessment of seagrass ecosystem services in Kotania Bay waters (Table 2). The existence of seagrass ecosystem services at the research sites, enjoyed by various layers of society or stakeholders, among others, managers of natural tourism parks, government, fishers, seaweed farmers and academician/researchers.

Perceptions of different stakeholders on the existence or function of the seagrass ecosystem (seagrass ecosystem services) vary. Seagrass ecosystem services in Kotania Bay areas production/provider services (habitat of sea organism), administrative/regulation services (carbon storage, current absorber) and cultural services (Tourism, research sites). The seagrass ecosystem services cannot be doubted as revealed by [14].

Table 2. Ecosystem services matrix

Seagrass ecosystem services matrix		Ecosystem services matrix				
		Habitat of sea organism	Current absorber	Tourism	Research sites	Carbon storage
Stakeholder	Managers of marine nature parks	0	0	+	0	0
	Government	0	0	+	0	0
	Fisherman	-	0	0	0	0
	Seaweed farmers	-	0	-	0	-
	Academician / Researcher	0	0	0	+	0

The presence of seaweed cultivators has an adverse impact on seagrass ecosystem services (habitat of sea organism, tourism and carbon storage). Similarly, the presence of fishermen to sea organism habitat services. This is supported by the results of research conducted in the Kotania bay that the bay is rich in sea organisms such as seaweed, sea cucumbers, mollusks, fish, and dugong. But now the existence of this sea organism is increasingly reduced and become scarce [15]. Managers of marine nature parks and Government are related or have a positive impact on tourism services. Similarly, the presence of academician/researcher on seagrass ecosystem services as a research site. According to [9], ecosystem services have a real relationship with the physical elements of the ecosystem, such as the number of service providers (food/organism) or regulatory services (control of erosion/current absorber).

In the descriptor matrix (Table 3), stakeholders utilizing seagrass ecosystems in the Kotania Bay describe the observable characteristics of the seagrass ecosystem itself. The observable characteristics associated with seagrass ecosystem services are the percentage of seagrass cover, seagrass species density, fish abundance, association organism, seagrass biomass, and beauty. Features described by stakeholders as a whole have a positive impact on seagrass ecosystem services, unless the characteristics of beauty are neutral to the service of the current absorber and the carbon storage. Similarly, the same value to tourism services

Table 3. Descriptor matrix

Descriptor matrix		Descriptor Stakeholder					
		% of Seagrass cover	Seagrass species density	Fish abundance	Association organism	Seagrass biomass	Beauty
Ecosystem services	Habitat of sea organism	+	+	+	+	+	+
	Current absorber	+	+	+	+	+	0
	Tourism	+	0	+	+	+	+
	Research sites	+	+	+	+	+	+
	Carbon storage	+	+	+	+	+	0

The ecosystem property intended for the following matrix is the typical or functional component of the seagrass ecosystem (Table 4). This matrix links the biophysical description to the specific features of the seagrass ecosystem as a service provider. Overall, ecosystem property and descriptor stakeholder have an active relation to each other, except for beauty aspect and fishery production have a negative relationship.

Table 4. Ecosystem Property Matrix

Ecosystem Property Matrix		Ecosystem Property		
		Important value index	Water quality	Fishery production
Descriptor Stakeholder	% of seagrass cover	+	+	+
	Seagrass species density	+	+	0
	Fish abundance	+	+	+
	Association organism	+	+	+
	Seagrass biomass	+	+	0
	Beauty	0	+	-

The matrix of Utilization Attributes correlates between descriptor utilization with the utilization quality. The use attribute is a description of the seagrass ecosystem management practices on the characteristics that characterize the ecosystem (Table 5). In the matrix shows that the activity of seagrass ecosystem utilization in the form of conservation is very closely related to attribute of seagrass cover, association organism, and beauty. This assessment is in contrast to fisheries activities on the seagrass ecosystem.

Table 5. Matrix of Utilization Attributes

Matrix of Utilization Attributes		Utilization Attributes		
		Seagrass cover	Association organism	Beauty.
Descriptor utilization	Conservation	+	+	+
	Fisheries	-	-	-

The matrix that links the functional trait with the ecosystem property is called the functional character matrix (Table 6). The relationship described in this matrix is a direct correlation between the ecosystem ownership and the benefits received by society. The functional trait characteristics of the seagrass ecosystem can be interpreted as a function or benefit that can be accepted and done by the community from the seagrass ecosystem property. Elements of major value indexes, water quality, and fishery production are either active or in other words closely related to people's income and carbon existence.

Table 6. Matrix functional trait

Matrix functional trait		Functional trait	
		Community Revenue	Carbon Existence
Ecosystem Property	Important value index	+	+
	Water quality	+	+
	Fishery production	+	+

The linkage matrix used in assessing the vulnerability of SES to this seagrass ecosystem, making general seagrass ecosystem services, is slightly easy to quantify and be measured. The integration of social values into ecosystem services taking into account the ecological and economic aspects of natural resource assessment will result in a fair and efficient policy that can enhance the resilience of the socio-ecological system [12, 16].

IOP Conf. Series: Earth and Environmental Science **89** (2017) 012023 doi:10.1088/1755-1315/89/1/012023

4. Conclusion

1. The socio-ecological system of the seagrass ecosystem in the Kotania Bay waters is evidenced by the significant dependence of coastal communities/fishermen on the seagrass ecosystem.
2. Social values are rooted in the public perception of ecosystem goods and services, which are rarely considered. The social value that exists in natural resources is highly recognized in conservation.
3. The ecological and economic value of natural resources is increasingly being used to determine the priority areas in the planning and management of coastal zones.

References

[1] Hutomo M. 2009. Kebijakan, Strategi dan Rencana Aksi Pengelolaan Ekosistem Lamun di Indonesia. Makalah disampaikan pada Lokakarya Nasional I Pengelolaan Ekosistem Lamun, 18 Nopember 2009, Jakarta: Sheraton Media.

[2] Anderies JM, Walker BH , Kinzig AP. 2006. Fifteen Weddings and A Funeral: Case Studies and Resilience-based Management. *Ecol Soc* 11:21.

[3] de la Torre-Castro M, Ronnback P. 2004. Links Between Humans and Seagrasses-An Example from Tropical East Africa. *Ocean & Coastal Management* 47:361–387.

[4] Patterson TM, Coelho DL. 2009. Ecosystem Services: Foundations, Opportunities, and Challenges for the Forest Products Sector. *Forest Ecology and Management* 257:1637–1646.

[5] Newcome J, Provins A, Johns H, Ozdemiroglu E, Ghazoul J, Burgess D, Turner K. 2005. *The Economic, Social and Ecological Value of Ecosystem Services: A literature Review.* Economic for The Environment Consultancy (EFTEC).16 Percy St, London W1T 1DT.

[6] Supriyadi IH. 2009. Pemetaan Lamun dan Biota Asosiasi untuk Identifikasi Daerah Perlindungan Lamun di Teluk Kotania dan Pelitajaya. *Oseanologi dan Limnologi di Indonesia* 35:167-183.

[7] Supriyadi IH. 2000. Mangrove resource management planning in coastal areas of the District Piru, West Seram regency. Master Thesis, Bogor Agricultural University.

[8] Brown K, Tompkins E, Adger WN, 2001. Trade-off analysis for participatory coastal zone decision-making. Norwich : Overseas Development Group, University of East Anglia.

[9] de Chazal J, Quetier F, Lavorel S, Doorn A van. 2008. Including Multiple Differing Stakeholder Value Into Vulnerability Assessments of Socio-Ecological Systems. *Global Environmental Change* 18:508-520.

[10] Wawo, M., Y. Wardiatno., L. Adrianto., and D.G. Bengen., 2014. Carbon stored on seagrass community in Marine Nature Tourism Park of Kotania Bay, Western Seram, Indonesia. *Journal of Tropical Forest Management 20*(1):51–57. EISSN: 2089-2063. DOI: 10.7226/jtfm.20.1.51.

[11] Olsson P. 2003. Building capacity for resilience in social-ecological systems [Doctoral Thesis]. Stockholm. Stockholm University.

[12] Marshall NA, Fenton DM, Marshall PA, Sutton SG. 2007. How resource dependency can influence social resilience within a primary resource industry. *Rural Sociology*. 72(3): 359-390.

[13] Costanza R, d'Arge R, de Groot R, Farbek S, Grasso M, Hannon B, Limburg K, Naeem S, O'Neill R, Paruelo J, Raskin R, Suttonkk P, van den Belt M. 1997. The value of the world's ecosystem services and natural capital. *Nature 387:253–260.*

[14] Green EP, Short FT. 2003. *World Atlas of Seagrasses.* Los Angeles: University of California Pr. 298 pp.

[15] Kuriandewa T. 2009. Tinjauan Tentang Lamun Indonesia. Makalah disampaikan pada Lokakarya Nasional I Pengelolaan Ekosistem Lamun, 18 Nopember 2009, Jakarta: Sheraton Media.

[16] Berkes F, Folke C. 2000. *Linking Social and Ecological Systems: Management Practices and Social Mechanisms for Building Resilience.* Cambridge Univ Pr.

IOP Conf. Series: Earth and Environmental Science **89** (2017) 012024 doi:10.1088/1755-1315/89/1/012024

Efficiency and economic benefits of skipjack pole and line (*huhate*) in central Moluccas, Indonesia

Stevanus M Siahainenia[1], Johanis Hiariey[2], Mulyono S Baskoro[3], Wellem Waeleruny[2)]

[1] Graduate Student of Pattimura University, Ambon, Indonesia
[2] Faculty of Fisheries and Marine Science, Pattimura University, Ambon, Indonesia
[3] Faculty of Fisheries and Marine Science, Bogor Agriculture University, Indonesia

Email: johanishiariey6@gmail.com

Abstract. Excess fishing capacity is a crucial problem in marine capture fisheries. This phenomenon needed to be investigated regarding sustainability and development of the fishery. This research was aimed at analyzing technical efficiency (TE) and computing financial aspects of the skipjack pole and line. Primary data were collected from the owners of the fishing units at the different size of gross boat tonnage (GT), while secondary data were gathered from official publications relating to this research. Data envelopment analysis (DEA) approach was applied to estimate technical efficiency whereas a selected financial analysis was utilized to calculate economic benefits of the skipjack pole and line business. The fishing units with a size of 26-30 GT provided a higher TE value, and also achieved larger economic benefit values than that of the other fishing units. The empirical results indicate that skipjack pole and line in the size of 26-30 GT is a good fishing gear for the business development in central Moluccas.

1. Introduction

The potential of fisheries resources can be used as an economic resource for the development of coastal areas. The sustainable use of fish resources will contribute to the increase of the country's foreign exchange, fishermen's income, and employment and business opportunity. The trends of poor management will lead to a massive pressure on fish resources. For example, the use of very excessive fishing inputs in the territorial waters will influence the reduction of quality and quantity of fish resources.

Fish resources as a natural capital need to be optimally managed for the sustainability of fisheries business [1]. The more fishing gears operating on the regulated open-access fisheries conditions regardless of fishing capacity will cause inefficient fishing activities. Also, low control on the resource will trigger the fishermen to utilize the fish resources in excess, resulting in inefficient resource utilization [2]. This condition will cause the fishermen's inability to maximize profits from fishing business [3,4]. The control of input use is one of the sustainability-based approaches to fisheries management which aims to improve the performance of the fishing industry through the reduction of excessive fishing capacity [5].

In fishery business, boats and other inputs are directly controlled by fishermen except for the production. Due to the amount of production which depends on the level of fishing effort and the size of the fish population. The size of the fish population varies and is determined by fishing intensity [3]. Theoretically, these conditions can be explained by the model of Gordon-Schaefer, to examine the economic aspects with the constraint of the fish resources biology, to what level of fisheries inputs (GT of boats, trip, etc.) should be controlled to generate maximum economic benefits [6].

In open-access fisheries, fishery production function is assumed to be in a state of biological balance, so the value of net income (π) from fishing activities is the difference between total revenue

(TR) and total cost (TC). At the condition where the marine fishery is open-accessin nature, and the fishing industry is competitive, the fishery will expand to point where sustainable resource rent is completely dissipated. And if the TR = TC, a balance production occurs, so that economic rent of fish resources is zero, $(\pi) = 0$ [7]. Furthermore, in a certain period of equilibrium, fishing effort will be gradually decreased until zero when TR ≤ TC, which in turn, has an impact on the fishing company.

It is necessary to have fish stocks balance with fishing inputs to achieve fishery business sustainability by a fishing company. If the fisheries, in fact, are well managed, the resource will stabilize at Maximum Economic Yield, where sustainable resource rent is maximized [7]. Umar [8] suggests the ideal principle for the company to take advantage of natural resources is a policy to control expenditure so that the marginal revenue is greater than or at least equal to the marginal cost.

The continuation of exploitation of fish resources needs to be managed effectively to control 'race to fish' which leads to overcapacity and overfishing [9]. Excess capacity and overfishing that occur in the exploitation of skipjack tuna resources will cause a massive pressure on the resources. This condition has implications on the social and economic aspects such as the decline in fishery production and income and other socio-economic issues [10,11]. Whether the exploitation of fish resources in a certain waters is efficient regarding economic overfishing, production decreases due to an imbalance of fishing inputs and fish stocks can be determined by data envelopment analysis (DEA) approach. This is a non-parametric method that uses mathematical programming to determine an optimal solution that is subject to existing constraints.

The potential of large-pelagic fish resources, i.e., tunas and skipjack in Moluccas contained in the Republic of Indonesia's Fisheries Management Area (FMA)-714, FMA-715, and FMA-718 were estimated at 261.5 ton/year, while total production reached 101.2 ton/year [12]. These figures indicate that large-pelagic resources still have a chance for future development. If the potential is not managed optimally, it will lead to over-exploited and pressure on tuna skipjack resources. This research was aimed at analyzing the technical efficiency and financial index of skipjack pole and line in Central Maluku, in the perspective of the sustainable development of pole and line (*Huhate*) fisheries business.

2. Research Methods

2.1 Methods of data collection

Sampling was conducted randomly with GT size stratification of the pole and line boats at the capture fishery centers. It covered the skipjack pole and line units that were still active in fishing during the research. Primary data were collected from the boat owner which includes the main size of the vessels and machinery, fishing gear, production, the price of fish, the cost of input, duration of the fishing trip and investment value. Secondary data were obtained from the official publications that relate to the research.

2.2 Data Analysis

Measurement of technical efficiency (TE) was conducted by using the technique of DEA (Data Envelopment Analysis). In the context of capture fisheries, the transformation of inputs to outputs could be either variable returns to scale, VRS, where the production function is decreasing returns to scale, so the DEA model was modified according to the model developed by Banker-Charnes-Cooper (BCC) [13]. It could be explained that the use of the BCC model is possible for the analysis of VRS economic activities. The use of the specific model is based on assumptions that (i) there is still budget constraint faced by pole and line units in fishing skipjack at the waters of central Moluccas; (ii) not all pole and line units are operating in optimum scale. Consequently, the BCC model of DEA can be utilized to analyze technical efficiency. Therefore, the analysis of technical efficiency with the VRS and input orientation was used to determine input utilization rates in producing outputs, in addition to describing the technical efficiency of each DMU.

The DEA model of input minimization with VRS assumption is formulated as follows,

$$TE = \text{Max } \varnothing \tag{1}$$

Subject to the following restrictions,

$$\theta u_j \leq \sum_{j=1}^{J} z_j x_{jn} \leq x_{jn'}, n \in \alpha$$

$$\sum_{j=1}^{J} z_j x_{jn} = Y_j x_{jn'}, \ n\square a$$

$$Z_j \geq 0, Y_{jn'} \geq j = 1,2...J, n = 1,2,...N$$

It was assumed that j = 1, 2, ..., J is the number of observations on 20 skipjack pole and line (*huhate*) boats or J = 20 which is also the DMU (Decision Making Unit). Inputs used are the duration of the fishing trip and operating costs so that n = 1, 2, N, input (n = 2), where:

TE = technical efficiency of fishing gear to-j^{th}

θ = measurement value of each observation (≥1)

Uj = output for fishing gear to-j^{th} , which is 1 output (catch fish)

x_{jn} = used inputs to-n which are the duration of fishing trip and operating costs

Y_j = level of the variable input use to-n

z_j = utilization rate of variables.

The determination of financial feasibility index of fishery business was conducted by using investment criteria of NPV, IRR, and Net B/C with formula, as follows.

(a) Net Present Value (NPV)

$$NPV = \sum_{t=1}^{n} \frac{B_t - C_t}{(1+i)^t} \tag{2}$$

Where:

B_t = the benefit of *i* year

C_t = the cost of *i year*

r = interest rate per year

i = time of calculation

Decision criteria: NPV ≥ 0 means *go project;* NPV <0 means *no go project.*

(b) Internal Rate of Return (IRR)

$$IRR = P_1 - C_1 \times \left(\frac{P_2 - P_1}{C_2 - C_1}\right) \tag{3}$$

Where:

P_1 = the 1st interest rate

P_2 = the 2nd interest rate

C_1 = the 1st NPV

C_2 = the 2nd NPV

Decision criteria: if IRR ≥ *social discount rate* means *"go" project;* IRR < *social discount rate* means *"no go" project.*

(c) Net Benefit Cost Ratio (Net BCR)

$$\text{Net B/C} = \frac{\sum_{t=0}^{n} \frac{B_t}{(1+i)^t}}{\sum_{t=0}^{n} \frac{C_t}{(1+i)^t}} \quad (4)$$

Decision criteria: Net BCR ≥ 1 then investment activity is feasible to be implemented and if Net BCR <1, investment activity is not feasible to be implemented.

3. Results and Discussion

3.1 Technical description of skipjack pole and line (huhate)

Pole and line as a fishing gear consist of fishing rods, fishing lines, unrelated fishing rods and baits operated by fishers together on board. The fishing rods are made of quite an old bamboo or synthetic materials such as plastics or fibers having a good level of elasticity. The length of the rod is 2 -2.5 m with diameter 3-4 cm at the base and the tip 1-1.5 cm. The line is made from monofilament materials such as a white fishing line or wire leader with a length of 20 cm. The hook size used is 2.5 - 2.8.

The dimension of skipjack pole and line (*huhate*) boat size ranged from 13.6 to 32 GT, powered by the main engine of 105 to 250 HP and equipped with GPS to find out the fishing ground of skipjack. The duration of fishing time ranged from 1 to 6 days/trip, and 17 to 27 fishermen employed in one boat for the fishing operation.

3.2 Technical efficiency

The measurement of the technical efficiency of the skipjack pole and line boat in size of 20-32 GT utilized the technique of DEA with a variable return to scale (VRS). This scale was applied to assess changes in production levels due to the changes of input use. The results of the analysis are shown in Figure 1.

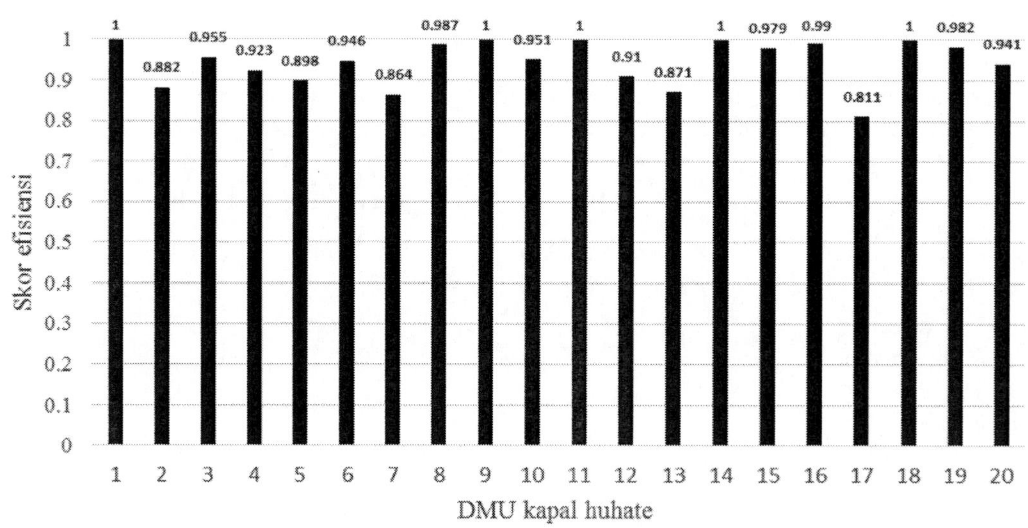

Figure 1. Technical efficiency scores of the skipjack pole and line units

The technical efficiency of skipjack pole and line boat tended to be fluctuated by its size. This result indicated that the use of inputs per vessel was not proportional to the outputs, except for DMU with efficiency score equal to 1 that is an optimal input use which is a base value in determining relative efficiency. Empirical results indicate that 5 out of 20 DMUs had the most efficient scores. And 10 DMUs had scores above 0.9, while the remaining 5 DMUs had efficiency scores above 0.8. Assumed that the efficiency score equals to 0.8 as a basis value, the entire DMUs find to be efficient. Based on this assumption, the skipjack fishing activities in the Maluku Sea indicated excess capacity. Pascoe et al. [14] state that the excess capacity is more in short-term and can be gradually changed to optimum input use. According to Cooper et al. [15], DEA application is used for calculating the efficiency by reducing input or output of vessels individually or in total. The highest input use was found in the DMU-17 which is indicated by the lowest score of technical efficiency compared with other DMUs. To achieve input utilization optimally, each input of the duration of the fishing trip and the operation cost should be decreased by 35.5% and 18.8% respectively. To improve the efficiency score of all the DMUs or the 20 units of skipjack pole and line it could be conducted by reducing 58% of the duration of the fishing trip and 42% of the operating cost. However, the input of operating cost is difficult to control except for the length of the fishing trip. Hence, the duration of the fishing trip can be used as an instrument for capacity control in skipjack fishery.

3.3 Financial analysis

3.3.1 Investment. Investment is a permanent working cost to purchase capital goods that are expected to have long-term benefits. The types of capital goods of skipjack pole and line business by boat size are presented in Table 1.

Table 1. The capital goods and their values by boat size

Boat size (GT)	Type of equipment and investment value (in rupiah)					
	Boats	Main Engine	Auxiliary Engine	Navigation Equipment	Fishing Gear and Other	Amount
≤ 20	507,410,000	225,000,000	32,000,000	23,000,000	3,400,000	903,431.500
21-25	652,385,000	250,000,000	32,000,000	23,000,000	3,400,500	1,104,902,750
26-30	690,000,000	300,000,000	32,000,000	23,000,000	5,000,000	1,207,500,000
≥ 31	858,280,000	350,500,000	32,000,000	23,000,000	5,000,000	1,458,522,000

Source*:* Fiber vessel industry

The highest investment was on the skipjack pole and line boats in size ≥ 31 GT, and the lowest one was in size ≤ 20 GT. The average use of capital goods for skipjack pole and line company were 66.54% for boats, 27.64% for main engines, 3.15% for auxiliary engines, 2.26% for navigation equipment and the remaining 0.41% for fishing gear.

3.3.2. Variable costs. Variable costs are the costs incurred due to the use of variable inputs in which the changes are linear to the changes in production levels [16]. The variable costs of skipjack pole and line fishery consisted of the wage of fishermen 30.57%, fuel 27.12%, ice 17.41%, baits 13.87, and ration 11.08% (Figure 2). Also, the expenditure of operation costs by boat size per year is the following. The annual operation cost spent to the boat size of less than or equal 20 GT amounted to Rp202,850,000; the boat size of 21-25 GT Rp 330,509,074; boat size of 26-30 GT Rp 461,808.642; and boat size ≥ 31 GT Rp 532,420,560.

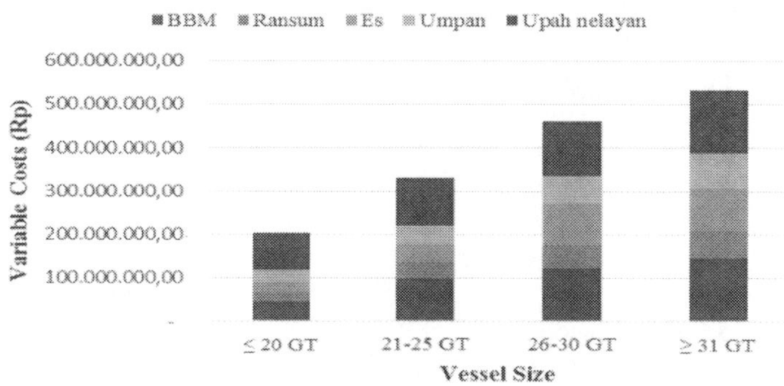

Figure 2. Variable cost components by boat size

3.3.3 Fixed cost. Fixed cost is the cost that companies or producers have to pay at any rate of output. The fixed costs of the skipjack pole and line company identified consisted of the costs of depreciation, repair and maintenance cost, and the cost of company license for fishing (Table 2).

Table 2. The expenditure of fixed cost of skipjack pole and line fishery

Boat size (GT)	Fixed cost (Rp/year)			Amount
	Depreciation	Repair and Maintenance	Fishing business license	
≤ 20	30,188,266	13,886,483	5,100,000	49,174,749
21-25	51,813,142	19,048,949	5,100,000	75,962,092
26-30	60,030,000	21,750,000	5,100,000	86,880,000
≥ 31	69,830,824	22,307,068	5,100,000	97,237,892

The calculation of depreciation cost was based on the straight-line method, in which the economic life of the boat was estimated 30 years, the main engine 15 years, the auxiliary machinery five years, the navigation equipment four years, and the fishing gear four months. Of the total fixed costs, expenditure for the depreciation component amounted to 68.51%, repair and maintenance 24.90%, fishing business license 6.60%. The highest fixed cost was in the depreciation component. In the period of economic life of fishing capital goods, the internal sources of funds should be well allocated to generate the opportunities for business development.

3.3.4 Analysis of revenue and earnings. Revenue is the product of multiplication of the fish production and its price. The revenue of the fishing business has a linear relationship to the production. It means that the higher production leads to greater revenue, and vice versa. The average price of skipjack was Rp 7,500/kg. Earning is the difference between total revenue and total cost. The earnings before tax are earning value minus taxes (10%), while earnings after tax or net revenue represent the difference between earnings before tax and with tax value (Table 3).

Table 3. Costs and earnings of skipjack pole and line fishery by boat size

Boat size (GT)	Revenue (Rp/Year)	Expenditure (Rp/Year)			Earning (Rp/Year)	
		Variable cost	Fixed cost	Total cost	Before tax	After tax
≤ 20	576,330,000	202,850,000	49,174,749	252,024,749	324,305,250	291,874,725
21-25	864,075,000	330,509,074	75.962,091	406,471,165	457,603,835	411,843,451
26-30	1,233,994,710	461,808,462	86,880,000	548,688,642	685,306,247	611,775,623
≥ 31	1,316,752,500	532,420,560	97,237,892	629,658,452	687,094,047	618,384,643

The empirical results of the research indicated that business of the skipjack fishing of all boat size (GT) observed was found to be profitable. Value of the net income varied due to the differences in revenues and expenditures among the vessel sizes. The total cost per kilogram of skipjack was found to be Rp. 3,451, variable cost Rp. 2,871, and net income reached Rp 3,643.

3.4 Business feasibility

Business feasibility aims to provide information about the feasibility of a business project, which is usually an investment project [8]. Business feasibility analysis was conducted by using investment criteria approach. According to Umar [17], the investment criteria are categorized as a criterion based on (1) income, and (2) cash flow which does not consider the time value of money (discounting factor) and considers the time value of money.

The business feasibility was analyzed using the investment criteria of NPV, IRR, and Net B/C for the skipjack fishing business. Results of the feasibility analysis are presented in Table 4.

Table 4. Index of investment criteria of skipjack pole and line by boat size

Boat size	Investment criteria		
(GT)	NPV (Rp)	IRR (%)	BC Ratio
≤ 20	297,111,637	8.34	1.35
21-25	471,734,375	9.04	1.76
26-30	1,284,091,063	13.61	1.91
≥ 31	855,990,866	8.81	1.67

Based on the empirical analysis of investment criteria, the skipjack pole and line fishery were financially feasible to be developed due to the higher financial index of B/C ratio as well as the NPV and IRR figures. Among the skipjack fishing boats, the boat in size of 26 to 30 GT was the priority to be developed. The priority is determined by using the decision-making method by Gray et al. [18]. Results of the empirical analysis for development priority are shown in Table 5. In the business perspective, the choice of investment for fishing business development should utilize the capital efficiently so it could provide a higher level of economic benefits.

Table 5. Development priority of skipjack pole and line boat

No.	Boat size (GT)	Net B/C Index	Ranking
1	≤ 20	1.35	4
2	21-25	1.76	2
3	26-30	1.91	1
4	≥ 31	1.67	3

The Net B/C equaled to 1.97 implied that the expenditure of Rp. 1.00 could generate a profit of Rp. 1.97. The IRR value of 13.85% indicated that discount rate was greater than the social discount rate of 8%, that is feasible. The NPV figure indicated that the profit to be gained is Rp. 1,236,422,956 for the next five years based on the present value of money, if the skipjack fishing business is managed effectively in term of capital input.

4. Conclusions

Skipjack pole and line (*Huhate*) fishery indicated excess capacity except for the boat in size of 26-30 GT which has already used input optimally. The highest excess capacity was found at the boat in size of 20 GT and below 20 GT. To achieve optimal inputs used, the duration of fishing trip needed to be reduced by 35.5% and the operation cost by 18.8% as well. The skipjack fishing boat in size of 26 to 30 GT had the highest values of the NPV, the IRR, and the BC ratio. The findings suggest that the development of skipjack pole and line business in Central Maluku should be focussed on the fishing boat in size of 26 to 30 GT because of the use of inputs efficient in fishing and having the highest scores in financial analysis.

References

[1] Yulianto G, Suwardi K, Adrianto L and Machfud 2016 Status Pengelolaan Ikan Demersal Sekitar Pantai di Kabupaten Indramayu, Jawa Barat Research Article: Scientific Communication in Fisheries and Marine Sciences *Omni Akuatika* Vol 12 No 3 pp 1-10

[2] Subade RF, and Abdullah NMR 1993 "Are Profit Maximizers? The case of Gillnetters in Negros Occidental and Loilo, Philippines" Asean Fisheries Science 6. Pp 39-49

[3] Anderson LG 1977 *The Economics of Fisheries Management* (London: The John Hopkins University Press, Baltimore) p 214

[4] Panayotou T 1985 *ProductionTechnology and Economic Efficiency: A Conceptal Framework* (Ottawa: Ont IDRC) pp 95-100

[5] Nikijuluw VPH 2002 *Rezim Pengelolaan Sumberdaya Perikanan.* (Jakarta: Kerjasama P3R dengan PT Pustaka Cidesindo) p 254

[6] Fauzi A 2004 *Ekonomi Sumberdaya Alam dan Lingkungan: Teori dan Aplikasi* (Jakarta: Penerbit Gramedia Pustaka Utama) p 259

[7] Gordon S 1954 The Economic Theory of a Common-Property Resource: The Fishery *J. of Political Econ.* Vol. 62 pp 124-42

[8] Umar H 1997 *Studi Kelayakan Bisnis; Manajemen, Metode dan Kasus* (Jakarta: Penerbit PT Gramedia Pustaka Utama) pp 7-8

[9] Rogoffs M. 2009 Overfishing and Sustainable Fisheries: A Program for Political Action *Ocean and Coastal Law J.* Vol.14:2

[10] Badington JR, Agnew DJ and Clark CW 2007 *Current Problems in the Management of Marine Fisheries*. Science 316 pp 1713-16

[11] Fauzi A 2010 *Ekonomi Perikanan. Teori, Kebijakan, dan Pengelolaan* (Jakarta: PT Gramedia Pustaka Utama. Kompas Gramedia) p 224

[12] DKP RI Departemen Kelautan dan Perikanan Republik Indonesia 2006 (Jakarta: Statistik Kelautan dan Perikanan Tahun 2005) p 314

[13] Fauzi A and Anna S 2005 *Pemodelan Sumberdaya Perikanan dan Kelautan untuk Analisis Kebijakan* (Jakarta: Penerbit Gramedia Pustaka Utama) p 343

[14] Pascoe S, Kirkley JE, Gebroval D and Paul CMJ 2003 Measuring and Assessing Capacity in Fisheries 2 Issue and Methods (UN Roma: FAO Fisheries Technical *Paper* 433/2 p 130

[15] Cooper WC, Seiford LM, Tone and Kaoru 2004 *Data Envelopment Analysis*. (Massachusets: Kluwer Academic Publisher)

[16] Sugiarto, Herlambang T, Brastoro, Sudjana R and Kelana S 2007 *Ekonomi Mikro: Sebuah Kajian Komprehensif* (Jakarta: Penerbit PT Gramedia Pustaka Utama) pp 248-251

[17] Umar H 2003 *Studi Kelayakan Bisnis* (Jakarta: Penerbit PT Gramedia Pustaka Utama) p 37

[18] Gray C, Simanjuntak P, Sabur LK, Maspaitela PFL and Varley RCG 1993 *Pengantar Evaluasi Proyek* (Jakarta: Penerbit Gramedia Pustaka Utama) pp 83-93

The Support of MPA (Marine Protected Area) in Coral Triangle Area: Evidence from Kei Islands, Indonesia

Syahibul K. Hamid[1], Wellem A. Teniwut[1], Roberto M.K. Teniwut[1], Meyske A. Rahantoknam[1], Cawalinya L. Hasyim[1], Marselus Hungan[1]

[1] Fisheries Agribusiness Study Program, Tual State Fisheries Polytechnic

Email: alkahfi.kahfi2@gmail.com

Abstract. Kei Islands located inside the coral triangle. Therefore, the biodiversity level on the sea in this area is considered high. United nation has proposed for water that included in the coral triangle has to apply marine protected area (MPA) to preserve the area. The main problem is most of the community especially in Kei Islands have depended on the sea as their sources of the economy even fisheries commodity like fish play a large part on the inflation rate and other prosperity indicators likes school and housing. Also, Kei Islands practice on form local wisdom for owning areal of the sea which calls "petuanan laut" by certain of villages or group of villages in one area. This study aimed to map the cluster of catching fisheries area based on the quantity of fish supply on a local market in Kei Islands and measure each cluster on their support and perspective on Marine Protected Area (MPA). We conducted a focus group discussion and collecting additional data by questionnaires with descriptive and quantitative analysis with logistic regression. The implication of this study can provide a clear view of coastal communities view on MPA program also to identify an area that has marine resources, human resources, and equipment to provide government an empirical view on catching fisheries in Kei Islands to issued better policy to develop fishing industry in Kei Islands.

1. Introduction

Coral triangle is sea area with world richest and highest marine biodiversity located at Asia Pacific's region consist of six countries namely Indonesia, Malaysia, Philippine, Timor Leste, Papua New Guinea, and Solomon Islands also known as CT6 (Coral Triangle 6). The areas are a home of over 120 million people who depend on their live fisheries and marine activities from fishing, fish culture and mariculture and marine ecotourism [1, 10]. The coral triangle area contains over 70% of coral species and more than 35% reef fishes of world supply [9]. Therefore it is important to conserve the area since it also relates to future food security. In 2008 inter-governments initiative between countries on coral triangle area to preserve the marine resources with agreement and policy known as CTI (Coral Triangle Initiative). The initiative proposed by President Yudhoyono of Indonesia, where it covers an area of 5.7 million km^2, biogeographically delineated by high coral diversity, one of the goals of CTI was to establish MPA (Marine Protected Area).

Furthermore, the purposed of Marine Protected Area (MPA) is to conserve and maintain the sustainability of marine resources [2]. Only 2.1% part of the sea in the world to conduct MPA program, in another side United Nation has targeted by 2020 at least 10% or sea around the world will apply for MPA Program. The resistance of many countries and coastal community on the application of MPA arise cause of the signification of sea to each local community on their welfare [4]. Thus, the most efficient way to manage well-received MPA is involving the coastal community regarding their interests in crafting this program.

In a rural region in particular developing country like Indonesia, barriers to face in the matter of conservation on land and sea is socio-economy factors. Where it starts from the social gap, education, job opportunity and welfare comparison between rural and urban area, also in recent years is climate change and political stability [7-8]. There are only a few successful cases for the application of MPA in the coastal area. Aside from economic factor, in Fiji, the local coastal community willing to accept MPA with the environmental approach, where local community approved MPA program to preserve their area from climate change which has taken great influence on their region [5].

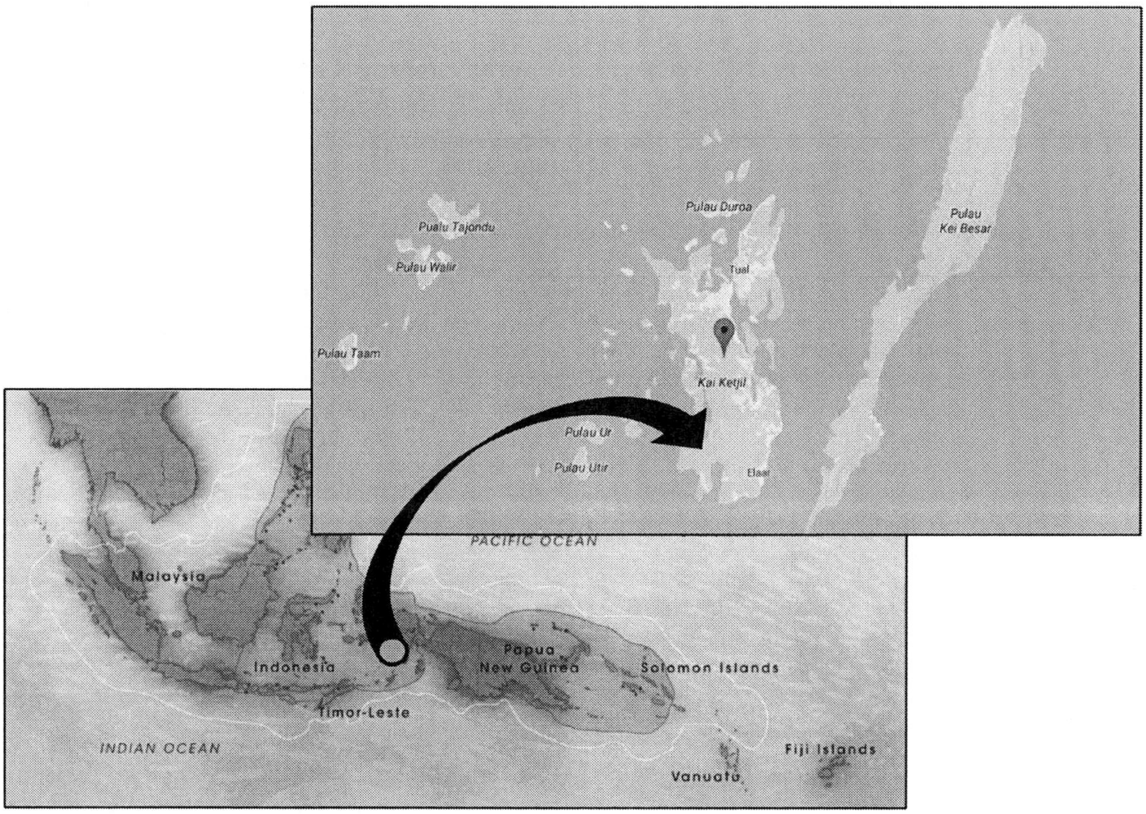

Figure 1. Kei Islands in Coral Triangle Area

Kei Islands located in the coral triangle area, which means this area has abundant of biodiversity resources. Therefore it is necessary to preserve it by applied MPA program. As an archipelagic region, Kei Islands needs to maintain their marine resources in the long run since it is the only sources of revenue for this area. The potential of catching fisheries of this region is high, although based on the study conducted by [6], if the trend of fishermen behavior, fleet, fishing technology includes illegal fishing on the same course then in 8-10 years from now the number of marine resources in this region, in general, will drop significantly. This current condition makes the application of Marine Protected Area (MPA) is greatly needed because it is not only will have a significant impact on future of world marine biodiversity but also the future welfare of Kei Islands community. The aimed of this study was to measure the acceptance of MPA by catching fisheries community in Kei Islands.

2. Methodology

To get sufficient data to obtain a better result on this study, we divided step into data collecting into two approaches. First, we conducted a preliminary survey to determine active fishermen in the region along with the fishing ground based on supply-side information from active fishermen. With the information that we have gathered, we divided fishermen villages into clusters as baseline source for the second step of our data collecting. Where we conducted Focus Group Discussion (FGD) on fishermen on each cluster on their on the MPA related to their knowledge, their view and their input on the application of MPA in the region. Also with a total of 255 respondents which were fishermen that didn't include on FGD to measure their social-economy condition and their support on the concept of MPA applied in Kei Islands Sea. This study conducted from May to August 2017 on Kei Islands, Indonesia. For data analysis, we used descriptive analysis and logistic regression.

Table 1. Clusters Fishermen and Fishing Ground in Kei Islands

No	Clusters	Villages	Fishing Ground
1	Dullah Laut	• Dullah Laut • Duroa • Tamedan • Labetawi	Water of Dullah Laut Islands, Mas island and Rumadan islands
2	Ohoirdetutu	• Ohoidertutu • Ohoiren • Ohoira • Wab	Water of west bank Kei Kecil Island and Tanimbar Kei island
3	Selayar	• Selayar • Sitniohoi • Dunhawan	Water of dua island, sepuluh island and Ngidiun cape
4	Sathean	• Sathean • Ibra	Nerong strait
5	Kei Besar	• Ngafan • Sungai	Kei Besar island water

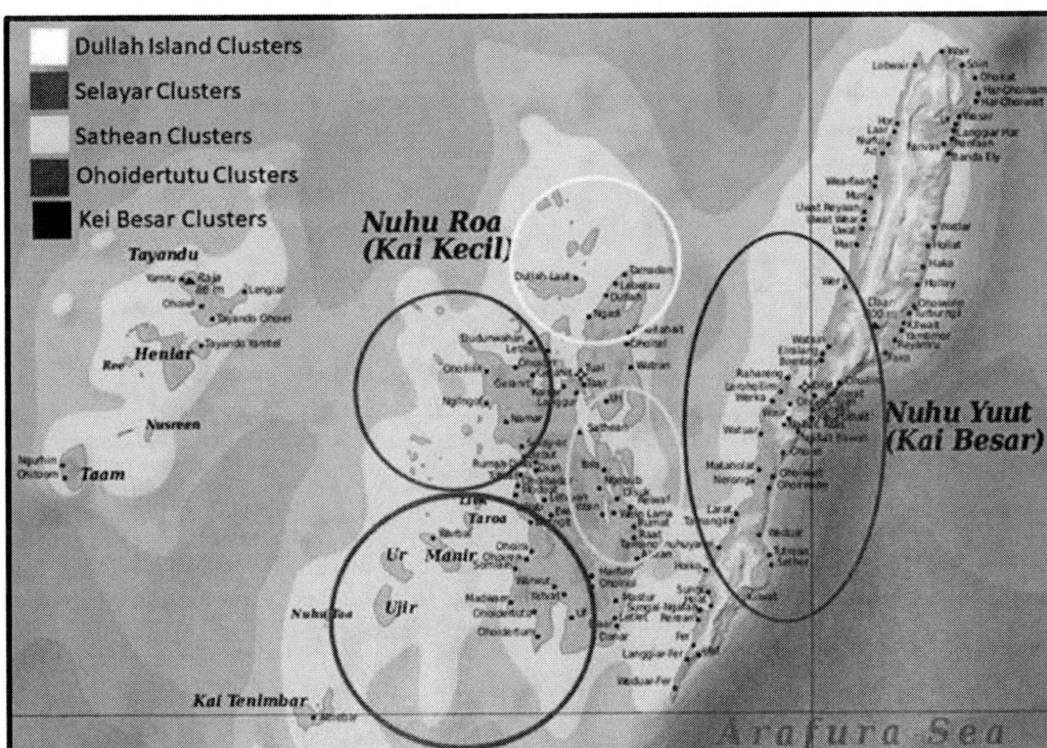

Figure 2. Village Clusters based fishing activity in Kei Island

On table 1 as we can see, we have five groups with total 15 villages in Kei Islands. On each cluster consist of fishermen who used fishing as their source of revenue to support their family, also fishermen that conducted fishing activity as their primary job. These five clusters also consist of high productivity of fishermen in the region. Therefore these were the locations to collect the data on measuring the support on the application of MPA in Kei Islands.

3. Results and discussion

The variables in this study consisted of 1 dependent variable namely support of MPA and seven independent variables, ie role variables in capture fisheries, fishing gear, fishing catching location, length of time to go to sea, income, the combination of local wisdom and reason to support MPA. The result showed that of the seven there were 24 indicators, with only 12 independent indicators of 6 variables that statistically significant to the support of MPA this can be seen in table 2.

Table 2. Significant Variables Result

	B	Sig.	Exp(B)
Role of fishing activities			
Fisherman (Own all of their fishing equipment)	-1.252	.072*	.286
Lift nets owner	-2.046	.035**	.129
Fish seller	-3.010	.044**	.049
Fishing gear			
Fishing net and *bubu*	2.340	.049**	10.382
Fishing area			
On sea own by their village	1.511	.035**	4.533
On sea own by other village but has "*pela*" relation	-2.387	.097*	.092
Duration			
Duration 4 – 6 hours	1.088	.072*	2.968
Combine with local wisdom			
Sasi	5.346	.000**	209.865
Sustainability reason			
Economic	3.075	.010**	21.653
Environment	6.294	.001**	541.258
Social	2.633	.084*	13.913

N = 255, Significant at **5% and *10%

3.1. Role of Fishing Activities and Fishing Gear

On table 2, based on the role in fishing activities for fishermen who own their fishing gear and equipment was statistically significant to affect their perception on support MPA, where they have tendencies not to support MPA. It happens because compared to the fisherman who doesn't own their fishing gear and equipment, those who have their gear are having a broad range of fishing activity and liberty to conducted fishing activity anytime. Within said they feel the application of MPA this region will limit their range and therefore will eventually affect their revenue. Same goes to lift net owner, where they also tend to support MPA, since lift net has a limited range of operation, with MPA it will impact their productivity. A fish seller in the local market also tends not to support MPA, because they said it is related to the supply of fish which also will impact the price of fish they buy from fishermen which can decrease their margin profit.

For fishing net user tend to support MPA. Fishermen who use the fishing net on their take on MPA is tended to be positive since it will help maintain the carrying capacity and resources of the shallow marine environment. The fishing net user often operates in shallow water which makes the application of MPA tend to not significantly affect their productivity in a way that decreases their catching result.

3.2. Fishing Area and Duration
On fishing area variables, fishermen who are conducting their fishing activity in the sea that own by their village tend to support MPA but those who usually did their fishing activity on sea own by their neighbor but have *pela* relation, where *pela* is local wisdom form of bonding between villages, tend to not to support MPA. For fishermen that used to conducted fishing activity in believing that if MPA applied in their water will mean good for future. Therefore they tend to support MPA but on other side, fishermen who used to conduct fishing activity in water own by other village did it out of the reason that, water own by their village is not high on marine resources as water own by other villages especially when they have *pela* relation because they will have a permit to conduct fishing activity.

Based on the duration of fishing activity, fishermen who used to spend 4-6 hours each day to conducting fishing activity tend to support MPA. The positive tendencies occur because most of the fishermen in Kei Islands did fishing activity in their water, thus for a short period it only cover shallow water makes the application of MPA not affect their fishing activity.

3.3. Combine with local wisdom and Sustainability reasons
Most fishermen in Kei Islands based on our filed study sounded skeptic on their take of the application of MPA mainly because it will limit their fishing activity since more than half of coastal community in Kei Islands depend their life on marine resources. The idea of MPA tend to have a negative perception of fishermen, but on the notion of combining with a form of local wisdom called *sasi* then MPA get more positive responses. *Sasi* is the spatial and temporal closure of fields, forests, reefs and fishing grounds, tidal zone (*meti*) or village-controlled sea (*petuanan laut*), is a conspicuous feature of Maluku region [11], with *sasi* will make sure that nobody takes what does not belong to them.

Based on sustainability reasons, measured by indicator namely environment, social and economy, fishers tend to have a positive perception on supporting MPA. This result indicated that fishermen on Kai Islands are understood the importance of preserving the sea for better future, but based on our field survey, they hesitated to showed directly support MPA because of they afraid it may negatively affect their lives. Since there are no other natural resources in this region, aside of agriculture and newly emerged marine ecotourism, marine resources are the only source of their revenue. Therefore in order to make MPA program work in the region, the local and central government have to make sure that another sector is also alive to support their daily need, in addition to that combining with a form of local wisdom also can help the application of MPA in this region.

4. Conclusion
Most fishermen tend to support MPA model in a way where there are strict rule and law by the government to prevent to conduct fishing activity in a certain sea in Kei Islands, but if it combines with a form of local wisdom called *sasi*, then MPA can work in this region. Fishermen in the region have understood the important maintain the marine resources for the future, but they tend to be practical in short-term because the sea is where they go to support their family. The idea of MPA may work in Kei Islands if central and local government find another sector aside of agriculture as new sources of revenue while at the same time modified the form MPA with local wisdom to make it works.

References
[1] Cros, A., Fatan, N. A., White, A., Teoh, S. J., Tan, S., Handayani, C., ... and Fitriana, R 2014 The Coral Triangle Atlas: an integrated online spatial database system for improving coral reef management. *PloS one*, 9(6), e96332.

[2] e Costa, B. H., Claudet, J., Franco, G., Erzini, K., Caro, A., dan Gonçalves, E. J 2016 A regulation-based classification system for Marine Protected Areas (MPAs). *Marine Policy*, 72, 192-198.

[3] Sala, E., Costello, C., Parme, J. D. B., Fiorese, M., Heal, G., Kelleher, K., ... and Rosenberg, A. A 2016 Fish banks: An economic model to scale marine conservation. *Marine Policy*, 73, 154-161.

[4] Robinson, E. J 2016 Resource-Dependent Livelihoods and the Natural Resource Base. *Annual Review of Resource Economics*, 8, 281-301.

[5] Takasaki, Y. 2016. Learning from disaster: community-based marine protected areas in Fiji. Environment and Development Economics, 21(01), 53-77.

[6] Teniwut, W. A 2016 For sustainable revenue of fisheries sector in small islands: evidence of Maluku, Indonesia. AACL Bioflux, 9(3).

[7] Neleman, S., and de Castro, F 2016 Between nature and the city: youth and ecotourism in an Amazonian 'forest town'on the Brazilian Atlantic Coast. *Journal of Ecotourism,* 15(3), 261-284.

[8] Rosegrant, M. W., Dey, M. M., Valmonte-Santos, R., and Chen, O. L 2016 Economic impacts of climate change and climate change adaptation strategies in Vanuatu and Timor-Leste. *Marine Policy*, 67, 179-188.

[9] Green A, Petersen N, Cross A, MacLeod E 2008 Coral Triangle facts, figures and calculations. PartII: Patterns of biodiversity and endemism. Brisbane: The Nature Conservancy.

[10] Coral Triangle Initiative 2009 Regional Plan of Action, Coral Triangle Initiative on Coral Reefs, Fisheriesand Food Security (CTI-CFF). Manado: Coral Triangle Initiatlive p.42.

[11] Thorburn, C.C 2000 Changing Customary Marine Resource Management Practice and Institutions: The Case of Sasi Lola in the Kei Islands, Indonesia. *World Development* Vol. 28, No. 8, pp. 1461±1479

Management and Feasibility Analysis of Smoked Fish Business in Ambon

Hellen Nanlohy[1], Yolanda M T N Apituley[1], Alfonsina M Tapotubun[2] Frederik Reiuwpassa[2], Theodora E A A Matrutty[2]

[1] Fisheries Agrobusiness Study Program, Faculty of Fisheries and Marine Science, Pattimura University
[2] Fisheries Technology Study Program, Faculty of Fisheries and Marine Science, Pattimura University

Email: nanlohy_hellen@yahoo.com

Abstract This research aims to examine management and feasibility aspects of smoked fish business in Ambon. By using survey, this research focused on smoked fish business in Negeri Hative Kecil and Silale, known as dried fish producer villages.Primary and secondary data collected by using interview, observation, and recording. Analysis methods used are qualitative descriptive analysis and business feasibility analysis covers NPV, Payback Period (PP), and Break Even Point (BEP). The result shows that most of the smoked fish businessmen do not apply the proper management in their business. Two to three people do all the works (from production to marketing) without a clear job description. Feasibility analysis for smoked fish business in Negeri Hative Kecil shows that NPV is 21.501.053,- PP is 58 days, and Benefit Cost Ratio (B/C) is 1,06. BEP Production is 1,455 kg, and BEP Price is IDR 19,941, while the feasibility analysis for smoked fish business in Desa Silale shows that NPV is 30.745.837,-, PP is 24 days and Benefit Cost Ratio (B/C) is 1,41. BEP Production is 988 kg, and BEP Price is IDR 7,966. Based on the result, the two smoked fish business in Ambon is feasible to be developed. However, good management with clear job description should be applied to improve the business.

1. Introduction

Maluku is a well-known province of fish and non-fish resources that are potential to be managed and developed as income and foreign exchange for the country and the region itself. The resources will not produce benefits or high economic value if not followed by proper processing and good marketing [13]. [14] Stated that processing is needed to increase the economic value of fish resources that could lead to optimal utilization. Processing development is also expected to increase the value added and create a variation of products to broaden the market and absorb labor. Fish processing can be done in various ways: salting, drying, smoking, fermentation, processing in low and high temperature and by-products processing [1][14]. Smoked fish processing is a famous fishery business in Ambon, which is still operated in small scale (household scale) and traditional manner. Most of the home industries are still being undertaken traditionally. Fish smoked is a typical product of Ambon which is already famous in the domestic, national and foreign markets. Kind of fish commonly smoked are skipjack (*Katsuwonus pelamis*) that found mostly in the waters around Ambon Island and Banda Sea.

IOP Conf. Series: Earth and Environmental Science **89** (2017) 012026 doi:10.1088/1755-1315/89/1/012026

Traditional smoked fish processing business contributes economy to the businessmen and the development of the region. Stated that fish processing business can drive the regional economy development [7]. However, this kind of business is very dependent on weather condition and lead to the unstable price of products. Difficulties of getting capital also influence this business, so the development of it sometimes is risky. Therefore, in developing this business, a good planning is crucial. To be developed, each business needs evaluation, especially financial evaluation with many criteria or aspects. This assessment will determine whether the business is experiencing a profit or loss. Risk can be minimized or even avoided through good planning [8][9]. This research aims to analyze business management and feasibility of smoked fish household industry in Ambon. Financial analysis criteria in this study are Net Present Value (NPV), Payback Period (PP) and Break Event Point (BEP).

2. Methodology

The basic research methodology used in this research surveys [10], which consists of observation, interview by using questionnaire and documentation.

2.1. *Population and Samples*

Population in this research is smoked fish producers in Ambon, and samples were determined by using purposive technique [10]. Samples taken were smoked fish producers in Negeri Hative Kecil and Silale who actively produce smoked fish to be sold. The sample in each village amounted to four people, so the total sample amounted to eight people.

2.2. *Data Analysis*

This research used qualitative and quantitative descriptive analysis method. The qualitative descriptive analysis method is used to describe the current facts carefully [10][2] while quantitative descriptive analysis used to calculate and explain the financial aspects such Net Present Value (NPV), Net Benefit/Cost, Payback Period (PP) and Break Even Point (BEP) [12].

2.3. *Location and Time*

This research was conducted in Negeri Hative Kecil and Silale, Ambon, on June – July 2016.

3. Result and Discussion

There are three smoked fish processing centers in Ambon, namely Negeri Hative Kecil, Silale village and Laha village. The first two villages are the greatest productivity of smoked fish in Ambon. Traditional smoked fish processing requires several production factors, such as capital, raw material, labor and adequate infrastructures. Experienced human resources are necessary to produce specific smoked fish products in Ambon.

3.1. *Business Management Aspect*

The management aspect of this traditional smoked fish processing business is focused on internal conditions. Management aspects analyzed in this research are planning of the raw materials, labor, capital and equipment used [4].

Raw materials used in smoked fish processing is skipjack (*Katsuwonus pelamis*). The smoked fish raw material obtained in two ways. First, the processors buy the raw materials directly from the fishermen. This could happen in a fish season. Second, the processors buy the raw materials from cold storage, and it occurs in no fish season. However, raw materials used in smoked fish processing in Ambon are usually purchased from fishermen [7]. The producers are not able to plan the raw materials purchased well, due to the uncertainty of raw materials availability.

All the smoked fish business in Negeri Hative Kecil and Silale are individual businesses and household scaled only. Such kind of business does not need many labors, and those who work in the smoked fish processing business in the two villages are family members. The processors are also functioned as marketers, and therefore, there are only two to three people in that business with the unclear job description. Initial capital used by the traditionally smoked fish processors comes from their capital. As theirs are very limited, the processors often borrowing money for the continuation of their business. The limited capital of the processors causes this business are constrained to be developed. Instruments used in smoked fish processing are still traditional, simple and using manpower. Most of the equipment used in smoked fish processing is impropery.

3.2. Financial Aspect

The financial aspect of this business covers investment, expenditure, revenue, and income. [12][3]. Stated that financial aspect is crucial in a business activity. Investment capital is an initial cost/venture capital incurred at the time the business is running. The average investment of smoked fish business in Negeri Hative Kecil is IDR 13.325.000 whereas in Silale is IDR 8.193.750,- Investment capital used by each smoked fish processors in the two villages can be seen in the following table.

Table 1. Investment of Smoked Fish Processing Business in Ambon

Villages	Respondents	Total Investment (IDR)
Hative Kecil	1	15.000.000
	2	11.250.000
	3	17.500.000
	4	9.550.000
	Total	53.300.000
	Averages	13.325.000
Silale	1	8.900.000
	2	9.250.000
	3	8.950.000
	4	5.675.000
	Total	32.775.000
	Averages	8.193.750

Costs incurred by this business are fixed and variable costs. Fixed costs cover depreciation and maintenance costs, while the variable costs are incurred to purchase the raw material, salt, kerosene, coconut oil, coconut shell, yarn, bamboo, detergent, plastic bags, transportation, and consumption.

Table 2. The Average Costs Incurred by Smoked Fish Business in Ambon

Villages	Respondents	Total variable cost (Rp)	Total Fixed Cost (Rp)	Total Cost (Rp)
Hative Kecil	1	62,197,250,-	4.413.333,-	66.610.583
	2	71.994.417	2.775.000	74.769.417
	3	67.329.333	3.721.666	71.050.999
	4	79.000.000	3.926.667	82.926.667
	Total	280.521.000	14.836.666	141.380.000
	Averages	70.130.250	3.709.167	35.345.000
Silale	1	12.126.857	212.917	12.339.774
	2	17.307.000	185.417	17.492.417
	3	14.620.500	222.083	14.842.583
	4	15.313.998	253.000	15.567.000
	Total	59.368.355	873.417	60.241.774
	Averages	14.842.089	218.354	15.060.444

Table 2 shows total variable costs, fixed costs and total costs incurred by the smoked fish business. It can be seen that the average total cost in Negeri Hative Kecil was amounting to IDR 35,345,000,- and in Silale amounting to IDR 15,060,444,-.

Table 3. Revenue and Income of Smoked Fish Business in Ambon

Villages	Resp.	Prod (Kg)	Revenue (IDR)	Total Cost (IDR)	Income (IDR/Mo)	Income (IDR/Year)
Hative Kecil	1	3.077	71.500.000	66.610.583	4.889.417	58.673.004
	2	3.485	80.990.000	74.769.417	6.220.583	74.647.000
	3	3.048	74.818.333	71.050.999	3.767.334	45.208.008
	4	3.598	86.020.833	82.926.667	3.094.166	37.129.992
	Total	13.208	313.329.166	295.357.666	17.971.500	215.658.004
	Averages	3.302	78.332.292	73.839.416	4.492.875	53.914.501
Silale	1	1.080	18.900.000	12.339.774	6.560.226	78.722.712
	2	1.620	24.300.000	17.492.417	6.807.583	81.690.996
	3	1.350	20.250.000	14.842.583	5.407.417	64.889.004
	4	1.450	21.320.000	15.567.000	5.753.000	69.036.000
	Total	5.500	84.770.000	60.241.774	24.528.226	294338712
	Average	1.375	21.192.500	15.060.444	6.132.057	73.584.678

Table 3 shows that the average revenue received by the smoked fish business in Negeri Hative Kecil is IDR 78,332,292,-, per month while those in Silale received IDR 21,192,500,- per month. The total expenditure of smoked fish business per month in Negeri Hative Kecil is IDR 73,839,416,- and in Silale is IDR15,060,444,-. Based on the revenue and expense issued, the income of the business can be calculated. The average income of the smoked fish business in Negeri Hative Kecil is IDR 4,492,875,- per month or IDR 53,914,501,- per year, and the average income of the similar business in Silale is IDR 6,132,057,- per month or IDR 73, 584,678,- per year.

3.3 Business Feasibility Analysis

Business feasibility of smoked fish home industries in NegeriHative Kecil and Silale are analyzed by using several criteria of financial analysis., Net Present Value (NPV), Payback Period (PP), and Break Even Point (BEP).

Table 4. Business Feasibility Analysis of Smoked Fish Business in Ambon

Villages	Resp.	NPV	PP	B/C Ratio	BEP Prod.	BEP Price	Feasibility Criteria
Hative Kecil	1	22,186,394,-	0.5	1,07	1,416	11,025	Feasible
	2	31,473,734,-	0.27	1,08	1,037	7,169	Feasible
	3	18,060,413,-	0.45	1,05	1,446	12,200	Feasible
	4	14,283,672,-	1.08	1,04	1,919	13,370	Feasible
	Total	86,004,213,-	2.30	4,25	5,818	43,764	
	Average	21,501,053,-	0.58	1,06	1,455	19,941	
Silale	1	33,780,487,-	0.23	1,53	705	7,860	Feasible
	2	35,071,316,-	0.22	1,39	1,166	6,825	Feasible
	3	27,581,546,-	0.27	1,36	990	8,189	Feasible
	4	26,550,000,-	0.25	1,37	1,090	8,990	Feasible

Total	122,983,349,-	0.97	5,65	3,951	31,864
Average	30,745,837,-	0.24	1,41	988	7,966

Smoked fish production depends on the season and production per month. If there is no fish season, the production will decline, and vice versa. The fish season usually takes place in January, February, March, and December, during the no fish season last from April to November. The production in every month ranges from 25-28 days. Raw materials taken from the cold storage in Ambon priced IDR15,000,- per kilogram in the fish season, and IDR 24,000,- in no fish season.

The result of NPV analysis shows that NPV value of the two villages is different. The average value of NPV in Negeri Hative Kecil 21,501,053,-, while in Silale30,745,837,-. Those values are > 0 and means that investment in traditional smoked fish processing businesses in both villages can be accepted. The NPV value indicates that this business in Ambon is estimated able to turn a profit. Stated that positive value of NPV indicates that the running businesses will be able to turn a profit [11]. The value of Payback Period for traditional smoked fish business in Negeri Hative Kecil 0.58 (58 days) and Silale 0. 24 (24 days), which indicates that business able to restore the investment for less than one year. The shorter the time to return the investment, the better the business could be developed [11]. Benefit-Cost Ratio (B/C) in Negeri Hative Kecil is 1,06 and Benefit Cost Ratio (B/C) in Silale is 1,41.

Analysis of Break Even Point (BEP) shows that the average of BEP production in Negeri Hative Kecil is 1.455kg, while the average of BEP price is IDR 19,941,-. The average production in Negeri Hative Kecil when the research was undertaken 3.302 kg and average selling price IDR 20,000, -. The average production in Silale 1,455 kg, and averages BEP price is IDR 7,966,-. And the average production of fish smoked in Silale was 1,375 kg with selling price IDR. 20.000,-. Analysis of business feasibility is critical to undertake by the smoked fish processor in Ambon. By that means, the processors can improve and increase the advantages of business [12]). Investment available at a venture is also determined the advantage gained [3][14]. A good investment will benefit the business the processing of this traditional.

4. Conclusion

1. Management aspects of the smoked fish business were feasible but still, needs to be improved regarding the technical and operational aspects.
2. The smoked fish business in Ambon is financially feasible and able to be developed.

Acknowledgements
Thanks to Government of the Republic of Indonesia through the Directorate General of Higher Education on MP3EI funds for implementation of this research.

References

[1] Adawyah, R. 2007. *Processing and Preserving of Fish*. Jakarta: Pt. Earth Alphabet.
[2] Arikunto, S., 2010. *Research Prosedure, An Practice*. Revised Edition. Rineka Cipta Press. Pp 413.
[3] Halim, A., 2005. *Analisis investasi*. Four Salemba, Jakarta.
[4] Handoko, T. H., 2001. *Personnel Management and Human Resources*. Issue 2. University Of Gajah Mada, Yogyakarta.
[5] Karimah, A., I. Gumilar., and Z. Hasan, 2012. A Prospective Analysis of Cultivation in Freshwater Aquarium Fish Freshwater Aquaculture Park (Obedience) and Taman Mini Indonesia Jakarta Indah (TMII), Journal 3 (3): 145-156.
[6] Lumi, K. Mantjoro, E. Dan Wagiu, M., 2013. Economics Value of Resources in North Sulawesi. (Case Study of *Katsuwonus pelamis*). *Jurnal Ilmiah Platax,* Vol 1-2.
[7] Mangantar, Adolfina, M., and D. N. Baramuli, 2015. Merchantibility of *Katsuwonus pelamis* in the Town of Bitung. *Journal EkoSosBudKum Fields*, LPPM 2 (2): 73-84.
[8] Munawaroh, M., H. Miniartidan L. Hidasah. *BisnisPlan for Graduate Programme 1*. LP3M UMY, Yogyakarta.

[9] Nanlohy, H., Y. M. T. N., Apituley, M. Tapotubun, and F. Reiuwpassa, 2016. Traditional fish Processing Business profile in the Moluccas. *Proceeding of National Seminar MPHPI 2016 and the Eifght of Scientific Conferencce.*
[10] Nazir, M. 2003. *Research Metod.* Ghalia Indonesia Press. Jakarta.
[11] Pasaribu, A. M., Joseph, d., and Amiluddin, 2005. *Planning and Project Evaluation Fisheries.* Hasanuddin University Press, Macassar.
[12] Rahardi, A., and Hartono, 2003. *Agribusiness Farms,* Revised Edition. The Diffuser Is Self-help, Jakrta.
[13] Retraubun, A. W. S. , J. Abrahamsz, Y. Lopulalan, S Tubalawony, P. Usmany, H. Nanlohy, and F. Ayal, 2016. Research Reports Development Planning Provincial Maluku Based Maritime. Cooperation Regional Development Planning Board Maluku with the Faculty Fisheries and of Marine Science Pattimura University.
[14] Sormin R., B., D., M. Ch. A. Latumahina, T. H. E. A. Matrutty, A. M. Tapotubun and H. Nanlohy, 2015. Quality Profile and Economic Analisis of Dried Smoked Julung in Banggoi Vilage, District of Eastern Seram, Malucas. *International J. Of Current Research,* 7 (02) pp 12775-12778.
[15] Sujarno, 2008. Analysis Factors Affecting Income Fishermen in Kabupaten Langkat. The Thesis. A Graduate School University North Sumatra, Medan.
[16] Umar, H. , 2003. *A Feasibility Study on Business, Technique Analyze Feasibility Business Plan Comprehensively.* Edition second. Gramedia Main Library Press, Jakarta.

Mapping Marine Resources Utilization Based on Seascapes Area: A Study on Gender Comparison

Simon Picaulima, Wellem A. Teniwut, Syahibul Kahfi, Roberto M.K Teniwut, Ida I Dewa Ayu Raka Susanti, Marselus Hungan, Meyske Rahantoknam, Cawalinya Hasyim, Siska D. Rahakbauw, M.R. Renjaan, Anna M. Ngabalin, Tati A. Ngangun1, Frischila Pentury, K.D Betaubun, A.K Ngamel, E.C Ohoiwutun

Fisheries Agribusiness Study Program, Tual State Fisheries Polytechnic, Langgur-Sathean Ringroad Km.6, Langgur, Southeast Maluku, Indonesia 97611

Email: spicaulima@yahoo.com

Abstract. Each section of coastal area from village area to deep sea has particular fisheries and marine activities. Therefore requirements for knowledge, skill, and risk are distinct between each scape. Based on this notion we divided seascapes based on three different area, first is village area to mangrove and coconut trees before coast area, second is area from beach to shallow water and the third is deep sea. We conducted our data collection in Kei Islands, Indonesia, with purposive sampling to target fishermen and aquaculture farmers in the coastal area. Purposes of this research are to analyze the role and the contribution of male and female on each area to formulate the best way to improve and maintain the sustainability of microenterprises and coastal community welfare; we used logistic regression to analyze the data. The result demonstrated each fisheries activity in each seascape with based on gender on each area and activity. Recommendation of this study especially for the government as empirical guidance to improve the economic welfare of the coastal community in Kei Islands and small islands area in general.

1. Introduction

The tropical coastal seascape is one of the locations that contain high of marine biodiversity which also place of many natural resources and ecosystem services that are significant for coastal community and people in general's daily need. Researchers used many different terms of seascapes, [1, 2] proposed seascapes as areal where combination between mangrove forests, seagrass beds, and coral reef ecosystems exist. Each section of seascapes requires distinct knowledge, and skill even risk to utilize, because it's a delicate situation to handle since every part of seascapes depends and supports each other. Therefore, in Caribbean coastal ecosystems there're trying to distinguish the importance of seagrass and mangrove habitats to populations of coral reef fishes [4]. Countries in the world especially developed countries face whole new challenge compare to undeveloped countries or country with high gap on infrastructure between big cities and rural area. In country such as Australia [3], the main challenge is to preserve the seascapes, but for countries like in Africas, the main challenges are to optimum utilize since most of the coastal community still live under poverty.

With each section of seascapes has its distinct features, then, the role of gender in the utilization on seascapes become clearly different especially related to the productivity and risk. Male and female use coastal area differently and will produce differential economic and environmental impacts. Many types of research had been conducted to view role of gender on utilization of coastal area, [5] Showed the different on knowledge poses based on gender, [6] Studied on the contribution of female on coastal area utilization, [7] Stated to gender should be integrated to maximize the productivity of economic resources and maintain environmental existence of coastal area.

Indonesia is an archipelagic nation that covers 5.8 million km^2 of marine area and has approximately 17,504 islands, a coast line of 95,181 km [8] and 8,090 coastal villages [9]. Millions of Indonesia's poor belong to small-scale coastal fishing households in the eastern region of Indonesia and are heavily reliant on fish for daily food and to generate income to meet daily living costs as well as other basic needs such as education. Kei Islands are located in Eastern Indonesia, to be precise is in Maluku Province. Thus many of coastal community in Maluku in general and Maluku, in particular, rely their daily life on fisheries sector, either from fishing, mariculture, food processing and marine ecotourism. Despite high potential on marine and fisheries resources, [10] stated that in Maluku if the trend stands still then in 8-10 years number of fish in Maluku sea, in general, includes Kei Islands will drop gradually and significantly. Therefore, the ability to utilize other section of sea is important for the sustainability of fisheries productivity, especially on mariculture and food processing based fisheries resources.

For fishing activities, male as dominant gender, but for another sub sector of fisheries, its supposed to be balanced on the number between male and female, where female in particular play huge role on selling fish and other fisheries product compare to male [11]. As a coastal area, the ability on the utilization of fisheries resources in Kei Islands is important to increase community well-being and welfare, therefore, purposes of this research were to analyze the role and contribution of gender on each seascapes section in coastal area. On this study we used three seascapes area. First, village and mangrove area, the second is beach and shallow water and the third one is deep sea and coral reefs as it shows in figure 1.

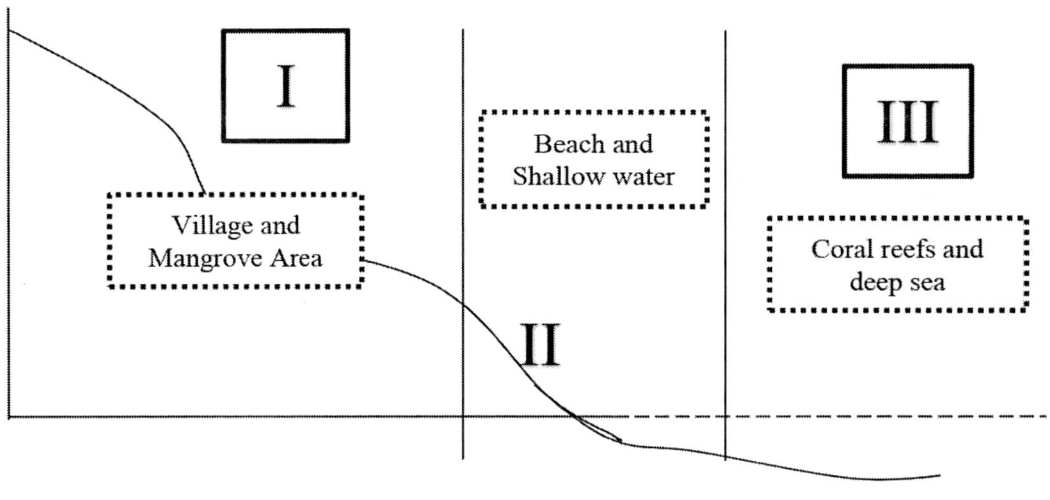

Figure 1. Seascapes Section of This Study

2. Methodology

The study was conducted from May to July 2017, a preliminary survey conducted during two weeks period to determine the best research location that meets the criteria. Primary data collected by purposive sampling using questionnaire and focus group discussion (FGD). The sample size of 122 men and women 122 people who work mariculture farmers, fishermen and micro enterprises of food processing and marine ecotourism in 10 (ten) villages (Ohoi in local language) in Kei Islands which consists of two administrative government namely Southeast Maluku Regency and Tual City. Name of villages on this research as follows Ohoi ngilngof, Ohoi evu, Ohoi letvuan, Ohoi loon, Ohoi kelanit, Duluh laut village, Ohoi sathean, Ohoi ngafan, Dulah darat and lupus hamlet (as seen of figure 2). The methodology of analysis using Descriptive and logistic regression analysis.

We conducted our study by adopting the previous study conducted by [12] but we also did some adjustment on the research, for instance, we focus only fisheries sector, and also we choose villages with same particular trait as seascapes that we have divided. Based on our preliminary survey there are more than 15 villages that meet the criteria and we choose 10 villages where they have active villages for processing, they also have mangrove that contains many fisheries resources and have beach and fishing activity on deep sea and coral reefs. Aside from FGD we also collected the data by questionnaire to have comprehensive result on the comparison on the contribution of gender in fisheries resources utilization in Kei Islands.

The seascape section divided based on data and information obtained by discussion during the preliminary study. In section 1 usually, has mangrove ecosystem still affected by splashing sea water or tidal, plantation, get sun exposure and village area with the risk that is considered low. In section 2 where coastal and shallow seas have characteristics of tidal, waves, coral reef and seagrass ecosystems, species of biotic and abiotic resources, especially fisheries, aquaculture, resource utilization properties, some private and some common properties, have physical and moderate risks. The deep sea region has high salinity characteristics, has variations in surface temperature with depth, species of biotic and abiotic resources, considered high risks and physical.

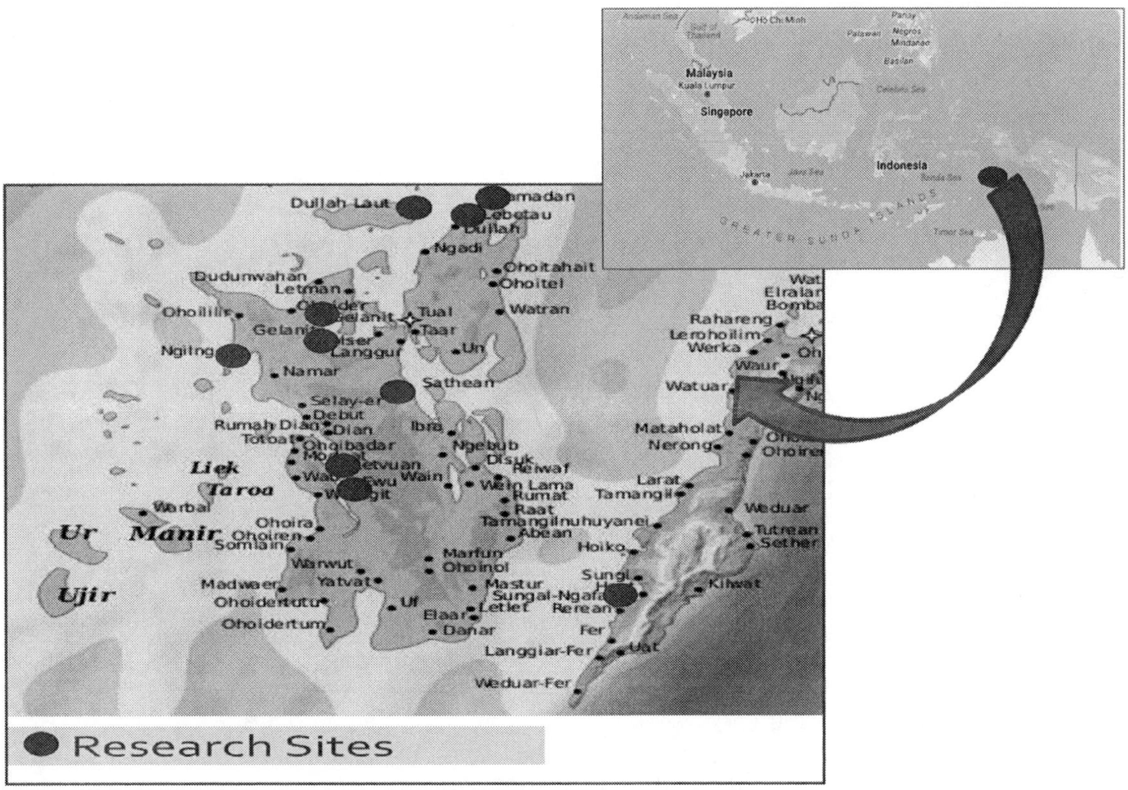

Figure 2. Research sites

On table 1 can be seen that number of fishermen in this study is the highest on fishing activities category 43.9% followed respectively by mariculture farmers 32.4%, marine ecotourism 11.1%, a combination of several activities 9.4% and fisheries processing 3.3%. Based on seascapes area more than half of respondent of study operate on section 2 of seascapes 59.4% followed by section 3 14.8% and section 1 7% also result showed that respondents also operate on the combination section 18.9%. Most of the respondent generate their income between IDR 500.000 to IDR 2.000.000,- also more than half of respondent work around 1 to 6 hours a day.

IOP Conf. Series: Earth and Environmental Science **89** (2017) 012027 doi:10.1088/1755-1315/89/1/012027

Table 1. Social economy characteristic of respondents

Fisheries Activities		
	Frequency	Percent
Catching fisheries	107	43.9
Fisheries processing	8	3.3
Mariculture	79	32.4
Marine Ecotourism	27	11.1
Combination	23	9.4
Seascapes Activity		
	Frequency	Percent
Section 1	17	7.0
Section 2	145	59.4
Section 3	36	14.8
Combination	46	18.9
Income		
	Frequency	Percent
<Rp.500.000	35	14.3
Rp.500.000-Rp.999.999	89	36.5
Rp.1.000.000 Rp.1.999.999	93	38.1
Rp.2.000.000-Rp.2.999.999	18	7.4
>Rp 3.000.000	9	3.7
Hours of work		
	Frequency	Percent
< 2 hours	44	18.0
2-3 hours	56	23.0
4-5 hours	87	35.7
>6 hours	57	23.4
< 2 hours	44	18.0

Female and Male: 122; N: 244

3. Result and Discussion

3.1. Income generated by gender based on seascapes section

In section 1 female's income was greater than male's because in this area the dominant activities are fishing processing and ecotourism, but the income is not too significantly large because the processing product price has not been maximized also due to the influence of bad weather when we conducted the research. The number of visitors to the tourist sites for ecotourism activities decreased due in our research period, but this is an empirical condition on Kei islands where income from fisheries and marine resources utilization often fluctuate during the time. In section 2, female's income is greater than male's because, in this area, the highest activity are mariculture, ecotourism, and fishing where female conducted more mariculture and ecotourism with an addition of low on productivity on fishing activities in the shallow area makes female has more advantage in section 2. In section 3, male's incomes are larger than women because the third area of dominant activity is fishing and ecotourism with high risk and physical requirements make male because of dominant forces on this section (Table 2).

Table 2. Income generated by gender based on seascapes section

Section	Male	Female	Total
Combination Location	35.150.000	8.820.000	43.970.000
Section 3	32.800.000	12.950.000	45.750.000
Section 2	50.600.000	85.850.000	136.450.000
Section 1	8.000.000	8.900.000	16.900.000

3.2. Income generated by gender based on seascapes activity

In section 1 and section 2 was dominant performed by female the highest income fisheries activity for the female is on marine ecotourism activities, fishing processing, mariculture, where it's the total income from the utilization of all section combine for female. Especially the fisheries processing where male did not contribute to income generated. The highest income fisheries activity for the male is in catching fisheries activities because most of the fishermen either in section 2 and section 3 are male, so it's also understandable for the male to have more income generated in fishing activity than female. For combination on the section of seascapes to operate, male generate more income than female. This also understandable since male has the main job to provide economic support for each household where female is actually has double job on house job and also to support their family in fisheries activity, therefore male has more time to conducted more fisheries activity in various of seascapes sections, this aside from physical advantage of male from female (Table 3).

Table 3. Income generated by gender based on seascapes activity

Section	Male	Female	Total
Mariculture	27.500.000	40.200.000	67.700.000
Combination	23.100.000	8.000.000	31.100.000
Fisheries processing		3.720.000	3.720.000
Catching fisheries	73.950.000	34.000.000	107.950.000
Marine Ecotourism	2.000.000	30.600.000	32.600.000

3.3. Logistic regression result

Table 4. Significant results of logistic regression

Logistic Results		
	Sig.	Exp(B)
Marine Ecotourism	.008**	.070
Villages, Mangrove and Coconut areal	.072*	.264
Beach and Shallow water	.014**	.269
Work < 2 Hours	.046**	.258

*Significant in 10%; **significant in 5%

Based on the result showed on table 4, out of 8 variables with total 17 indicators only 4 indicators that have significant effect on gender contribution. Where it can be seen on result of logistic regression, showed that the male tends to prefer conducted marine ecotourism activity, in section 1 and section 2 also the result showed that male also not to prefer operating in this section. These results support the other result on this study where female generated more income in section 1 and section 2, where based on our discussion with coastal community showed that male seems to have thought about high potential on economic in sea especially on fishing activity in deep water that's why they prefer to go fishing. This argument also supports the result of logistic regression where male did not prefer to work less than 2 hours since they mostly go fishing in deep water which took more than half a day.

4. Conclusion

Male generated IDR 126,550,000.00 where, female create IDR 116,520,000.00 each per month, although, Female has a great potential for generated more income if provided with sufficient knowledge and skill related. In general, both gender still underperformed compared to marine and fisheries resources availability in the region, therefore the role of higher education institutions and research organization in significant to increase coastal community skill and capability on utilizing and maximize the marine resources in Kei Islands.

References

[1] Duke, N. C., and Wolanski, E 2001 Muddy coastal waters and depleted mangrove coastlines–depleted seagrass and coral reefs. Oceanographic processes of coral reefs: Physical and biological links in the Great Barrier Reef, 77-91.

[2] Moberg, F., and Rönnbäck, P 2003 Ecosystem services of the tropical seascape: interactions, substitutions and restoration. *Ocean & Coastal Management*, 46(1), 27-46.More references

[3] McCartney, A 2006 The social value of seascapes in the Jurien Bay Marine Park: an assessment of positive and negative preferences for change. *Journal of Agricultural Economics*, 57(3), 577-594.

[4] Mumby, P.J., Edwards, A.J., Arias-Gonzalez, J.E., Lindeman, K.C., Blackwell, P.G., Gall, A., Gorczynska, M.I., Harborne, A.R., Pescod, C.L., Renken, H., Wabnitz, C.C.C., Llewellyn, G 2004 Mangroves enhance the biomass of coral reef fish communities in the Caribbean. *Nature* 427, 533–536.

[5] Iniesta-Arandia, I., Del Amo, D. G., García-Nieto, A. P., Pineiro, C., Montes, C., and Martín-López, B 2015 Factors influencing local ecological knowledge maintenance in Mediterranean watersheds: insights for environmental policies. *Ambio*, 44(4), 285-296.

[6] Kleiber, D., Harris, L. M., and Vincent, A. C. (2015). Gender and small-scale fisheries: a case for counting women and beyond. *Fish and Fisheries*, 16(4), 547-562.

[7] Diamond, N. K., Squillante, L., and Hale, L. Z 2003 Cross currents: navigating gender and population linkages for integrated coastal management. *Marine Policy*, 27(4), 325-331.

[8] Ministry of Marine Affairs and Fisheries. 2009a. Kelautan dan Perikanan dalam Angka. [Marine Affairs and Fisheries in numbers]154 pp.

[9] Ministry of Marine Affairs and Fisheries.2009b. Jumlah nelayan menurut kategori nelayan 2009. Statistik Perikanan Tangkap. The Ministry of Marine Affairs and Fisheries. Jakarta. Viewed 26 July 2011 http://www.statistik.kkp.go.id[Number of fishers based on fisher category. The Statistic of capture fisheries]. 1p.

[10] Teniwut, W. A. (2016). For sustainable revenue of fisheries sector in small islands: evidence of Maluku, Indonesia. *Aquaculture, Aquarium, Conservation & Legislation-International Journal of the Bioflux Society (AACL Bioflux)*, 9(3).

[11] Salagrama, V. and A. Salka 2010 A study of the fisheries post harvest and market supply chains in Nias Island, North Sumatra Province, Indonesia. FAO Nias Information Bulletin 1.*Food and Agriculture Organization of the United Nations*. 128 pp.

[12] de la Torre-Castro, M., Fröcklin, S., Börjesson, S., Okupnik, J., & Jiddawi, N. S. (2017). Gender analysis for better coastal management–Increasing our understanding of social-ecological seascapes. *Marine Policy*, 83, 62-74.

Characterization of chitin extracted from fish scales of marine fish species purchased from local markets in North Sulawesi, Indonesia

I.F.M. Rumengan[1], P. Suptijah[2], S. Wullur[1], and A. Talumepa[1]
Faculty of Fisheries and Marine Science, Sam Ratulangi University, Manado
Faculty of Fisheries and Marine Science, Bogor Institute of Agriculture, Bogor

Email: innekerumengan@unsrat.ac.id

Abstract Chitin is a biodegradable biopolymer with a variety of commercial applications, including in the food food-supplement industries as a marine-derived nutraceutical. The purpose of this study was to characterize the molecular structure of chitin extracted from fish scales of important marine fish purchased from local markets in North Sulawesi. Chitin compound material was obtained from a specific fish scale, and then sequentially carrying out a boiling treatment to separate it from a complex with collagen. From the scales of two fish species, parrotfish (*Chlorurus sordidus*) and red snapper (*Lutjanus argentimaculatus*), the rendemen of chitin obtained were 45 % and 33%, respectively. Structural characteristics of the chitin were discussed by FTIR (Fourier Transform Infrared) analysis data. FTIR analysis was done using infrared spectroscopy, which is the resulting spectrum represents the molecular absorption and transmission, creating a molecular fingerprint of the sample. The molecular structure of chitin, $C_{18}H_{26}N_2O_{10}$, where the hydroxyl group on the second carbon replaced by acetyl amide, was shown by the infrared spectra. In the infrared spectra, chitin from parrot fish scales indicated the amide band at 1627.13 cm^{-1}, and chitin from red snapper fish scales the amide band at 1648.09 cm^{-1} which are a typical one for marine chitin. The hydroxyl and amino bands at the ranged spectra up to 3500 cm^{-1}. The yields of chitin isolated from fish scale were relatively huge. Some treatments are necessary to confirm the molecular conformation and deacetylation behavior. All products from the extraction of fish scales could be more accessible for structural modifications to develop biocompatible materials for pharmaceutical purposes.

1. Introduction

Chitin is potential biomaterial for biotechnological industries and tissue engineering, due to some characteristics, including their polyelectrolyte and cationic nature, the presence of reactive groups, high adsorption capacities, bacteriostatic and fungistatic influences [1,2,3,4]. In Indonesia, an interest of the potential of marine-derived chitin has been mainly due to the concern on the environmental problems regarding the disposal of marine processing shellfish wastes consisting of crustacean exoskeletons.

Depending on the source, chitin can occur in the α-, β- and γ-forms. The differences among them depend on the arrangement of chains of the crystalline regions [5]. Because of these differences, each chitin polymorph has different properties specific to it. In most cases, crystallinity index provides information about the crystal state, but it is also very useful for distinguishing α-chitin from β-chitin. The crystallinity index (CI) can also be calculated by X-ray diffractograms. The molecular structure could be characterized by infrared spectrum, in which different bonds or functional groups absorb a different range of infrared with a certain wavelength. The spectra are associated with the different vibration that occurs after deacetylation process, and therefore can be used for calculation of the degree of deacetylation (DD) as an important property that affects the biodegradability and immunological activity [6], because this parameter has effects on solubility, chemical reactivity, and biodegradability. Depending on the source and preparation procedure, DD may range from 30% to 95% [7].

Fish scales are wastes of seafood restaurants and fish markets. Fish scales could be a better source of chitin production. The extracted chitin can be used to produce chitin-derived products, such as chitosan and glucosamine. The purpose of this study was to characterize the molecular structure of chitin extracted from marine fish scales by using FTIR data. This information would be useful for further developing it as a nutraceutical compound for food industries, in particular as natural preservatives.

2. Material and Methods

The fish scales of two marine fish species, parrotfish (*Chlorurus sordidus*) and red snapper (*Lutjanus argentimaculatus*) were collected from local markets in North Sulawesi. The fish scales were washed and sun-dried for two days. The procedure of chitin extraction was adopted for collagen extraction with modification. The pre-treatment step was carried out using NaOH 0.5 M solution for 10 hours, following with hydrolysis with HCl 0.75 M solution for 24 hours. The solution was then neutralized with destilled water, and heated at 40ºC for 2 hours. The residue was separated from solution, washed with distilled water and then re-heated at 80ºC for 2 hours. The materials remaining after physical process were chitin.

Infrared spectroscopic analysis was adopted [8, 9]. Chitin samples were prepared in a potassium bromide (KBr) disk and film. Approximately 40-60 mg of chitosan powder and 120 mg of KBr were blended and triturated with agate mortar and pestle for 10 min. Approximately 40 mg of the mixture were compacted using a IR hydraulic press at a pressure of 8 tons for 60 seconds. The disk was conditioned in a desiccator placed in an oven at 80^0C for 16 hr before analysis. Sample was then inserted into ZnSe ATR cell.

3. Result and Discussion

Fish scale chitin was typically associated with a complex matrix with collagen. This collagen was not easily removed during deproteinization step. To provide a method separating a collagen, chitin compound material was separated during hydro-extraction with distilled water at 80^0C after repeatedly washed with distilled water for neutralization.

Christalinity of chitin and chitosan was generated from hydrogen bond between corresponding hydroxyl and N-acetyl groups [10]. Each crystalline peak characterizes the crystallographic the structure, which is generated from parallel and antiparallel alignments of polymeric chains or sheets. Semi-crystalline chitin and chitosan have amorphous and crystalline regions [11].

The results of FTIR spectra of fish scales are shown in Fig. 1. The IR spectra of the fish scales chitin was compared to the IR spectrum of commercial as shown in Table 1. The IR spectra of chitin of *C. sordidus* and *L. argentimaculatus* fish scales displayed similar IR spectra to that of the commercial chitin.

Table 1. Functional groups of two fish scale samples, *C. sordidus* and *L. argentimaculatus* compared to a commercial standard chitin [12].

Group	Wave length (cm^{-1}) Standard chitin	Parrot fish *C. sordidus*	Red snapper *L. argentimaculatus*
OH	3450	3437.67	3436.31
N-H stretching	3300-3250		
C-H stretching	2891,1	2925.19	2145.36
/C=O stretching	1680-1660	1627.13	1648.09
N-H bending	1560-1530	1558.41	
CH_3	1419,5	1380.63	
C-O-C	1072,3	1075.51	1082.35
N-H	750-650	602.14	606.11

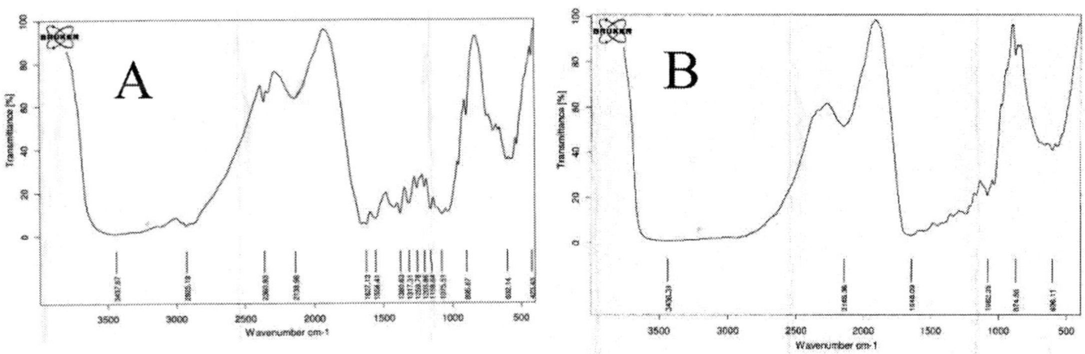

Figure.1. Infrared spectra as a wave length function displayed by fish scale chitin of parrot fish (*C. sordidus*) (A) and red snapper (*L. argentimaculatus* (B)

The FTIR spectra of both fish scale chitin exhibited a characteristic band at around 3436-3438 cm^{-1} is attributed to –NH and –OH groups stretching vibration and the band 2145-2826 cm^{-1} were an aliphatic C-H stretching bands that converges to OH stretching with N-H. The characteristic carbonyl C=O stretching of chitin at 1627 cm^{-1} are attributed to the vibrations of the amide I band. The sharp band at 1380 cm^{-1} from *C. sordidus* chitin corresponds to a symmetrical deformation of the CH_3 group. The vibrations bands at 1075-1082 cm^{-1} showed C-O-C vibration inside chitin ring and produced many peaks caused by the presence of hydroxide from chitin which contains a single bond C=O [12].
Chitin $C_{18}H_{26}N_2O_{10}$, like cellulose is a β-(1,4)-linked glycan, but is composed of 2-acetamida-2-deoxy-D-glucose (N-acetylglucosamine), while chitosan is a low acetyl form of chitin and is composed primarily of glucosamine, 2-amino-2-deoxy-D-glucose [13]. In the infrared spectra, the fish scales chitin indicated the amide II band at 1558.41 cm^{-1}, a typical one for marine chitin. A typical amide I band was found at 1627-1648 cm^{-1}.

4. Conclusion

The present findings prove that molecular structure of this marine-derived chitin as a high molecular weight polymer has a linear polyamine whose amino groups are readily available for chemical reactions [13]. For future works, it remains to explore its physical and chemical properties to develop an applicable nutraceutical biomaterial for food industries.

Acknowledgements
This work was a part of the research project sponsored by the Research Technology and Higher Education Ministry with a research scheme, MP3EI for fiscal year 2016-2017. The authors thank the Directorate Research and Community Service staff and the involved board of the ministry for financial support.

References
[1] Islam, Md. M., Shah. Md. M., Rahman, M, M., Molla, Md.a. I., Shaikh, A. A., Roy, S.K 2011 Preparation of Chitosan from Shrimp Shell and Investigation of Its Properties. International *Journal of Basic & Applied Sciences*. 11(1). 116-130 pp.
[2] Hasri 2010 Prospek kitosan dan kitosan termodifikasi sebagai biopolimer alami yang menjanjikan. *Jurnal Chemical*, 11(2). Hal 1-10.
[3] Khoushab, F. and Yamabhai, M 2010 Chitin Research Revisited. *Marine Drugs*, 8(7). 1988-2012 pp.
[4] Aranaz, I., Mengíbar, M., Harris, R., Paños, I., Miralles, B., Acosta, N., Galed, G. and Heras, H 2009 Functional characterization of chitin and chitosan. *Chemical Biology*, 3. 203-230 pp.
[5] Jang, M.K., Kong, B.G., Jeong, Y.I., Lee, C.H., Nah, J.W 2004 Physicochemical characterization of α-chitin, β-chitin and γ-chitin separated from natural resources. J. Polym. Sci. Part A Polym. *Chem.* 42, 3423–3432 pp
[6] Fernandez-Kim 2004 Physicochemical and functional propertiesof crawfish chitosan as affected by different processing protocols. Unpublished thesis (MSc), Louisiana State University.
[7] Synowiecki, J., Al-Khateeb, N.A 2003 Production, Properties, and Some New Applications of Chitin and Its Derivatives. *Crit. Rev. Food Sci. Nutr.* 43, 145–17 pp.
[8] Khan. T. A., Peh, K. K., Ch'ng. H. S 2002 Reporting degree of deacetylation values of chitosan: the influence of analytical methods. *Journal Pharmaceutical Science*. 5(3), 205-212 pp.
[9] Xu Y.X., K.M. Kim, M.A. Hanna, and D. Nag 2005 Chitosan–starch composite film:preparation and characterization. *Industrial Crops and Products* 21. 185–192 pp.
[10] Bartnicki-Garcia, S 1988 The biochemical cytology of chitin and chitosan synthesis in fungi. In: Chitin and chitosan (G. Skjå-Bræk, T. Amthonsen, and P. Sandford,eds). Proceedings from the 4th International Conference on Chitin and Chitosan held in Trondheoim, Norway, August 1988. Elsevier Application Science. 23-35 pp.
[11] Jung, J 2013 New Development of β-Chitosan from Jumbo Squid Pens (Dosidicus gigas) and its Structural, Physicochemical, and Biological Properties. Oregon State University. Dissertation. 1-210 pp.
[12] Puspawati, N.M. dan Simpen, I. N 2010 Optimasi deasetilasi kitin dari kulit udang dan cangkang kepiting limbah restoran seafood menjadi kitosan melalui variasi konsentrasi NaOH. *Jurnal Kimia*, 4 (1). Hal 79-90.
[13] Sanford, P.A 1988. Chitosan: commercial uses and potential applications. Proceedings, Elsevier Appl. Sc. London. *In*: Chtin and chitosan by G. Skjåk-Bræk , T. Anthonsen and P. Sanford (eds). 51-69 pp

IOP Conf. Series: Earth and Environmental Science **89** (2017) 012029 doi:10.1088/1755-1315/89/1/012029

Sound characteristics of *Terapon jorbua* as a response to temperature changes

Amron[1,2], I Jaya[2], T Hestirianoto[2], K v Juterzenka[2]

[1]Marine Science Department, Jenderal Soedirman University, Kampus Unsoed Karangwangkal, Purwokerto 53122 Indonesia
[2]Marine Science and Technology Department, Bogor Agricultural University, Kampus IPB Darmaga, Bogor 16610 Indonesia

Abstract. The change of water temperature has potential impact on the behavior of aquatic animal including fish which generated by their sound productivity and characteristics. This research aimed to study the response of sound productivity and characteristics of *Terapon jorbua* to temperature change. As a response to temperature increase, *T. jorbua* to have decreased the number of sound productivity. Two characteristic parameters of fish sound, i.e. intensity and frequency as were quadratic increased during the water temperature rises. In contrast, pulse duration was quadratic decreased.

1. Introduction

Water condition has an important role for aquatic organism. The change of water parameter may influence the existence of the biota. Temperature is one of the main parameters that could potentially elevate during the global climate change. The global impact of the climate change is increasing in global mean temperature (GMT), which approximately increased 3–4°C has impact to many aspects such as sea level rise, agriculture, water resources, human health, energy, terrestrial ecosystems productivity, forestry, biodiversity, and marine ecosystems productivity [1]. Potential impacts of global warming to the water ecosystem and resources have been studied. [2-6] are investigated the global warming impact to aquatic ecosystem, from the polar terrestrial to marine ecosystem. The other study, [7-12] are studied the impact of global warming on the benthic, fish, and other organism communities. The most important factor that contributes to ecosystem changes is the rise of sea temperature as global warming consequences.

Almost of freshwater and marine fishes are exotherms that cannot regulate their body temperature through physiological means and whose body temperatures are virtually identical to their environmental temperatures [13]. Much of the research on the behavior of fish as a response to temperature changes have tended to focus on generally fixed patterns of behavior [14]. However, most water environment characteristics especially temperature tends to be an important factor to change the fish behavior. [15] Analyzed that dial temperature could be changes the migration behavior and orientation of chum salmon. [16] Founded that vertical movement and behavior of Pacific bluefin tuna are depended by temperatures gradient in the vertical water column. [17] Studied and concluded that increases in stream temperatures in Arizona had reduced the habitat available for native fishes and therefore had favor those non-native species with higher thermal tolerances.

The alteration of fish behavior may associate with a variety of fish production. During reproductive, territorial, agonistic, aggressive, social and feeding, the sound production by fish may have different characteristics [18]. Sound production of fish plays an important role in studying fish behavior because the performance of fish is difficult to be conducted and studied in the ocean. Therefore, analyzing fish sound might be a possible way to understand their behavior when some

IOP Conf. Series: Earth and Environmental Science **89** (2017) 012029 doi:10.1088/1755-1315/89/1/012029

fishes could emit specific sound in frequency, amplitudes, and other acoustic characteristics. Study on acoustic behavior of marine fishes due to the rise of temperature could be developed, especially to stenothermal (narrow thermal range) species, like terapon. The species that included in the family of Teraponidae or terapons (thornfishes, grunters, or tiger perches) has tolerance range in sea temperature 25.86-29.04°C and salinity 31.94-35.18 psu (www.fishbase.org).

2. Material and method

2.1 Materials

Terapon jorbua harvested fishermen in Pelabuhan Ratu Bay, Indian Ocean was transported and acclimated in laboratory. Fish was fed by artificial feed (Super-Vit®), with a chemical composition are moisture 12%; protein 30%; fat 6%; and dietary fiber 6%. During the experimental realization, water was circulated 30 ml/s using external water pumps and filtered by external filters.

2.2 Experimental setting

Two specimens of *T. jorbua* in 80-100 mm size ($1B_1$ and $1B_2$) was placed in each of small aquarium (50 × 25 × 30 cm; length × width × height; water depth 25 cm). In the other similar size of aquarium, two specimen fishes ($2B_1$ and $2B_2$) with similar size were placed. In this study, two replications were conducted to minimalize the random error (Figure 1). Water temperature is increased ± 2°C every 2 days by using the water heater. It was 29 ± 1°C in the starting period, and 33 ± 1°C in the last period. A light condition in the laboratory reis adjusted to the environmental conditions during the day and night period without additional lighting.

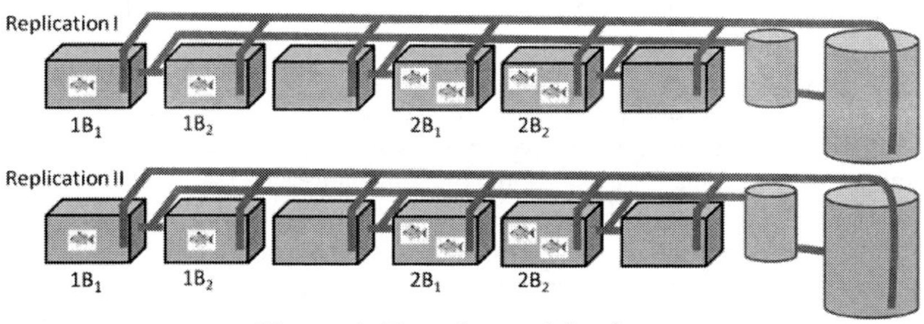

Figure 1. Experimental Setting

2.3 Data acquisition

The video system is used to takes data of fish behavior during resting, territorial, aggressive, and social behavior. The video system was placed in front of the aquarium. The passive acoustic technique was used to measures sound production of fish as long as the underwater camera recorded. Commonly instrumentation was used to convert the production of the sound from fishes into a voltage that can recored and analyzed is a hydrophone [19]. The hydrophone was positioned in the center of aquarium. Both the video camera and hydrophone was connected to personal computer to record audio and visual signal synchronously.

2.4 Data analysis

Sound productivity based on fish behavior in every observation time was analyzed descriptively to explain the relationship between sound productivity to water temperature. Variability of sound characteristics during the rise of water temperature was analyzed by curve fitting.

3. Results and Discussion

3.1 Results

3.1.1 Response of sound productivity of fish to temperature change. One of sound parameters which can be measured to representing the changes of fish behavior as an impact of climate changes is sound productivity. The optimum temperature of *T jorbua* in nature habitat is 26 – 29°C, however, that was increased during the climate changes and has influenced the sound productivity of fish. Simulation results showed the sound productivity of fish was significantly decreased during the level of water temperature increased after the optimum temperature (Figure 1). Based on two sound types produced by *T. jorbua*, the productivity of "click" sound was changed compared to "frog" sound.

The alteration of click sound productivity happened in all observation period, dawn (05 - 07 h), peak of the day (11 – 13 h), dusk (17 – 19 h), even in peak of the night period (23 – 01 h). In the dawn period, sound productivity at temperature 29 – 30°C rose to the peak when it neared at 06 - 07 h (Figure 2a).

In the same way when the temperature was 31 – 32°C, sound productivity achieved the peak at 06 h. On the other hand, sound productivity dropped at nearly 07 h. The changes of sound productivity were significantly shown when the temperature rose up to 33 – 34°C. This situation appeared when the fish experienced similar sound productivity in this period. Therefore, there were almost none of sound productivity peak at the water temperature 29 – 30°C and 31 – 32°C. In the peak of the day (temperature 29 – 30°C), the sound productivity was relatively high during this period (Figure 2b). Sound productivity decreased was displayed when the temperature rose up to 31 – 32°C, it was relatively low before achieved the peak of the day (12 h), and rose afterward. Contrast sound productivity revealed when the temperature achieved 33 - 34°C, the sound productivity decreased during this period. The tendency of sound productivity pattern that decreased during the water temperature increases also happened at dusk and the peak of night period. In the other hand, sound productivity was relatively high at temperature 31 – 32°C during at dawn and peak of the day period (Figure 2c, and 2d). Sound productivity was decreased when the temperature climbs up at 31 - 32°C, and it was continuously decreased at temperature 33 -34°C.

3.1.2 Response of sound characteristics of fish to temperature changes. As similar with *T. jorbua* sound productivity, sound characteristics (intensity, frequency, and duration) were also changed along the rise of water temperature (Figure 3). Sound characteristics either click or frog sound had a quadratic increase during the increase of water temperature. Even if the intensity, frequency, and duration of frog sound were higher than click sound, both of the sound types had almost similar increased pattern.

The intensity of fish sound either click or frog sound had quadratic elevation along the rise of water temperature (Figure 3a). The increase of intensity average was first shown when the temperature elevates from 29 – 30°C to 31 – 32°C, and variability of intensity revealed when water temperature elevates up to 33 – 34°C. Simulation results indicated that fish sound intensity would simultaneously escalate along with the elevation pattern of intensity when the water temperature increased. The intensity of click sound which produced by the fish had sharper elevation than frog sound, especially at water temperature below 34°C.

Another sound characteristic which also had similar elevation pattern along the rise of water temperature was frequency (Figure 3b). Fish sound frequency either click or frog sound had quadratic increased along with the escalation of water temperature. A sharp increase in frequency exhibited when the water temperature rose up to 33 – 34°C, and continuously showed considerable elevation along with the addition of water temperature. On the other hand, the intensity and frequency of sound either click or frog had relatively similar pattern.

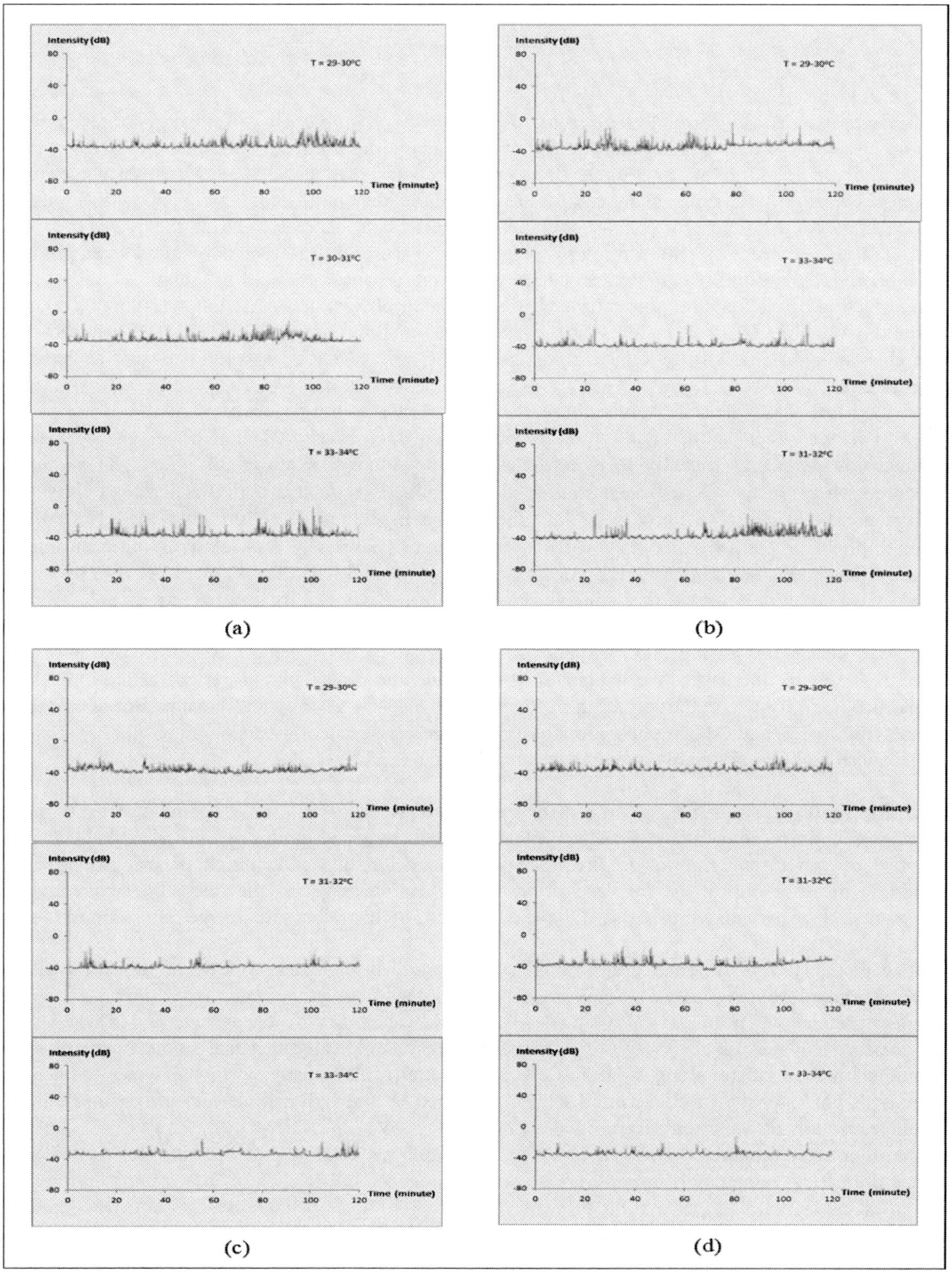

Figure 2 Sound productivity of *Terapon jorbua* recorded in temperature level: 29-30, 31-32, and 33-34°C. (a), (b), (c), and (d) respectively represents recording time at 05-07, 11-13, 17-19, dan 23-01 h

There were significant differences eein intensity and frequency, meanwhile pulse duration had quadratic decreased along with water temperature increased (3c). Pulse duration either click or frog sound was dropped when the water temperature rose up to 33 – 34°C. Pulse duration continuously depleted along with the increase of water temperature. A different pattern of depletion of pulse duration revealed between click and frog sound, whereas click sound had sharper decrease pattern than frog sound. The larger range of pulse duration of frog sound tends to cause it sharper elevation than the pulse duration of a click sound.

3.2 Discussion

The decrease of sound productivity of *T. jorbua* during the rise of water temperature was highly related to changes in their behavior. One reason that possible to describe this phenomenon is the influence of temperature on the muscle ontogeny of the fish [20]. Different temperatures lead to variation in body muscle mass (regarofding size and number of muscle fibers) that directly affect the swimming ability. In the case of zebrafish, Jhonston *at al.*, (2009) revealed that the optimal embryonic temperature for fast muscle fiber recruitment (hyperplasia) is at 26°C, as in that temperature there were 18.8% more fast fibers than at 22°C and 13.7% more fibers than at 31°C. The Increase of swimming ability changed the territorial behavior to aggressive behavior [20]. As the consequences, this change affected the decrease of productivity of click sound.

Besides changing the sound productivity, the rise in water temperature also has impacted the sound characteristics of *T. jorbua*. The elevation of water temperature aeffects to raise the intensity, frequency, and to decrease the pulse duration. This condition is similar to Connaugton *et al.*, (2000) investigation, reported that sound pressure level, repetition rate, and dominant frequency of weakfish are increase, but pulse duration is decrease during water temperature increases [21]. With an elevation in water temperature, the rate of exchange of calcium ions in tympanic muscles responsible for sound generation increases [22], and, as a consequence, the schedule of their work changes substantially, which, in turn, is reflected in most parameters of sounds produced by fish. In *Polimyrus marianne* and *P. adspersus*, *Q*10 coefficient is close to 2.07 for the amplitude of the sound of the moan type and for the duration and frequency of a sequence of pulses in grunt sounds [23]. A decrease in the duration of acoustic pulses, an increase in the rate of their generation, and an increase in the frequency of the sound and the level of acoustic pressure that accompany water temperature increase were found in many fish [24, 21]. The nature of these changes is related to temperature effects on the work of the center regulating acoustic activity and is common for all fish [24]. The intensification of sound generation with water temperature increase takes place parallel to an increase in the behavioral activity of fish [25].

A distinctly pronounced positive linear dependence on temperature is also exhibited by the level of acoustic pressure of anxiety sounds and their dominant frequency while the duration of pulses with an increase in temperature successively decreases [21]. The dominant (basic) frequency of sounds of males of plainfin midshipman *P. notatus*, oyster toadfish *O. tau*, and Lusitanian toadfish *H. didactylus* depends on water temperature similarly [26, 27]. In *Prionotus carolinus*, basic frequency increases by 43 Hz with an increase in temperature by 2.5°C [28]. In *Eutrigla gurnardus*, the duration of pulses in sounds of the grunt type noticeably decreases with a seasonal increase in water temperature [29].

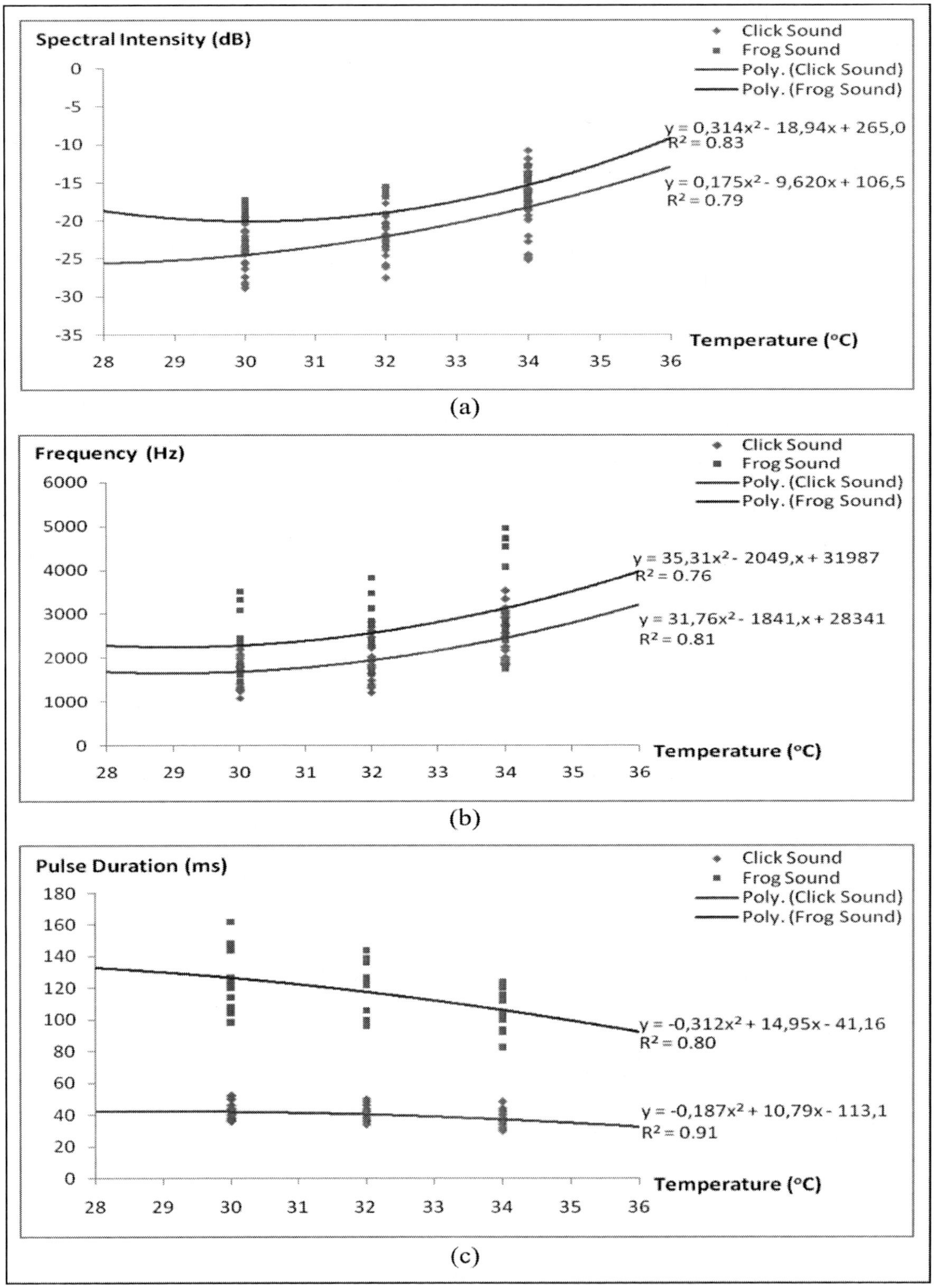

Figure 3 Sound characteristics of *Terapon jorbua* recorded in temperature level: 29-30, 31-32, and 33-34°C. (a) Intensity, (b) frequency, and (c), pulse duration

4. Conclusion

Increases in water temperature have an impact on the existence of *Terapon jorbua* which can be represented by the changes in productivity and characteristics of sound that they produced. The sound productivity of fish decreased during the increase of the water temperature. The sound productivity of fish decreased during the rise of water temperature. Sound characteristics that have been quadratic increases along the increase of water temperature were intensity and frequency. Meanwhile, another characteristic, pulse duration has quadratic decreased due to the rise of water temperature. Temperature range of this study still very limited caused it was difficult to predict the stress threshold of fish. Therefore further investigation by giving preferential treatment in wider temperature range is needed.

Acknowledgement

Authors gratefully acknowledge to Directorate General of Higher Education – Indonesian Ministry of Education and Cultural (DIKTI – KEMENDIKBUD) for financial support of this research (Hibah Doktor).

References

[1] Hitz, S., and J. Smith, 2004. Estimating Gobal Impacts from Climate Change. *Global Environmental Change*, 14:201-218.

[2] Meyer, J.L., M.J. Sale, P.J. Mulholand, and N.L. Poff, 1999. Impacts of Climate Change on Aquatic Ecosystem Functioning and Health. *Journal of the American Water Resources Association*, 35(6):1373-1386.

[3] Walther, G.R., E. Post, P. Convey, A. Menzel, C. Parmesank, T.J.C. Beebee, J.M. Fromentin, O. Hoegh-Guldberg, and F. Bairlein, 2002. Ecological Responses to Recent Climate Change. *Nature,* 416:389–395.

[4] Soto, C.G., 2002. The potential Impacts of Global Climate Change on Marine Protected Area. *Reviews in Fish Biology and Fisheries*, 11:181-195.

[5] Beuhler, M., 2003. Potential Impacts of Global Warming on Water Resources in Southern California. *Water Science and Technology*, 47: 165-168.

[6] Hawkins, S.J., A.J. Southward, and M.J. Genner, 2003. Detection of Environmental Change in a Marine Ecosystem-evidence from the Western English Channel. *The Science of the Total Environment*, 310:245-256.

[7] Schiel, D.R., J.R. Steinbeck, and M.S. Foster, 2004. Ten Years of Induced Ocean Warming Causes Comprehensive Changes in Marine Benthic Communities. *Ecology*, 85(7):1833-1839.

[8] Coyle, K.O, B. Konar, A. Blanchard, R.C. Highsmith, J. Carroll, M. Carroll, S.G. Denisenko, andB.I. Sirenko, 2007. Potential Effects of Temperature on the Benthic Infaunal Community on the Southeastern Bering Sea shelf: Possible Impacts of Climate Change. *Deep-Sea Research*, 54:2885–2905.

[9] Occhipinti-Ambrogi, A., 2007. Global Change and Marine Communities: Alien Species and Climate Change. *Marine Pollution Bulletin*, 55:342-352.

[10] Ficke, A.D., C.A. Myrick, and L.J. Hansen, 2007. Potential Impacts of Global Climate Change on Freshwater Fisheries. *Rev. Fish. Biol. Fisheries*, 17: 581-613.

[11] Brander, K., 2009. Impacts of Climate Change on Marine Ecosystems and Fisheries. *J. Mar. Biol. Ass. India*, 51(1):1-13.

[12] Planque, B., J. Fromentin, P. Cury, K.F. Drinkwater, S. Jennings, R.I. Perry, and S. Kifani, 2010. How does Fishing Alter Marine Populations and Ecosystems Sensitivity to Climate? *Journal of Marine Systems*, 79:403-417.

[13] Moyle, P.B., and J.J. Cech, 2004. *Fishes: an Introduction to Ichthyology, 5th edition.* Prentice Hall, Englewood Cliffs, New Jersey.

[14] Oddling-Smee, L., and V.A. Braitwite, 2003. The Role of Learning in Fish Orientation. *Fish and Fisheries* 4(2): 235-246.

[15] Friedland K.D. R.V. Walker, N.D. Davis, K.W. Myers, G.W. Boehlert, S Urawa, and Y. Ueno, 2001. Open-ocean Orientation and Return Migration Routes of Chum Salmon Based on Temperature Data from Data Storage Tags. *Mar ecol prog ser.*, 216: 235-252.

[16] Kitagawa T., H. Nakata, S. Kimura, T. Itoh, S. Tsuji, and A. Nitta, 2000. Effect of Ambient Temperature on the Vertical Distribution and Movement of Pacific Bluefin Tuna *Thunnus orientalis. Mar Ecol Prog Ser.*, 206:251–260

[17] Carveth, C.J., A.M. Widmer, and S.A. Bonar, 2006. Comparison of Upper Thermal Tolerances of Native and Nonnative Fish Species in Arizona. *Transactions of the American Fisheries Society*, 135:1433-1440.

[18] Kasumyan, A.O., 2008. Sounds and sound production in fishes. *J. Ichthyol.*, 48(11):981-1030.

[19] Mann D.A., P.A. Cott, B.W. Hanna, and A.N. Popper, 2007. Hearing in Eight Species of Northern Canadian Freshwater Fishes. *J. fish biol.*, 70(1): 109-120.

[20] Johnston, I.A., H.T. Lee, D.J. Macqueen, K. Paaranthaman, C. Kawashima, A. Anwar, Jr. Kinghorn, and T. Dalmay, 2009. Embrionic Temperature Affects Muscle Fiber Recuitment in Adult Zebrafish: Genome-wide Changes in Gene and MicroRNA Expression Assosiated with the Transition from Hyperplastic to Hypertrophic Growth Phenotypes. *J. Exp. Biol*, 212(Pt12):1781-1793.

[21] Connaughton M.A., M.H. Taylor, and M.L. Fine, 2000. Effects of Fish Size and Temperature on Weakfish Disturbance Calls: Implications for the Mechanism of Sound Generation. *J. Exp. Biol.*, 203:1503–1512.

[22] Feher, J.J., T.D. Waybright, and M.L. Fine, 1998. Comparison of Sarcoplasmic Reticulum Capabilities in Toadfish (*Opsanus tau*) Sonic Muscle and Rat Fast Twitch Muscle. *J. Muscle Res. Cell Motil.*, 19:661–674.

[23] Crawford, J.D., 1997. Feature-detecting Auditory Neurons in the Brain of a Sound-producing Fish. *J. Comp. Physiol.*, 180:439–450.

[24] Bass, A.H., and R. Baker, 1991. Evolution of Homologous Vocal Control Traits. *Brain Behav. Evol.*, 38:240–254.

[25] Ladich, F., 1988. Sound Production by the Dudgeon *Gobio gobio* (a Common European Freshwater Fish (Cyprinidae, Teleostei). *J. Fish Biol.*, 32:707–715.

[26] Brantley, R.K., and A.H. Bass, 1994. Alternative Male Spawning Tactics an Acoustic Signals in the Plainfin Midshipman Fish *Porichthys notatus* Girard (Teleostei, Batrachoididae). *Ethology*, 96:213–232.

[27] Amorim, M.C.P., R.O. Vasconcelos, J.F. Marques, and F. Almada, 2006. Seasonal Variation of Sound Production in the Lusitanian Toadfish *Halobatrachus didactylus*. *J. Fish. Biol.*, 69:1892–1899.

[28] Connaughton, M.A., 2004. Sound Generation in the Sea Robin (*Prionotus carolinus*), a Fish with Alternate Sonic Muscle Contraction. *J. Exp. Biol.*, 207:1643–1654.

[29] Amorim, M.C.P., 2005. Dial and Seasonal Variations of Sound Production in Captivity Grey Gurnards *Eutrigla gurnardus*. *Acta Zool. Sinica*, 51(6):1013–1022.

Dispersion, Speciation, and Pollution Assessment of Heavy Metals Pb and Zn in Surface Sediment from Disturbed Ecosystem of Jeneberang Waters

Najamuddin[1], Surahman[2]

[1]Marine Science Study Program, Khairun University, Ternate 97719, Indonesia
[2]Fisheries Resources Management Study Program, Khairun University, Ternate 97719, Indonesia

Email: najamuddin313@gmail.com

Abstract Surface sediments were collected from seventeen stations in Jeneberang waters (riverine, estuarine, and marine). Lead (Pb) and zinc (Zn) concentrations were determined by atomic absorption spectrometry, and the speciation of metals was obtained by a sequential extraction procedure. Dispersion of Pb and Zn were found higher in the riverine and marine samples than the estuarine samples. Following speciation, the metals were found similar composition of fraction in the riverine and estuarine samples but any different in the marine samples. The results indicated that there is a change of dispersion pattern and speciation composition of metals due to the presence of the dam that lies at the boundary between the estuary and the river. The toxicity unit was indicated low toxicity level; pollution level was in weakly to moderately polluted while the aquatic environment risk attributed were no risky to light risk.

1. Introduction

The aquatic environmental contamination by heavy metals is a serious and worldwide problem due to their high toxicity, persistence, and bioaccumulation [1]; [2]. Despite the advances in sewage treatment technologies, water pollution caused by sewage discharge has still been a serious problem in many developing countries due to the high operating cost of sewage treatment facilities as well as lack of effective management and enforcement (e.g. [3]; [4]; [5].

Aquatic environment particularly estuarine and coastal is important zone due to its close relationship with human activities. These locations, the ultimate destination of terrestrial runoff, have become increasingly contaminated by the heavy metals in recent decades. The toxic effects of metals preserved in sediments seriously threaten the coastal ecosystems. Accordingly, determination of concentration and distribution of heavy metals deposited on the seabed is of important [6].

Sediments in the aquatic environment are sensitive indicators for assessment and monitoring heavy metal contaminant. Sediments not only serve as reservoirs for deleterious chemical species which cause negative effects on aquatic system and human health but also act as a source of metals for aquatic organisms via several pathways, including disposal of liquid effluents, terrestrial runoff, and leachates carrying chemicals originating from numerous urban, industrial and agricultural activities, as well as atmospheric deposition. Therefore, sediments reflect the current quality of the aquatic environment as well as provide information on the impact of pollution sources [1].

Concentration and speciation assessments of heavy metals in sediment are useful to study the pollution level in the aquatic environment and to provide basic information for determination of aquatic environmental health risks [1]. Consequently, the concentrations of heavy metals in sediment are often monitored to provide basic information for the environmental risk assessment [7]. Determination of the speciation of heavy metals in sediments is important to assess their potential toxicity, a threat to the ecosystem, and source (natural processes or anthropogenic sources) [8].

The number of studies on dispersion, speciation, and pollution assessment of heavy metals in sediment have been published [9]; [10]; [11]; [12]; [13] however, all the studies carried out at naturally (normally) ecosystem and none information is available at disturbed ecosystem. The change of ecosystem will drive the change physic-chemical parameters within the environment and they could influence process and dynamics of heavy metal caused by the change of interaction pattern of metals and environment. A seasonal variable can stimulate the change of distribution and geochemical fractionation of heavy metals [14]. Following this research, analyze the effect of ecosystem changes due to the presence of the dam in an estuary on dispersion and speciation of heavy metals lead (Pb) and zinc (Zn).

2. Materials and Methods

2.1. Study area

Makassar, as the largest city in eastern Indonesia, is undergoing rapid economic growth and industrialization. The Jeneberang River flows through Makassar City, including the upstream area in the Bawakaraeng Mountains, Gowa district and the downstream area in the Makassar Strait. The Jeneberang River flows through village and city regions and therefore has the potential for contamination by various metals from agricultural, domestic, and transportation sources. There is also potential intake of metals from marine sources such as the Makassar international seaport, ship docking and manufacturing, hospital, and industrial sectors [14].

Jeneberang estuary is a disturbed ecosystem where around 3 km before the estuary there is a dam utilized by local water company for the material source. The dam changed flows pattern and mixing between fresh water and seawater in the estuary zone. The change will influence the physic-chemical parameters within the environment such as pH, temperature, and redox potential (Eh), could influence metal release from the solid to liquid phases that transformed sediments from the main metal sink to the sources for the overlying waters [2]. Moreover, there is a transport reduction of some suspended particles because of low water flow. Consequently, the flow of disposal and suspended particle from the rivers would be deposited along the dam.

2.2. Sediment sampling and pretreatment

A total of 17 surface sediment samples were collected from riverine, estuarine, and marine waters. The riverine water zone contained sampling stations R1–R3, the estuarine zone contained sampling stations E1–E7, and the coastal (marine) zone contained sampling stations CS1–CS7. Sample locations were determined by a global positioning system (Fig. 1).

The surface sediments were sampled using an Ekman grab sampler. At each sampling site, the top 0–10 cm of sediment were collected using a plastic spatula, and each sample was packed in a new polyethylene zipper bag. After collection, the samples were stored and transported to the laboratory in a cooler box with ice packs and then stored in a fridge at 4 °C before analysis. In the laboratory, samples were first sieved through a 2 mm nylon sieve to remove coarse debris and shell fragments, ground using an agate mortar and pestle in a fume cupboard, and then sieved through a 0.5 mm sieve to remove large particles and homogenize the mixture.

2.3. *Analysis of sediment properties and suspended particulate matter*

The texture of dry sediment samples was analyzed using a pipet procedure (after passing through a 63 µm sieve). Grain sizes from 2000 µm to 62.5 µm were classified as sand, those from 62.5 µm to 3.9 µm as silt, and those <3.9 µm as clay. The organic carbon (OC) content was analyzed using the Walkley-Black procedure and expressed as a weight percentage following sample combustion in a Loss on Ignition test. The suspended particulate matter (SPM) was measured using a gravimetric procedure, and the redox potential (Eh) of the sediment was measured using a portable Eh meter.

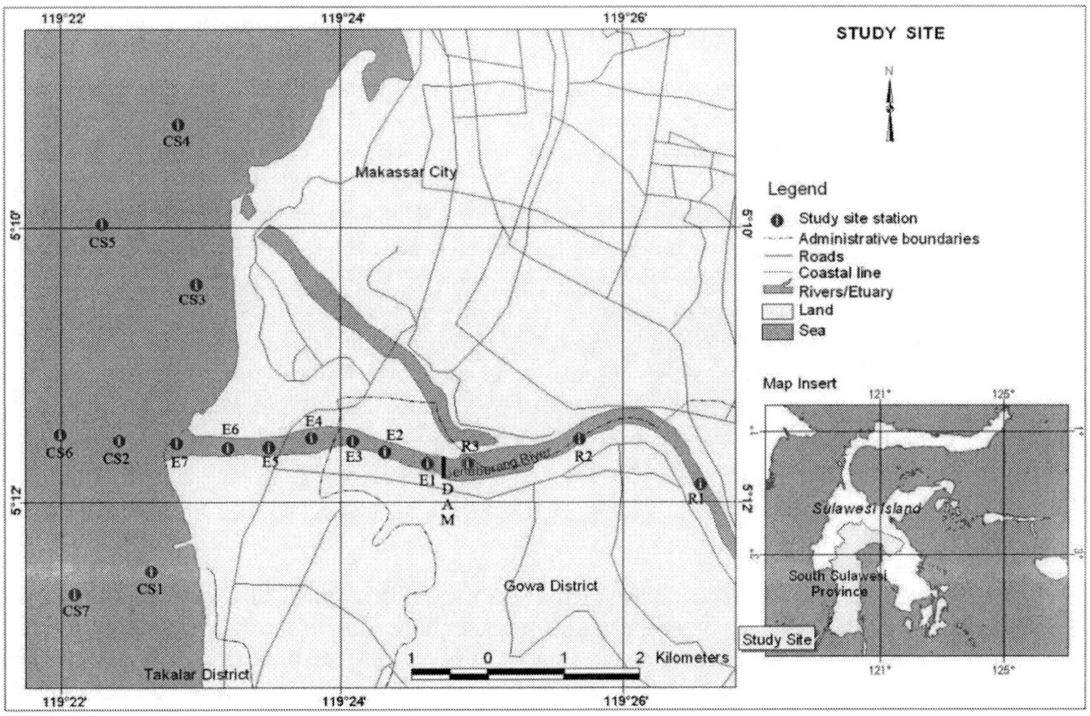

Figure 1. Location of sample sites

2.4. *Heavy metals analysis and geochemical fractionation*

The geochemical fraction of heavy metals was determined using sequential extraction procedure [15]. This method provides the exchangeable, bound to carbonates, bound to Fe-Mn oxides, bound to organic matter and residual fractions with the following procedures;

1. Exchangeable (F1). The sediment was extracted at room temperature for one h with 8 mL of either magnesium chloride solution (1 M $MgCl_2$, pH 7.0) or sodium acetate solution (1 M NaOAc, pH 8.2) with continuous agitation.
2. Bound to carbonates (F2). The residue from (i) was leached at room temperature with 8 mL of 1 M NaOAc adjusted to pH 5.0 with acetic acid (HOAc). Continuous agitation was maintained and the time necessary for complete extraction was evaluated.
3. Bound to Fe-Mn oxides (F3). The residue from (ii) was extracted with 20 mL of either 0.3 M $Na_2S_2O_4$ + 0.175 M Na-citrate + 0.025 M H-citrate or 0.04 M NH_2OH-HCl in 25% (v/v) HOAc. The latter experiments were performed at 96 ± 3 °C with occasional agitation and the time needed for complete dissolution of the free iron oxides was evaluated.
4. Bound to organic matter (F4). The residues from (iii) were added 3 mL of 0.02 M HNO3 and 5 mL of 30% H_2O_2 adjusted to pH 2 with HNO_3, and the mixture was heated to 85 ± 2 °C for 2 h

IOP Conf. Series: Earth and Environmental Science **89** (2017) 012030 doi:10.1088/1755-1315/89/1/012030

with occasional agitation. A second 3 mL aliquot of 30% H_2O_2 (pH 2 with HNO_3) was then added, and the sample was heated again to 85 ± 2 °C for 3 h with intermittent agitation. After cooling, 5 mL of 3.2 M NH_4OAc in 20% (v/v) HNO_3 was added, and the sample was diluted to 20 mL and agitated continuously for 30 min. The addition of NH_4OAc is designed to prevent adsorption of extracted metals onto the oxidized sediment.

5. Residual (F5). The residue from (iv) was digested with a HF-$HClO_4$ mixture according to the procedure described below for total metal analysis.

The "selective" extractions were conducted in centrifuge tubes (polypropylene, 50 mL) to minimize losses of solid material. Between each successive extraction, the separation was effected by centrifuging at 10,000 rpm (12,000 g) for 30 min. The supernatant was removed with a pipet and analyzed for trace metals, whereas the residue was washed with 8 mL of deionized water; after centrifugation for 30 min, this second supernatant was discarded. The volume of rinse water used was kept to a minimum to avoid excessive solubilization of solid material, particularly organic matter. Deionized water used in preparing stock solutions and in each step of the leaching procedure was obtained from a Millipore Milli-Q3RO/Milli-Q2 system. All glassware used for the experiments was previously soaked in 14% HNO_3 (v/v) and rinsed with deionized water.

For total or residual metal analysis, the solid was digested with a 5:l mixture of hydrofluoric and perchloric acids. For a 1 g (dry weight) sample, the sediment was first digested in a platinum crucible with a solution of concentrated $HClO_4$ (2 mL) and HF (10 mL) to near dryness; subsequently the second addition of $HClO_4$ (1 mL) and HF (10 mL) was made and again the mixture was evaporated to near dryness. Finally, $HClO_4$ (1 mL) alone was added, and the sample was evaporated until the appearance of white fumes. The residue was dissolved in 12 N HCl and diluted to 25 mL. All of the above metal-containing solutions were subsequently analyzed by graphite furnace atomic absorption spectrometry (Shimudzu-AA 7000).

2.5. *Quality control and assurance*

To guarantee the analytical data quality, laboratory quality control and quality assurance were implemented by utilizing standard operating procedures, calibration with standards, analysis of reagent blanks, recovery of spiked samples and analysis of replicates. The precision and accuracy of the analytical procedures were checked by recovery tests on spiked sediment samples. Adopted analytical procedures indicated 91% to 110% recoveries of the spiked metals and marine sediment reference material (NMIJ CRM 7302-a) were achieved from the tested samples, which agreed with certified values. The average differences between the certified and measured results were within 11% variability. Background correction and matrix interference were monitored through the analysis calibration checks. All analyses were duplicated to ensure the representativeness of the quantitative results, and the ultimate results were expressed as the mean and standard deviation. All reagents used in the experiment were an analytical-reagent grade or better. All glassware and plastic were pre-cleaned by soaking in 10% HNO_3 (v/v) for overnight.

2.6. *Data analysis*

Toxicity unit (TU), defined as the ratio of heavy metal concentration to probable effect level (PEL), was calculated to evaluate the toxicity of each heavy metals [16]. The potential acute toxicity of heavy metals was estimated by the sum of individual TU. Assessment of heavy metals pollution was analyzed by geoaccumulation index and risk assessment code. The geoaccumulation index (Igeo) is expressed as the following equation [17]:

$$Igeo = \log2\ [C_n/(1{,}5\ Bn)]$$

where Cn is the measured concentration of metal n in the sediment, and Bn is the geochemical background value of the metal n. Factor 1.5 compensates for possible fluctuations in the background values for a given substance in the environment, as well as very small anthropogenic influences. Müller defined seven classes in the geoaccumulation index:

Igeo ≤ 0 is class 0 (unpolluted);
0 < Igeo ≤ 1 is class 1 (weakly polluted);
1 < Igeo ≤ 2 is class 2 (moderately polluted);
2 < Igeo ≤ 3 is class 3 (moderately to strongly polluted);
3 < Igeo ≤ 4 is class 4 (strongly polluted);
4 < Igeo ≤ 5 is class 5 (strongly to extremely polluted);
Igeo > 5 is class 6 and extremely polluted.

Risk Assessment Code (RAC) is an assessment index of the risk of heavy metal existence on the aquatic environment based on the geochemical fraction of heavy metal interaction with sediment particle which is potentially absorbed by the organisms. Risk assessment index was obtained by calculating the percentage of exchangeable and carbonate fraction in sediment [18]; [19]. The index can be defined in five classes:

RAC<1 (not risky);
1<RAC≤10 (light risk);
11<RAC≤30 (medium risk);
31<RAC≤50 (high risk),
RAC>50 (very high risk).

3. Results and Discussion

3.1. Sediment properties and dispersion of heavy metals

Sediment texture was dominated by sand in the river and estuary zone, but in the marine zone, it was dominated by clay. There was a high water flow in the river and estuary from the upstream region which led to turbulence in the estuary, where larger particles (e.g., sand) deposited on the bottom. The increasing concentration of SPM in the water column was influenced by the high levels of erosion and water runoff. SPM binds to organic matter, and with greater levels of erosion, the OC concentration in sediments increases. A seaward decrease in Eh was caused by an increase in pH and the weak flow of the water mass. The decrease in Eh caused the number of electrons in sediment to decrease. The number of electrons was directly proportional to the Eh. This is following [20], who reported a strong correlation between pH and Eh. Based on the Eh, the study area was determined to be a reducing zone (anaerobic conditions). The sediment texture (clay, silt, and sand), OC, Eh, and SPM properties are shown in Table 1.

The dispersion of Pb and Zn were the highest concentrations found in the riverine and marine zones, while lowest concentrations were found in the estuarine zone The concentrations of Pb and Zn in sediment around the estuary of the Jeneberang waters are shown in Table 1. Higher Pb and Zn concentration of riverine and marine sediment indicated that the pollutant source of sediment was derived from mainland and marine. Pollutant source of elements derived from the mainland was in the particulate forms entering river basin and it settled and accumulated as sediment in water bottom.

Higher Pb and Zn concentration in the riverine zone were presumably due to a higher concentration of the particulates derived from soil erosion entering the river basin. The particulate that characteristically concentrates the heavy metals retained at the dam deposited to the water bottom, and, due to lower water flow, accumulated in the sediment. Beside rivers, the pollutant source of Pb and Zn can be found in the coastal area as indicated by a higher Pb and Zn concentration in the coastal zone compared to the estuary. The main source of Pb and Zn was

IOP Conf. Series: Earth and Environmental Science **89** (2017) 012030 doi:10.1088/1755-1315/89/1/012030

presumably originated from the seaport activities and traffic transportation in the shore of northern Jeneberang Estuary [21]; [22]; [23].

Heavy metal dispersion of sediment was found higher in estuary than that in riverine. Estuary commonly plays a role as natural pollutant filtration since there is smaller solubility change of heavy metals entering estuary waters. The reduced solubility was attributed to enhanced salinity and pH level leading heavy metals easily absorbed by particle then settled as sediment in water bottom [24]; [25]. However, the heavy metal concentration of sediment found in this study was at low concentration.

As different pattern found in this study that the heavy metal concentration was lower in estuary was presumably attributed to two factors; 1) the dam restricted the riverine flow leading to a reduction of water mass flow, therefore most particulate heavy metals were deposited along the dam, and there were only small amount entering the estuary; 2) there was a strong turbulence of water flow in the estuary thus it decreased the deposition rate of particle.

Table 1. Pb and Zn concentration from surface sediment and sediment properties in Jeneberang waters.

Sampling points	Sites group	Pb ($mg.kg^{-1}$)	Zn ($mg.kg^{-1}$)	Clay (%)	Silt (%)	Sand (%)	OC (%)	Eh (mV)	SPM (mg/L)
R1	Riverine	3.330	65.926	17	24	59	2.155	-23.970	88.0
R2	Riverine	3.114	73.102	23	15	62	2.170	-28.065	84.0
R3	Riverine	3.263	79.275	13	13	74	2.140	-35.305	78.5
E1	Estuarine	0.449	48.287	26	15	59	2.475	-10.520	153.5
E2	Estuarine	0.573	53.737	25	16	59	1.680	-12.075	137.0
E3	Estuarine	0.995	50.647	15	6	79	1.795	-9.360	108.5
E4	Estuarine	0.973	55.176	17	6	77	1.920	-25.685	72.5
E5	Estuarine	1.400	55.047	14	5	81	1.630	-20.565	109.3
E6	Estuarine	1.488	52.021	13	6	81	1.755	-12.825	134.5
E7	Estuarine	1.596	55.093	23	5	72	1.970	-25.590	101.3
CS1	Marine	1.733	55.363	23	14	63	1.660	-34.700	60.0
CS2	Marine	2.001	52.495	18	13	69	1.885	-35.300	84.8
CS3	Marine	2.316	59.935	22	35	43	2.820	-47.200	81.3
CS4	Marine	3.130	72.326	23	56	21	3.250	-55.035	37.3
CS5	Marine	2.729	65.289	68	8	24	2.755	-47.190	22.8
CS6	Marine	2.435	59.634	68	9	23	2.540	-52.980	22.3
CS7	Marine	2.252	57.410	66	9	25	2.110	-52.430	21.3

The existence of a dam at the boundary between river and estuary though does not change the amount of water volume into the estuary as the concept of Continuity Law but the dam affects the pattern of mixing of river water and sea water in estuary. The water flowing to estuary falls from the surface of the dam into the estuary system and then mixed evenly because of the dominance of sea water mass and without plume of river water flow in the estuary to the coastal waters.

The dam causes a change in the mass velocity of the water at the upper boundary of the dam to be small enough that the particle deposition becomes high. As a result, the transport and flux of the heavy particulate metal into the estuary reduced due to deposit with the particles. This indication is seen in data obtained from heavy particulate metal of Pb and Zn that high enough in the particles [26]. The change in the distribution of heavy metals showed a significant decrease in the concentration of particulate heavy metals. This is consistent with [27] which states that the behavior of heavy metals in waters is strongly influenced by the interaction between dissolved and particulate phases as well as metals inputs. The same is revealed by [28] that the temporal and spatial distribution of heavy metal strongly determined by local hydrodynamics plays a role in the dispersion of metals.

Table 2. Pearson's correlation coefficient between Pb, Zn, clay, silt, sand, OC, Eh, and SPM in Jeneberang waters.

	Pb	Zn	Clay	Silt	Sand	OC	Eh	SPM
Pb	1							
Zn	.873**	1						
Clay	.209	.053	1					
Silt	.431	.441	-.144	1				
Sand	-.445	-.313	-.803**	-.475	1			
OC	.508*	.479	.390	.706**	-.772**	1		
Eh	-.659**	-.493*	-.618**	-.428	.807**	-.659**	1	
SPM	-.627**	-.465	-.689**	-.198	.732**	-.481	.909**	1

**. Correlation is significant at the 0.01 level (2-tailed).

*. Correlation is significant at the 0.05 level (2-tailed).

The turbulence led to sediment resuspension resulting metal desorption from the sediment to the water column. Consequently, there was an increased element concentration dissolved in the zone as compared to riverine. Sediment resuspension caused desorption or release of heavy metals from the particle surface to the water column [29]; [30].

The sediment resuspension effect depends on the condition of the water bottom. As an example, there was an increase of nitrogen level that abruptly emerged as a reaction of the trapped water in the sediment (interstitial water) with the water column. However, the amount released depended on the diffusion gradient of long-term period. Sediment resuspension with clay character can increase the oxidation in the sediment increasing the potential redox level and reduce the organic materials, however it does not expose to an increased nutrient and oxygen consumption significantly in the water column [29].

The pattern of Pb and Zn dispersion in the sediment was similar to that in the particulate form. The results showed that heavy metals in the sediment where the deposition results of a particulate element that was further accumulated on the sediment [21]. The higher Zn concentration compared with Pb was attributed to a naturally increased intake of sediment particles from soil erosion and rock weathering on the mainland as a consequence of Zn concentration higher in soil and rock naturally. The heavy metal dispersion concentration was influenced by the sediment characteristics and concentration of suspended particulate matter (SPM). This was demonstrated by a correlation analysis of metal concentrations in the sediment, sediment characteristics, and the SPM. A strong correlation was found, as shown in Table 2. Heavy metal concentrations in the sediment were not only dependent on rock weathering but were also influenced by sediment deposition, the redox status, the OC concentration, and sediment texture. The results showed that the levels of SPM had the strongest correlation with heavy metals dispersion concentrations than in order sand and Eh.

Pb and Zn concentrations in this study area were approximately lower than several locations of Indonesian waters as reported by [31]; [32]; [9]. However, the results found in this study were by that reported by [23] and the other studies [33]; [34]; [35]; [36]; [8]; and [37]. The comparison of heavy metal concentration in this study with other reports is shown in Table 3.

Table 3. Comparison of Pb and Zn concentration (mg.kg^{-1}) of surficial sediments from different locations in Indonesia and some part of the world.

References	Locations	Pb	Zn
The present study	**Jeneberang Estuary**	**0.24-4.33**	**38.73-113,19**
[23]	Coastal of Makassar City	0,38-2,58	0-107,49
[31]	Tallo Estuary	13.25-83,01	40.02-294,76
[9]	Kendary Bay	0.84-17.02	-
[32]	Coastal of Semarang	37.80	161
[42]	Bangka Island	5,79	16,70
[43]	Coastal of Gresik	1,74-12,70	77,3-405,0
[44]	Kelabat Bay	11,46	13,64
[45]	Angke Estuary of Jakarta	18,64-29,57	56,58-69,30
[33]	Coast of Bohai Bay	16- 49	75- 288
[34]	Guadalquivir River Estuary	30–54	69–190
[11]	Caspian Sea, Iran	7.13-13.00	26.95-28.00
[12]	Changjiang River	13.5–35.5	47.2–124
[35]	Yellow Sea	4.31–21.9	18.8–96.2
[36]	Quintero Bay Chile	17–36	65–167
[10]	Coastal of Abu Dhabi	1.9	8.2
[8]	Yangtze River Estuary	28	70
[37]	Near-shore, north Shandong Peninsula	11.0–26.3	40.5–89.8

3.2. *Speciation of heavy metals Pb and Zn*

Speciation of heavy metals in sediment is very important to analyze as geochemically sediment condition expresses the process occurred in water column. The speciation of element could also be used as a basis to identify the source of element in aquatic environment. The speciation of heavy metals Pb and Zn from surface sediments of Jeneberang waters expressed as mean percentage is shown in Fig. 2.

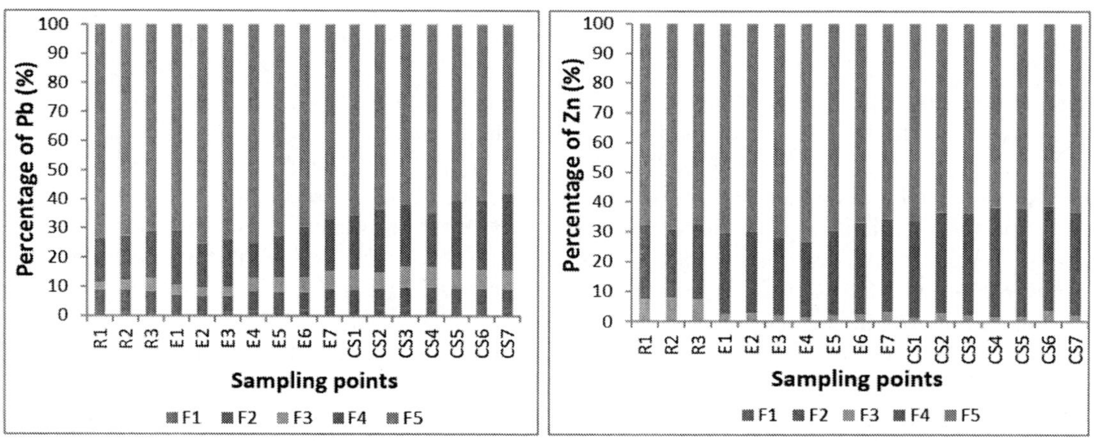

F1 (exchangeable), F2 (bound to carbonate), F3 (bound to Fe–Mn oxides), F4 (bound to organic), and F5 (residual)

Figure. 2 Percentage of speciation of Pb and Zn at Jeneberang waters

Measurement of total metal concentration is a fundamental aspect of sediment quality assessment. However, the metal fractionation in geochemical form is of crucial important that it had distinct mobility, migration ability, and chemical behavior. Therefore, sequential extraction procedure was proposed to study the strength and metal association with sediments [2]. The labile fraction of exchangeable are weak structures and could be easily broken down by ion exchange so that it is high reactivity and mobility leading to high bioavailability. The fractions of organic are medium to high mobility, while the fractions of carbonate and Mn-Fe oxides are medium mobility that the structure could be broken down as the alteration of redox and decomposition. The resistant components comprise the non-labile fraction that can only be disrupted by strong acids.

Most of the Pb and Zn in surface sediments was strongly retained in the residual fraction. Our results indicate that the metals were strongly bound to the crystal lattices of minerals and thus had relatively low mobility, bioavailability, and toxicity. Therefore, the existence of this fraction in sediments is not harmful to organisms. The heavy metal complexes can be broken down when the sediments are decomposed [38]; [2]. The resistant fractions of Pb and Zn were found to be dominant in the study area, indicating that the main sources of Pb and Zn were from natural processes, such as rock weathering or soil erosion on the mainland and runoff entering the waters. The resistant components comprise the non-labile fraction that can only be disrupted by strong acids [39].

The small percentage of labile materials in the carbonate and exchangeable fractions found in this study indicated that the pollutant intake derived from human activity (anthropogenic sources) was lower than that from natural sources. This is by [38] and [40], who reported that the labile fractions associated with sediments commonly originate from anthropogenic sources, while the non-labile fractions commonly originate from natural processes. Even though, in the study area found potential metal intake from anthropogenic sources (presumably originating from seaport activities, transportation, the hospital, and industrial sites along the northern shore of the Jeneberang River estuary), however the stronger currents around the coastal area, the metals also tended to be present in the water column rather than deposited on the water bottom. The metals were considered to have been desorbed from the surface of the particles, due to the strong water turbulence. This is following [29] and [30], who both reported that sediment resuspension leads to elemental desorption into the water column.

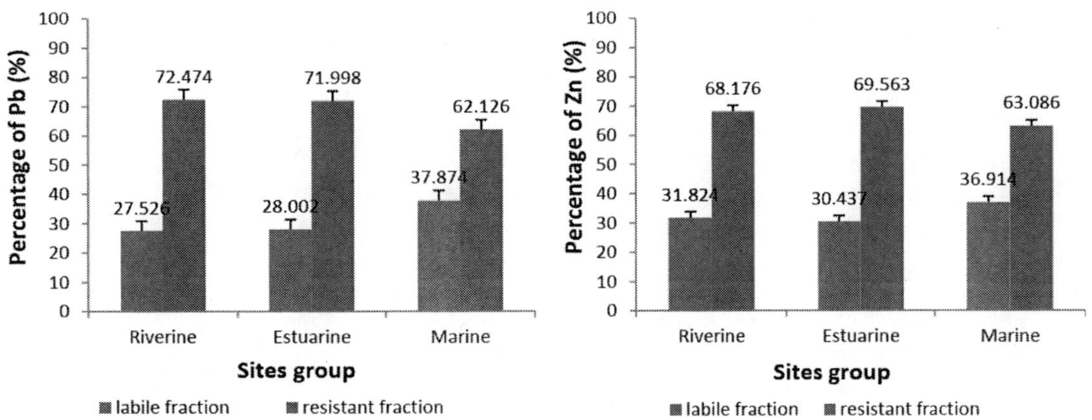

Figure. 3 The average percentage of geochemical fractionation (labile fraction and resistant fraction) of heavy metals Pb and Zn in based on sites group.

Most metals introduced by human activity are present in the exchangeable and carbonate fractions. The metals in the acid-soluble fraction are considered to be bound the weakest, and therefore the most mobile metals, in sediments and are defined as the exchangeable and carbonate fractions. Metals in these fractions have greater bioavailability. This situation poses a higher risk to the environment. The reducible (Fe-Mn oxide) and oxidizable (organic) phases act as a metal sink. The metals bound to these fractions can be mobilized by increasing/reducing oxidizing conditions in the environment [2].

Compare to the speciation in the natural ecosystem such as [23] classified the geochemical fractionation of samples around the Makassar coast as either acid soluble (exchangeable and carbonate fractions), reducible (Mn-Fe oxide fraction), or oxidizable (organic fraction). In their report, the mean percentages of Pb in the acid soluble, reducible, and oxidizable phases were 0–7, 64–78, and 16–33%, respectively. The mean percentages of the Zn fraction in the acid soluble, reducible, and oxidizable phases were 0–55, 39–100, and 0–25%, respectively. [46] reported that the Pb labile fraction from Jakarta Bay was dominated by the oxidizable (8.55–93.77), reducible (0.17–25.99%), and easily leachable and exchangeable phases (0–8.16%). Moreover, the percentage of Pb in the non-labile fraction ranged from 0 to 92.66%.

Table 4. Pearson's correlation coefficient between geochemical fractionation of Pb and Zn with sediment properties in Jeneberang waters.

		F1	F2	F3	F4	F5	Clay	Silt	Sand	OC	Eh	SPM
Pb	F1	1										
	F2	-.536*	1									
	F3	.906**	-.371	1								
	F4	.814**	-.535*	.638**	1							
	F5	-.929**	.434	-.841**	-.944**	1						
	Clay	.652**	-.473	.422	.796**	-.721**	1					
	Silt	.220	.133	.247	.046	-.162	-.144	1				
	Sand	-.712**	.341	-.524*	-.735**	.738**	-.803**	-.475	1			
	OC	.459	.013	.405	.433	-.500*	.390	.706**	-.772**	1		
	Eh	-.812**	.103	-.761**	-.680**	.820**	-.618**	-.428	.807**	-.659**	1	
	SPM	-.720**	.109	-.623**	-.588*	.700**	-.689**	-.198	.732**	-.481	.909**	1
Zn	F1	1										
	F2	.943**	1									
	F3	-.376	-.300	1								
	F4	.847**	.767**	-.625**	1							
	F5	-.842**	-.789**	.172	-.877**	1						
	Clay	.779**	.698**	-.219	.574*	-.595*	1					
	Silt	.071	.033	.000	.299	-.372	-.144	1				
	Sand	-.735**	-.640**	.194	-.690**	.753**	-.803**	-.475	1			
	OC	.373	.267	-.048	.491*	-.587*	.390	.706**	-.772**	1		
	Eh	-.801**	-.747**	.122	-.721**	.837**	-.618**	-.428	.807**	-.659**	1	
	SPM	-.775**	-.686**	.110	-.583*	.672**	-.689**	-.198	.732**	-.481	.909**	1

**. Correlation is significant at the 0.01 level (2-tailed).

*. Correlation is significant at the 0.05 level (2-tailed).

The speciation composition of the Mn-Fe oxide fraction in rivers is smaller (only 20%) than in estuaries with a composition of 80% found in waters without ecosystem disturbance [41]. In this study was found different, where the composition of Mn-Fe oxide fraction in rivers and estuaries is equal to heavy metal Pb, but on heavy metals Zn, the percentage of Mn-Fe oxide composition in rivers is greater than in estuaries. The high fractionation of Mn-Fe oxide in estuaries rather than in rivers because the estuary area is a high water mixing that increases manganese and iron oxides in water. The

Mn-Fe oxide fraction is higher than the organic fraction in rivers and will experience changes in composition when entering estuarine and coastal waters due to the complexity of the dynamics of the water masses in estuaries and beaches. Different things found in this study, the organic fraction has a higher composition than the Mn-Fe oxide fraction in the three aquatic regions (rivers, estuaries, and beaches). The presence of Mn-Fe oxides in sediments plays a very important role as it enhances the ability of sediments inbound heavy metals and naturally their content in high sediment/particle so that it is important to be fundamental in understanding the heavy metal geochemical processes in the sediments.

The correlations between the geochemical fractionation of Pb and Zn and the sediment properties are shown in Tables 4. The results of the correlation analysis indicated strong correlations between all the speciation and the sediment properties. The strongest correlation between speciation and sediment properties occurs in the sand, Eh, and SPM parameters. Analyzed by fraction was found the strong correlations between fractions F1 and F2 with clay, sand, Eh, and SPM. These fractions are weakly bound to surface of the sediments and depend on the physic-chemical parameters. The fraction F3 was strongly correlated with the Eh and SPM because this fraction was formed under reducing conditions. The fraction F4 was similar with fraction F1 and F2. This explains why OC which has not played a role in the formation of this fraction was distributed relatively homogenously throughout the study area. Fraction F5 was strongly correlated with the sediment texture and SPM because this component represents a chemical complex of metals in the crystalline mineral matrices within sediment particles.

3.3. Pollution assessment of heavy metals

Toxicity unit, geoaccumulation index (Igeo) and risk assessment code (RAC) were applied to assess the potential toxicity acute, pollution level and environmental health risk of heavy metals lead (Pb) and zinc (Zn). Compared to the guideline values in China (GB 18668-2002), heavy metals contamination of Pb dan Zn in Jeneberang waters still under by Grade I at all sampling points, and it was categorized no polluted. The calculation of toxicity unit was obtained $\Sigma TU < 5$ indicating low toxicity level as shown in Table 5.

The range of the Igoe of Pb was 0.007-0.053, which was classified as weak pollution. The range of the Igoe of Zn was 0.775-1.273, which was classified as weak to moderate pollution. The RAC of Pb was 6.707-9.664, which was classified as a light risk. The RAC of Zn was 0-0.223, which was classified as no risk. The higher RAC of Pb indicated that the anthropogenic source of this pollutant was larger than that of Zn. The RAC of Zn was not a threat to aquatic biota because the majority of the Zn was in non-labile or residual fractions.

Table 5. Toxicity, Geoaccumulation index (Igeo) and risk assessment code (RAC) of Pb and Zn from surface sediments at Jeneberang waters

Sampling points	Toxicity Unit (TU)		Geoaccumulation Index		Risk Assessment Code (RCA)	
	Pb	Zn	Pb	Zn	Pb	Zn
R1	0.030	0.243	0.053	1.058	8.872	0.000
R2	0.028	0.270	0.050	1.174	8.722	0.000
R3	0.029	0.293	0.052	1.273	8.390	0.000
E1	0.004	0.178	0.007	0.775	7.252	0.000
E2	0.005	0.198	0.009	0.863	6.707	0.000
E3	0.009	0.187	0.016	0.813	6.837	0.000
E4	0.009	0.204	0.016	0.886	8.574	0.000
E5	0.013	0.203	0.022	0.884	8.285	0.000
E6	0.013	0.192	0.024	0.835	8.016	0.000

E7	0.014	0.203	0.026	0.885	9.115	0.114
CS1	0.015	0.204	0.028	0.889	8.664	0.115
CS2	0.018	0.194	0.032	0.843	9.439	0.143
CS3	0.021	0.221	0.037	0.962	9.664	0.085
CS4	0.028	0.267	0.050	1.161	9.661	0.106
CS5	0.024	0.241	0.044	1.048	9.284	0.136
CS6	0.022	0.220	0.039	0.957	9.348	0.209
CS7	0.020	0.212	0.036	0.922	9.136	0.223
$\sum$	0.302	3.730				

4. Conclusions

Dispersion of metals was found higher in the riverine and marine samples than the estuarine samples. Following speciation, the metals were found similar composition of fraction in the riverine and estuarine samples but any different in the marine samples. These were indicated that the change of environment causes a change of dispersion pattern and speciation composition of metals due to the presence of the dam that lies at the boundary between the estuary and the river. There are potential hazards to environmental damage and marine biota caused by Pb due to the increased levels of this pollutant in the labile fraction in sediments along the coast of the Jeneberang waters, and this is a potential focus for future research and monitoring. This provides a strong indication that the source of anthropogenic heavy metal pollutants was mainly of the coastal waters and the potential level of pollutants of heavy metals Pb greater than heavy metals Zn.

Acknowledgements
This research was supported by General Directorate for Resources of Research-Technology and High Education (003/PEN-DD/PL/2015) with the grant program of research and public services.

References

[1] Hu, B.Q., Cui, R.Y., Li, J., Wei, H.L., Zhao, J.T., Bai, F.L., Song, W.Y., and Ding, X., 2013. Occurrence and distribution of heavy metals in surface sediments of the Changhua River Estuary and adjacent shelf (Hainan Island). *Mar. Pollut. Bull.* 76, 400–405.

[2] Wang, Z., Wang, Y., Chen, L., Yan,, C., Yan, Y., and Chi, Q., 2015. Assessment of metal contamination in coastal sediments of the Maluan Bay (China) using geochemical indices and multivariate statistical approaches. *Mar. Pollut. Bull.* 99, 43–53.

[3] Ikenaka, Y., Nakayama, S.M.M., Muzandu, K., Choongo, K., Teraoka, H., Mizuno, N., Ishizuka, M., 2010. Heavy metal contamination of soil and sediment in Zambia. *Afr. J. Environ. Sci. Technol.* 4, 729–739.

[4] Louhi, A., Hammadi,A., Achouri,M., 2012. Determination of some heavymetal pollutants in sediments of the Seybouse River in Annaba, Algeria. *Air Soil Water Res.* 5, 91–101.

[5] Chen, M., Zeng, G., Zhang, J., u, P., Chen, A., Lu, I., 2015. Global landscape of total organic carbon, nitrogen and phosphorus in lake water. *Sci. Rep.* 5, 15043.

[6] Li, T., Li, X., Sun, G., Zhou, Y., Chen, F., Zhong, H., Yang, C., Luo, W., Yao, Y., 2015. Assessment of the impacts of trace metals on benthic foraminifera in surface sediments from the northwestern Taiwan Strait. *Mar. Pollut. Bull.* 98, 78–94.

[7] SEPA (State Environmental Protection Administration of China), 2002. *Marine Sediment Quality*. Standards Press of China, Beijing.

[8] Liu, Z., Pan, S., Sun, Z., Ma, R., Chen, L., Wang, Y., and Wang, S., 2015. Heavy metal spatial variability and historical changes in the Yangtze River estuary and North Jiangsu tidal flat. *Mar. Pollut. Bull.* 98, 115–129.

[9] Armid, A., Shinjo, R., Zaeni, A., Sani, A., and Ruslan, R., 2014. The distribution of heavy metals including Pb, Cd and Cr in Kendari Bay surficial sediments. *Mar. Pollut. Bull.* 84, 373–378.

[10] Al-Rashdi, S., Arabi, A.A., Howari, F.M., and Siad, A., 2015. Distribution of heavy metals in the coastal area of Abu Dhabi in the United Arab Emirates. *Mar. Pollut. Bull.* 97, 494–498.

[11] Bastami, K.D., Bagheri, H., Kheirabadi, V., Zaferani, G.G., Teymori, M.B., Hamzehpoor, A., Soltani, F., Haghparast, S., Harami, S.R.M., Ghorghani, N.F., and Ganji, S., 2015. Distribution and ecological risk assessment of heavy metals in surface sediments along southeast coast of the Caspian Sea. *Mar. Pollut. Bull.* 81, 262–267.

[12] Hu, G., Bi, S., Xu, G., Zhang, Y., Mei, X., and Li, A., 2015. Distribution and assessment of heavy metals off the Changjiang River mouth and adjacent area during the past century and the relationship of the heavy metals with anthropogenic activity. *Mar. Pollut. Bull.* 96, 434–440.

[13] Wu, Q.H., Leung, J.Y.S., Tam, N.F.Y., Peng, Y.S., Guo, P.R., Zhou, S., Li, Q., Geng, X.H., Miao, S.Y., 2016. Contamination and distribution of heavy metals, polybrominated diphenyl ethers and alternative halogenated flame retardants in a pristine mangrove. *Mar. Pollut. Bull.* 103, 344–348.

[14] Najamuddin, Prartono, T., Sanusi, H.S., and Nurjaya, I.W., 2016a. Seasonal distribution and geochemical fractionation of heavymetals from surface sediment in a tropical estuary of Jeneberang River, Indonesia. *Mar. Pollut. Bull.* 111, 456–462

[15] Tessier, A., Campbell, P.G.C., and Bisson, M., 1979. Sequential extraction procedure for the speciation of particulate trace metals. *Anal. Chem.* 51, 844-851.

[16] Pedersen, F., Bjurnestad, E., Andersen, H.V., Kjholt, J., Poll, C., 1998. Characterization of sediments from Copenhagen Harbour by use of biotests. *Water Sci. Technol.* 37, 233–240.

[17] Müller, G., 1969. Index of geoaccumulation in sediments of the Rhine river. *Geo J.* 2,108-118.

[18] Honglei, L., L. Liqing, Y. Chengqing, and S. Baoqing, 2008. Fraction distribution and risk assessment of heavy metals in sediments of Moshui Lake. *J. Environ. Sci.* 20, 390–397.

[19] Sundaray, S. K., Nayak, B.B., Lina, S., and Bhatta, D., 2011. Geochemical speciation and risk assessment of heavy metals in the river estuarine sediments-A case study: Mahanadi basin, India. *J. Hazardous Materials* 186, 1837–1846.

[20] Rhoads, D.C, 1974. *Organism-Sediment Relations. In Oceanography and Marine Biology.* Barnes, H. (ed). George Allen and Unwin Ltd, London.

[21] Najamuddin, Prartono, T., Sanusi, H.S., and Nurjaya, I.W., 2016b. Distribution and behaviour of dissolved and particulate Pb and Zn in Jeneberang Estuary, Makassar. *J. Tropical Mar. Sci. and Tech.*8 (1), 11-28.

[22] Samawi, M. F., 2007. *Design System of Coastal City Waters Pollution Control. A Case Study in Coastal Waters of Makassar City*. Dissertation, Postgraduate of Bogor Agricultural Institute, Bogor. 58-59.

[23] Werorilangi, S., Tahir, A., Noor, A., and Samawi, M.F., 2013. Status of pollution and bioavailability potential of metals in sediment at coastal water of Makassar City. *Proc. National Symp. on Marine and Fishery*, Hasanuddin University, Makassar. 1-18.

[24] Burton, J.D and Liss, P.S., 1976. *Estuarine Chemistry*. Academic Press. New York. 131-158pp.

[25] Libes, S., 2009. *Introduction to Marine Biogeochemistry*. Academic Press. 561-571.

[26] Najamuddin, 2017. *Dynamic of Heavy Metals Pb and Zn in Jeneberang Estuary, Makassar*. Dissertation, Postgraduate of Bogor Agricultural Institute, Bogor. 88-95

[27] Chester, R., 1990. *Marine Geochemistry*. Unwin Hyman, London. 321-421pp.

[28] Turekian, K.K., 2010. *Marine Chemistry and Geochemistry. A Derivative of Encyclopedia of Ocean Sciences*, 2nd edition. Academic Press. 265-280.

[29] Lenzi, M. 2010. Resuspension of sediment as a method for managing eutrophic lagoons. *J. Ecol and Nat Environ* 2 (11), 230-234 .

[30] Tengeberg, A., Almroth, E., and Hall, P., 2003. Resuspention and its effect on organic carbon recycling and nutrient exchange in coastal sediment: in situ measurements using new experimental technology. *J. Experimental Mar. Biol. and Ecol.* 285, 119-142.

[31] Rastina, 2012. *Environmental Management Strategy Based on Water Quality Modelling in Tallo Estuary, South Sulawesi*. Dissertation, Postgraduate of Bogor Agricultural Institute, Bogor. 60-63.

[32] Takarina, N.D., Browne, D.R., and Risk, M.J., 2004. Speciation of heavy metals in coastal sediments of Semarang, Indonesia. *Mar. Pollut. Bull.* 49, 854–874.

[33] Wu, Q.H., Tam, N.F.Y., Leung, J.Y.S., Zhou, X.Z., Fu, J., Yao, B., Huang, X.X., Xia, L.H., 2014. Ecological risk and pollution history of heavy metals in Nansha mangrove, South China. *Ecotoxicol. Environ. Saf.* 104, 143–151.

[34] Tornero, V., Arias, A.M., and Blasco, J., 2014. Trace element contamination in the Guadalquivir River Estuary ten years after the Aznalcóllar mine spill. *Mar. Pollut. Bull.* 86, 349–360.

[35] Jiang, X., Teng, A., Xu, W., and Liu, X., 2015. Distribution and pollution assessment of heavy metals in surface sediments in the Yellow Sea. *Mar. Pollut. Bull.* 83, 366–375.

[36] Parra, S., Bravo, M.A., Quiroz, W., Querol, X., and Paipa, C., 2015. Distribution and pollution assessment of trace elements in marine sediments in the Quintero Bay (Chile). *Mar. Pollut. Bull.* 99, 256–263.

[37] Xu, G., Pei, S., Liu, J., Gao, M., Hu, G., and Kong, X., 2015. Surface sediment properties and heavy metal pollution assessment in the near-shore area, north Shandong Peninsula. *Mar. Pollut. Bull.* 95, 395–401.

[38] Teixeira, E.C., Rodrigues, M.L.K., Alves, M.F.C., and Barbosa, J.R., 2003. Study of geochemical distribution of heavy metals in sediments in areas impacted by coal mining. in J. Locat, R. Galvez-Cloutier, R. C. Chaney and K. Demars (eds.). *Contaminated Sediments: Characterization, Evaluation, Mitigation/Restoration, and Management Strategy Performance*, ASTM STP 1442, ASTM International, West Conshohocken, PA, USA. 72–86.

[39] Trujillo-Cárdenas, J.L, Saucedo-Torres, N.P., del Valle, P.F. Z., RíosDonato, N., Mendizábal, E., and Gómez-Salazar, S., 2010. Speciation and sources of toxic metals in sediments of Lake Chapala, Mexico. *J. Mex. Chem. Soc.* 54 (2), 79-87.

[40] Rodrigues, M.L.K., and Fomoso, M.L.L., 2006. Geochemical distribution of selected heavy metals in stream sediment affected by tannery activities. *Water Air and Soil Pollut.* 169, 167–184.

[41] Moore, J.W., Ramamoorthy. S., 1984. *Heavy Metal in Neutral Water*. New York (US): Springer Verlag.

[42] Ahmad, F., 2013. Distribution and prediction on heavy metals pollution level (Pb, Cd, Cu, Zn, and Ni) in sediment in Bangka Island waters using load pollution index and geoaccumulation index. *J. Tropical Mar. Sci. and Tech.* 5(1), 170-181

[43] Lestari, and Budiyanto, F., 2013. Concentration of Hg, Cd, Cu, Pb and Zn in sediment of Gresik waters. *J. Tropical Mar. Sci. and Tech.* 5 (1), 182-191.

[44] Arifin, Z., 2011. Heavy metals concentration in water, sediment and biota in Kalibata Bay, Bangka Island. *J. Tropical Mar. Sci. and Tech.* 3 (1), 104-114.

[45] Hamzah, F., and Setiawan, A., 2010. Accumulation of heavy metals Pb, Cu, and Zn in the mangrove forest of Muara Angke, North Jakarta. *J. Tropical Mar. Sci. and Tech.* 2 (2), 41-52.

[46] Arifin, Z., and Fadhlina, D., 2009. Fractionation of heavy metals Pb, Cd, Cu and Zn in sedimen and the bioavailability to biota at Jakarta Bay. *Mar. Sci.* 14 (1), 27-32.

IOP Conf. Series: Earth and Environmental Science **89** (2017) 012031 doi:10.1088/1755-1315/89/1/012031

Concentration of PSP (*Paralytic Shellfish Poisoning*) Toxin On Shellfish From Inner Ambon Bay and Kao Bay North Halmahera

F S Pello[1]; S Haumahu[1], N V. Huliselan[1] and M A. Tuapattinaja[1]

[1]Department of Aquatic Marine Resources, Fishery and Marine Sciences Faculty Pattimura University, Ambon. Jln. Ir. Chr. Soplanit, Kampus Poka-Ambon

Email: rikapello@yahoo.com

Abstract The Inner Ambon Bay and Kao Bay have potential on fisheries resources which one of them is molluscs. Molluscs especially for class bivalve have economical values and are consumed by coastal community. The research had been done to analyze saxitoxin (STX) concentration on bivalves from Kao Bay and Inner Ambon Bay. The Saxitoxin Elisa Test Kit Protocol was used to determine saxitoxin concentration. The measurement showed that the highest concentration of saxitoxin (392.42 µg STXeq/100g shellfish meat) was *Gafrarium tumidum* from Ambon Bay, whereas concentration of saxitoxin (321.83 µg STXeq/100g shellfish meat) was *Mactra mera* from Kao Bay

1. Introduction

The effects of *Pyrodinium* caused health problems to the coastal peoples in Inner Ambon bay and Kao bay, Indonesia. The worst was in 1994 has led to people mortality [12]. *Pyrodinium* natural habitat is close to mangrove ecosystem, where they receive almost continuously nutrients, therefore the achieve bloom concentrations. Waters of Inner Ambon bay and Kao bay are dominated by mangrove ecosystem and are influenced by fresh water. Those mangrove ecosystems fulfill an important role in marine ecosystem and fisheries, such as shrimp, baitfish (*Stolephorus* spp) and mollusk that provide a potential source to coastal peoples. In some part of Kao bay, experienced as red tide phenomena in 1930, caused by the blooming of toxic phytoplankton [11].

The bloom of toxic algal has threatened the human health through marine food chain. From those 5000 species of marine algal, about 300 species often has an extraordinary abundance and 40% of them reproduce toxin that can be accumulated to mollusks and fish and is harmful to human who consumes the organisms [3].

Commonly, toxin from algal can be divided into five groups based on the symptom, e.g., Paralytic Shellfish Poisoning (PSP), Diarrhetic Shellfish Poisoning (DSP), Amnesic Shellfish Poisoning (ASP), Ciguatera Shellfish Poisoning (CSP) and Neurotoxic Shellfish Poisoning (NSP) [2]. Paralytic Shellfish Poisoning (PSP) is famous for saxitoxin (STX) that produced by toxic algal, e.g., *Alexandrium tamarense, Pyrodinium bahamense* var *compressed, Gymnodinium catenatum* and other dinoflagellates [1].

The main goal of this study is to determine saxitoxin concentrations from several species of mollusks, which is consumed by people and to analysis the toxic Dinoflagellates from Inner Ambon bay and Kao bay.

2. Material and Method

The study was conducted in June 2016, in the inner Ambon bay, Province of Maluku and Kao bay of Halmahera, North Maluku. Geographically, sampling sites in inner Ambon bay is at 06,66° 39'29" – 03,63° 30'30"S and 128°19'4,03" – 128°24'33"E, while in Kao bay is 1°8'1.41"LU -1° 8'49.77" N and 127°49'32.88" - 127°54'14.29" E 'Figure 1'.

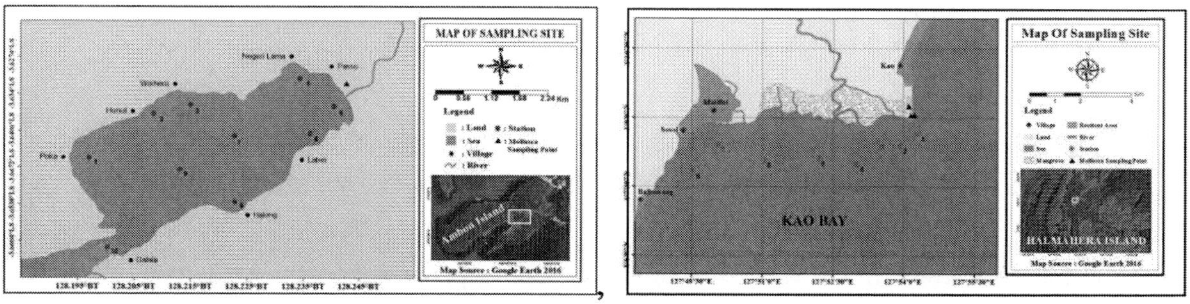

Figure 1. Map of sampling sites in inner Ambon bay, Region of Ambon and Kao bay, North Maluku.

Temperature, Salinity, and pH were measured. Data of saxitoxin concentration was detected by using Elisa test kit protocol. The phytoplankton was collected by hauling phytoplankton net of 48 µm mesh size from the depth of 5 meters to the water surface. Phytoplankton samples were preserved with formaldehyde of 4% immediately. Phytoplankton was identified according to [14], [8], and [10] to the lowest taxonomic as possible. The abundance of phytoplankton was calculated based on [9].

$$D = \frac{NfxVp}{V} \tag{1}$$

Where: D = phytoplankton abundance (ind/ m^3)
N_f = numbers of cell per 1 ml
V_p= volume of diluted sample
V = volume of filtered water (m^3)

Volume of filtered water was measured following [8].

$$V = \pi . r^2 .1 \tag{2}$$

Where:
V = volume of filtered water (m^3)
π = constant (3.14)
r = half of the mouth diameter of the net
l = the distance of net towing (m)

3. Results and Discussion

3.1. Paralytic Shellfish Poisoning (PSP) concentration

Four species of molluscs have been identified from inner Ambon bay, and Kao bay (Figure. 2, Figure 3 and Table 1) and the high concentration of saxitoxin in inner Ambon bay was found from *Gafrarium tumidum* with the value of 392.42 µg STXeg/100 g of mollusc meat, whilst the lowest concentration was from *Psammotea elongate*. From four species of mollusks collected from Kao bay, two of them have saxitoxin. The high concentration of saxitoxin was from *Mactra mera* (321.83 µg STXeg/100 g of meat and followed by *Polymesoda erosa*, while *Atactodea striata* and *Telescopium telescopium* have not saxitoxin.

Figure 2. Species of mollusks from inner Ambon bay

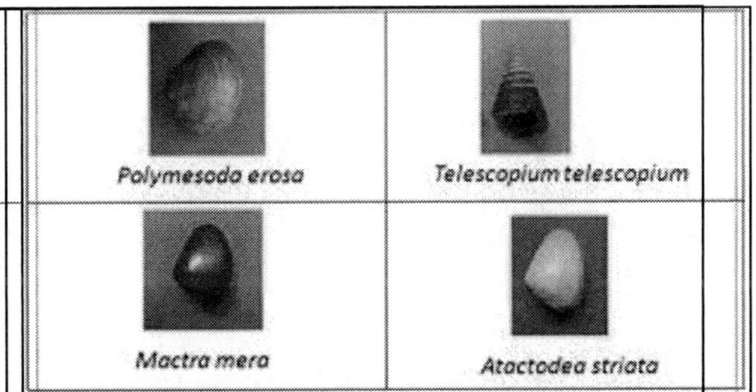

Figure 3. Species of mollusks from Kao bay

Table 1. Saxitoxin concentration of molluscs from inner Ambon bay and Kao bay

Locations	Species of Molluscs	Saxitoxin concentration (µg.100g^{-1})
Kao bay	*Polymesoda erosa*	196.11
	Atactodea striata	0
	Mactra mera	321.83
	Telescopium telescopium	0
Inner Ambon bay	*Gafrarium tumidum*	392.42
	Anadara antiquata	136.4
	Psammotea elongata	131.97
	Meretrix meretrix	142.53
	Limit of saxitoxin concentration (FAO,2004)	80

According to SNI 3460.1 year of 2009, saxitoxin allowed in frozen meat of mollusks is 80 µg/100 gr. The result of this study showed that the concentration was above of the limit. The concentration of saxitoxin in the meat of mollusks correlates with the abundance of toxic Dinoflagellates [4].

3.2. Composition and Abundance of phytoplankton

At least 64 species of phytoplankton belonging to 4 classes, i.e., Bacillariophyceae (Diatom), Dinophyceae, Cyanophyceae dan Chrysophyceae were found from 10 stations in inner Ambon bay. Bacillariophyceae has 36 species, and Dinophyceae has 26 species, while Cyanophyceae and Chrysophyceae only have one species. There were 72 species of phytoplankton from Kao bay belonging to 4 classes as recorded in inner Ambon bay. Bacillariophyceae consisted of 46 species, while, Dinophyceae has 23 species, Cyanophyceae and Chrysophyceae consisted of 1 and two species respectively. The abundant of Dinoflagellates in inner Ambon bay reached 22.71% of total number phytoplankton, while in Kao bay was 58.14% (Table 2).

IOP Conf. Series: Earth and Environmental Science **89** (2017) 012031 doi:10.1088/1755-1315/89/1/012031

Table 2. The abundance of phytoplankton and Dinoflagellates (cell/m^3) in inner Ambon bay and Kao bay

Inner Ambon Bay			Kao Bay		
Stations	Phytoplankton	Dinoflagellates	Stations	Phytoplankton	Dinoflagellates
1	141,847	44,459	1	53,651	33,810
2	150,000	62,389	2	30,698	14,698
3	130,987	37,898	3	39,476	8,238
4	147,994	27,898	4	31,587	22,857
5	494,331	116,369	5	81,810	43,206
6	346,051	62,532	6	32,190	15,746
7	215,924	25,318	7	61,143	46,794
8	771,847	132,293	8	49,175	35,429
9	61,576	38,169			
10	54,411	23,742			

Ten of 23 Dinoflagellates species found in Kao bay were present in all sampling sites, namely *Prorocentrum micans, Dinophysis caudata, Ceratium furca, Ceratium fucus, Ceratium lunula, Ceratium macroceros, Ceratium tripos, Ceratium trichoceros, Phyrophacus stenii* and *Protoperidinium depressum.* The most species which contributed to Dinoflagellates assemblage whereas *Dinophysis caudata* and *Ceratium trichoceros* (Table 3).

Table 3. Percentage of 5 dominant species of Dinoflagellates in Kao bay

Dinoflagellates	Stations							
	1	2	3	4	5	6	7	8
Dinophysis caudata	47.42	48.81	13.87	38.06	32.26	18.15	24.15	13.71
Ceratium trichoceros	33.52	30.67	39.88	26.94	24.76	44.05	46.27	40.50
Ceratium tripos	5.45	4.97		6.11		8.47	10.45	12.63
Ceratium furca	4.23		11.56	8.06	5.88	10.38	3.39	
Ceratium fucus	4.23		7.51	12.22	10.29	3.73	2.85	11.02
Ceratium lunula		6.05						
Ceratium macroceros		2.59	6.36					
Protoperidinium depressum					6.61			
Phyrophacus stenii								6.99

The result also showed that 13 species of 26 species of Dinoflagllates recorded in inner Ambon bay were found in all stations, i.e. *Dinophysis caudata, Dinophysis miles, Ceratium furca, Ceratium fucus, Ceratium lunula, Ceratium macroceros, Ceratium tripos, Ceratium trichoceros, Ceratium pulchellum, Gonyaulax spinifera, Pyrocystis noctiluca, Phyrophacus stenii* and *Protoperidinium depressum. Dinophysis miles* contributed the high percentage of the Dinoflagellates assemblage at all stations except for station ten which was dominated by *Dinophysis caudate* (Table 4).

Table 4. Percentage of 5 dominant species of Dinoflagellates in Inner Ambon Bay

Dinoflagellates	Stations									
	1	2	3	4	5	6	7	8	9	10
Dinophysis miles	61.89	49.69	38.15	53.88	70.99	57.45	37.23	70.68	38.30	17.30
Ceratium macroceros	14.61	10.34	8.99	14.84	9.20	8.25	11.07	10.64	13.14	
Phyrophacus stenii	7.02	8.04	11.76	10.27	10.62	12.38	19.12	5.10	12.01	7.85
Gonyaulax spinifera	3.94	9.42	11.76	4.45			4.53			4.63
Ceratium fucus	2.36					2.60	3.14			
Dinophysis caudata		5.28	8.40					2.79	4.63	43.86
Ceratium trichoceros				3.42	1.53	5.35		2.65	6.88	
Ceratium tripos					1.81					
Ceratium furca										8.45

Some species of Dinoflagellates, such as *Alexandrium, Gymnodinium* and *Pyrodinium* produce saxitoxin [6], and the concentration of the toxin correlate with the abundance of toxic dinoflagellates[4]. Based on this study, the abundance of *Alexandrium tamarense* and *Alexandrium affine* was only 3.64% of total dinoflagellates in inner Ambon bay, while the abundance of *Pyrodinium bahamense* and *Alexandrium affine* was about 0.37% of total the abundance of dinoflagellates in Kao bay. However, the concentration of saxitoxin in the meat of mollusks from inner Ambon bay and Kao bay were above the limit of concentration stated by [2]. It is argued that the high concentration of saxitoxin is caused by the ability of the mollusks to accumulate the toxin day by day. On the contrary, the study of [13] showed that concentration of saxitoxin was low even at the highest abundance of toxic dinoflagellate *Pyrodinium bahamense* in Lampung bay.

3.3. Hydrology parameters

The average of temperatures, pH and salinities in inner Ambon were nearly equal to those in Kao bay (Table 5). According to KEPMEN LH No.51/2004, the values of hydrology parameters in inner Ambon bay and Kao bay were in standard values.

Table 5. pH, Temperatures and salinities in inner Ambon bay and Kao bay

Stations	Inner Ambon bay			Kao bay		
	pH	Temp. (°C)	Salinities (°/₀₀)	pH	Temp. (°C)	Salinities (°/₀₀)
1	6	28	35	6	31	29
2	6	28	35	6	30	30
3	6	28	35	6	31	30
4	6	28	35	7	31	30
5	6	27	35	7	29	29
6	6	27	33	7	30	31
7	6	28	35	7	31	31
8	6	27	34	7	30	32
9	6	27	35			
10	6	27	34			
		Standart values		7-8,5	28-30	33-34

(*) Standard value according to KEPMEN LH No.51/2004

4. Conclusions

Saxitoxin concentration in the meat of mollusks from inner Ambon bay and Kao bay were above the limiting concentration according to SNI 3460.1. The abundance of *Alexandrium tamarense* and *Alexandrium affine* were only 3.64% of the total dinoflagellates in inner Ambon bay, while in Kao bay, the abundance of *Alexandrium affine* and *Pyrodinium bahamense* were only 0.37%. Moreover, the high concentration of saxitoxin in the meat of mollusks is accumulated for a long time.

Acknowledgement

The authors would like to thank the Research and Community Service of The Department of Education and Cultural of The Republic of Indonesia, who funded this research under MP3EI Research Scheme, 2016.

References

[1] Dam H.,S. Colin,S.Haley, D.Avery, L.Chen and H.Zhang 2009 Copoepod resistance to toxic Phytoplankton. http://www.pices.int/publications/presentation/Zoop%202007 /zoop% 202007%20S3/S3_Dam.pdf.

[2] FAO 2004 *Marine Toxin*. Food and agriculture organization of the United Nations. *http://www.fao.org/docrep/.*

[3] Hallegraeff G. M 1993 A review of harmfull algal blooms and their apparent global increase. Phycologia, 32 (2): 79-99.

[4] Hamasaki, K., T. Takahashi and S. Uye 2003 Accumulation of Paralytic Shellfish Poisoning Toxins in Planktonic Copepoda During a Bloom of the Toxic Dinoflagellate Alexandrium tamarense in HiroshimaBay, Western Japan. Mar.Biol. 143:981-988

[5] Hodgkiss, I.J. and S. Lu 2004 The Effects of Nutrients and their ratios on Phytoplankton abundance in Junk Bay, Hong Kong, Hydrobiologia. 512:215-229

[6] Lalli C.M, and T.R. Parsons. 1993. Biological Oceanography. An Introduction. Butterworth-Heinemann, Oxford.

[7] Menteri Kependudukan dan Lingkungan Hidup (MENKLH). 2004. Kepmen KLH No.51 Tahun 2004 Tentang Kriteria Baku Mutu Air Laut Untuk Biota Laut

[8] Newell GE, Newell RC 1977 Marine Plankton Practical Guide. Hutchinson-Co. (Pub.) Ltd., London.

[9] Perry, R. 2003 *A Guide To The Marine Plankton Os Southern California 3rd Edition*. UCLA Ocean Globe and Malibu High School.

[10] Tomas. CR, 1997 *Identifiying Marine Phytophlankton*. Academic Press.

[11] Wiadyana, N. N., 1996 Mikroalga berbahaya di Indonesia. Oseanologi dan Lymnologi di Indonesia, 29, 15-28.

[12] Wiadnyana, N. N., T. Sidabutar, K. Matsuoka, T. Ochi, and M. Kodama 1996 Note on the occurrence of Pyrodinium bahamense in eastern Indonesian waters. Dalam: Yasumoto, T., Y. Oshima, & Y. Fukuyo (eds.). 1996. Harmful and toxic algal blooms. IOC, UNESCO: 53-56.

[13] Widiarti, R 2004 Keberadaan Kista dari jenis mikroalga berbahaya di Teluk Lampung. *http://cdc.eng.ui.ac.id/article/articleprint/1505/-1/25*

[14] Yamaji, I., 1979 *Illusitrations Of The Marine Plankton Of Japan*. Hoikusha Publishing CO.,LTD. Japan.

Saponin Isolation as Main Ingredients of Insecticide and Collagen Type I From Crown of Thorn–Starfish (*Acanthaster planci*)

Anondho Wijanarko[1], Mikael Januardi Ginting[2], Muhamad Sahlan[1*], Imelda Krisanta Endah Savitri[3], Yunita Florensia[1], Maria Regina Sudiarta[1], Satria Pastika[1], Fakhri Rafiki[1], Heri Hermansyah[1]

[1]Department of Chemical Engineering, Universitas Indonesia, Kampus Baru UI Depok-16424, Indonesia
[2]Department of Biology, Universitas Indonesia, Kampus Baru UI Depok-16424, Indonesia
[3]Department of Aquatic Product Technology, Faculty of Fishery, Pattimura University, Kampus Unpatti-Poka, Ambon, Indonesia 97233

Email: sahlan@che.ui.ac.id

Abstract. The outbreaks of crown of thorns starfish (*Acanthaster planci*) resulted in the severe destruction of coral reefs in a large number of Indonesia's marine ecosystem, especially in the western part. At the moment, control efforts are proven to be ineffective because of its high cost and labor intensive. Recent research found that *A. planci* contain saponins that act as cytotoxic compound and can be used as an environment-friendly insecticide to eradicate *Kalotermitidae* pest. Saponins extracted by maceration using ethanol 96.0% with a total yield of saponins 9.04% and 4.66% for two test. Purification of saponin was achieved by utilization of activated carbon with a mass of carbon:volume sample 1:2 (w/v) and stirred for 20 minutes. Sapogenin can be isolated by hydrolyzing using hydrochloric acid, and thus 168.4 mg sapogenin is obtained. In addition to saponins, *A. planci* also contains collagen Type I. Collagen isolation by multistage extraction began with extracting the collagen with alkaline solvent, with water, NaOH 0.1 M, and Ca(OH)2 0.2 M as the solvent variations. The second step is acid-enzymatic extraction by pepsin digestion in 0.5 M acetic acid. Collagen extract will be further purified by salting out and dialysis method to obtain pure collagen yield called Pepsin Solubilized Collagens (PSC). Characterization of PSC consists of quantitative and qualitative analysis such as Lowry method, gel electrophoresis, UV spectroscopy, amino acid composition analysis, and Scanning Electron Microscopy (SEM). The result shows $Ca(OH)_2$ 0.2 M as the best extraction solvent with 2.26% yield of PSC.

1. Introduction

Acanthaster planci is a species of starfish from Echinodermata phylum, which feeds on live corals and thus possesses a great threat to corals ecosystem. Commonly known as crown of thorns starfish in English or Oni Hitode (means ogre starfish) in Japanese, *A. planci* is famous for its thorns throughout its body surface. Adult crown-of-thorns starfish normally range in size from 25 to 35 cm and the thorns are 5 to 6 cm long [1]. These starfish prefer to stay in more protected areas such as lagoons, and the deeper ocean along the reef front. As one of the main predators of coral reefs, *A. planci* can consume up to 5 – 6 m2 corals in a year [1]. However, we recently discovered that the consumption rate could reach 5 – 13 m2 per year [2]. Therefore, the population outbreak of *A. planci* will be resulted in corals destruction and thus give a negative impact on the ecosystem of the coastal region. Population outbreaks of *A. planci* are very dangerous because it damages coral reefs permanently, as well as disrupting the marine ecosystem since there are a lot of marine organisms live in the corals [1].

In Indonesia, *A. planci* is a species of starfish naturally found in the coral reefs in the Indo-Pacific Ocean. It acts as the control of coral reefs population because it is feeding on the corals, such as *Acropora* (small coral) [2]. However, in the last few years, there are many reports about *A. planci* population outbreaks, which damage coral reefs and marine ecosystem severely. Till now, *A. planci* population control in Indonesia, Japan, or Australia has been proven ineffective, such as manual removals, chemical injection into the starfish, provide fences alongside the corals, or cutting *A. planci* body into tiny pieces [2].

As an alternative of *A. planci's* population control, an integrated research for utilizing *A. planci* contained bioactive should be conducted. A. *planci* has numbers of proteins in the spines venom of their surface body. It was reported there are at least three kinds of proteins/peptides in that venoms, Plancinins that act as an anticoagulant factor, Phospolipase A2s (PLA2s) as an antibacterial agent, and Placitoxins that is similar to mammalian deoxyribonuleases II [4]. Besides the spines venom inside its thorns, *A. planci* body contains many other substances such as lipid, saponin, collagen, carotenoids, calcium, phosphor, and other minerals, which haven't been utilized effectively.

Saponins are natural substances that can act as secondary metabolites. Saponins consist of polycyclic aglycones attached to one or more sugar side chains. The aglycone (glycoside-free) portions of the saponins also called sapogenins [5]. However, the term asterosaponin has come to represent any saponin isolated from Asteroidea. Asterosaponin is famous for its toxicity and highly lethal for another organism, such as fish, Annelida, Mollusca, arthropods, and vertebrates. Because of its toxicity, asterosaponin is being utilized as a cytotoxic agent, antineoplastic, hemolysis agent, antitumor, antibacterial, antiviral, antifungal, and anti-inflammatory. There are several types of asterosaponin from *A. planci*, namely thornasteroid A [6] and Acanthaglycoside A, B, C, D, E, F [7].

Saponins are phytochemicals which can be found in most vegetables, beans, and herbs, such as tea, asparagus, ginseng, *Phaleria macrocarpa, Yucca sp., Polygala senega L.*, etc. As chemical compounds found abundantly in various plant species, saponins exhibit a variety of biological activities and have been investigated toward the development of new natural medicines because of their uses as an antibacterial agent, antiviral, anti-inflammatory, anticancer, and antitumor [8]. Besides its medical application, saponins are useful for pest control in agriculture or aquaculture because of its natural insecticide properties. In 2011, Agriculture Research Center in Bandung, studied the role of saponin from tea seed to exterminate shrimp pest. In 2007, Research and Development Center of Agriculture formulated a combination of tea saponin and *rerak* as a biopesticide for snail pest. Saponin utilization as a natural insecticide is highly potential because of its lethal effect on cold-blooded animals [9]. The deep experiment of this research and development center resulted that saponin utilization as an environmental safety pesticide is proven. Therefore, it's very beneficial to utilize saponins from *A. planci* which was known as a parasite and threat for corals and marine ecosystem. For increasing *A. planci*'s saponin utilization, preliminary research of its effectiveness against pest from *Kromoptera* family is conducted. This pest (termites) is very harmful to home appliances or furniture made of woods and bamboo. Termites booming caused financial loss up to Rp 250.000.000.000,- every year [10]. An easy and low-cost method for saponin extraction and purification must be performed as well as sapogenin isolation and in vitro test of its insecticide properties against termites.

As mentioned before, besides saponin, the whole body of *Acanthaster planci* highly contain type I collagen. Collagen is the most abundant protein in various connective tissues and was previously known only in animals. Therefore collagen for industrial purpose is mainly extracted from bovine,

IOP Conf. Series: Earth and Environmental Science **89** (2017) 012032 doi:10.1088/1755-1315/89/1/012032

porcine, and fowl origins. itiAlso, collagens are proven to be higher in mammals compared to marine animals [11-12]. However, the outbreaks of bovine spongiform encephalopathy and foot-and-mouth disease have caused concern enhancement regarding bovine collagen application. Furthermore, halal issue and H5N1 disease (avian flu) also influenced concern enhancement regarding collagen originated from porcine and fowl [13]. On the other hand, plant-derived collage, or phytocollagen, would not be as effective because it acts only as a precursor to boost collagen synthesis. These factors have led to extensive study on marine collagen as an alternative to another source of collagen. Most of the present researches on marine collagen are extracted from fish, jellyfish, sea urchin, sea cucumber, and starfish [14-17]. Moreover, marine collagen has been recognized as potential alternative sources, due to their stability and high biocompatibility with human tissues [18], as well as their low denaturation temperature [19]. Besides its superior properties, marine collagen from the under-utilized *A. planci* has high potential because it would not compete with food necessity like it does with fish, sea urchin, sea cucumber, squids, etc. Hereinafter, increase the utilization of *A. planci* to get collagen extract by multistage extraction method is also be performed. Multistage extraction method for the industrial purpose has been well developed, and thus it is very effective and feasible. Afterward, it provides some insights on the physicochemical properties of *A. planci*'s collagen despite the lack of information. Finally, it provides hopefully an optimum extraction method to obtain high collagen yields as an alternative source of commercial collagen for food, cosmetic, biomaterial, and biomedical industries.

2. Materials and Methods

2.1 *Saponin Isolation*

2.1.1 Solvent Selection for Saponin Extraction. Crown of thorns starfish (*Acanthaster planci*) was captured from Ambon, Island, Moluccas prefecture in February 2014. The starfish was then frozen and stored at -20°C until used. As pretreatment, this starfish was washed thoroughly with distilled water to remove salt or other materials. After that, all thorns on A. planci body surface were removed manually, and then the body was cut into small pieces (less than 1 cm). The sample was further treated with hexane 1:1 (w/v) for 1×24 hour to remove the non-polar substances. The next step is to treat the sample with ethyl acetate 1:1 (w/v) for 1×24 hour to remove the semi-polar substances. Afterward, the saponin was extracted by organic solvent with four variations, such as distilled water, 96% ethanol, n-butanol, and isopropyl alcohol for 3×24 hour by changing the solution daily. The sample and solvent ratio was 1:3 (w/v). The mixture was filtered with Whatman 42 and further thickened (to remove the solvent) by increasing the temperature until the solvent boiling point was reached. As a result, a viscous solution was obtained and subjected to UV spectroscopy at A = 255nm; E1% = 15 [6] to quantify the amount of saponin. On the other hand, qualitative analysis of saponin extraction was carried out by Liebermann Burchard and Froth test. The solvent with the highest yield of saponin will be selected for the scale-up process.

2.1.2 Froth Test. 0.5 ml crude extract / 0.05-gram pasta from extraction was diluted with 1 ml of water at 80°C and then shook vigorously. If foaming happens, wait until 30 minutes then add 1 drop of HCL 2N [20-21].

2.1.3 Liebermann Burchard Test. Liebermann Burchard (LB) reagent consisted of anhydrate acetic acid and concentrated sulfuric acid. Sample identification was done by adding LB reagent with the ratio of 5:5:3 (sample: acetic acid: sulfuric acid) [27].

2.1.4 UV Spectroscopy. Quantitative analysis of saponin extraction was achieved by UV spectroscopy at A = 255nm; $E^{1\%}_{1cm}$ = 15 [6]. Saponin concentration in the extract could be measured by Lambert-Beer Law.

$$A = E^{1\%}_{1cm} \times b \times c \quad (1)$$

A = absorbance
$E^{1\%}_{1cm}$ = specific absorptivity coefficient
b = cuvette width (1 cm)
c = saponin concentration in the sample (% or mg/100ml)

The absorption spectra were recorded by spectrophotometer Spectroquant Pharo 3000, Merck, Germany.

2.1.5 Scale up for Saponin Extraction. Scale up was achieved by using ethanol 96% as the extraction solvent (since it gave highest saponin yield from the selection process before). In scale-up, pretreatment was carried out similar to the pretreatment process in 2.1.1. Scale-up was carried out in duplicate, with the extract resulted from every step being filtered, thickened, and tested (Froth and Liebermann Burchard), as well as quantified by UV spectroscopy so that the daily profile can be obtained.

2.1.6 Saponin Purification with Activated Carbon. Saponin purification with activated carbon was repeated 3 times to find the optimum purification method. Commercial activated carbon with granular properties was used in this step. For every variation, the ratio of activated carbon mass and the sample solution is 1:1 or 1:2 (w/v). The selected solvent was ethanol 96% (sample:solution = 1:10 (v/v)) so that after purification sample can be further concentrated by evaporating the ethanol at 78.4 °C. The process was carried out in batch for 10 or 20 minutes with continuous stirring and several variations as follow:

1. Activated carbon and sample ratio is 1:1(w/v), retention time 10 minutes, batch process, and no repetition.
2. Activated carbon and sample ratio is 1:1(w/v), retention time 10 minutes, batch process, and repeated for 5 times.
3. Activated carbon and sample ratio is 1:2(w/v), retention time 20 minutes, batch process, and no repetition.

2.1.7 Sapogenin Isolation. Saponin extraction from method 2.1.6 (c) was further treated with 150 ml ethanol 50% and 50 ml HCl 1.85N to isolate the sapogenin. After that, the mixture was refluxed for 7 hours at 78.37°C [22]. The precipitate was filtered and analyzed by UV spectroscopy to find the λmax at EtOH=246.5 nm and E_{1cm} = 16[6]. The precipitate resulted will be further analyzed with LC-MS at Forensic Laboratory Centre – Indonesian Police Headquarter.

2.1.8 Saponin and Sapogenin Characterization. Saponin and sapogenin characterization by LC-MS is carried out on the sample before and after sapogenin isolation treatment. LC-MS was also measured at Forensic Laboratory Centre – Indonesian Police Headquarter.

2.1.9 In Vitro Test of Saponin Role Insecticide. In vitro test of saponin role as biopesticide was carried out by exterminating termites from a guava tree. Saponin extract with a concentration of 20, 10, 1, and 2 mg/ml were applied against 5 – 10 termites.

The death rate was noted for every concentration, and thus the concentration with precise results will be selected for next step. The selected concentration will be further analyzed with statistical method. 100 termites were divided into 20 groups inside a 10×10cm box. The first 10 groups were sprayed with water, and the other 10 groups were sprayed with the selected concentration of saponin. The death rate for every box was noted and further analyzed to find the efficiency of saponin as insecticide

2.2 Collagen Isolation

Collagen extraction from the body of *A. planci* was performed according to the method of extraction of pepsin solubilized collagen from the body wall of crown-of-thorns starfish by [23], modified with biochemical characterization of collagen from the starfish *Asterias amurensis* by [24].

2.2.1 Collagen Extraction. All procedures were performed at 4°C. First, the thorns on *A. planci* body surface was removed entirely; then the entire body was cut into small pieces (2 2 cm) and washed with distilled water. 15 g of sample was soaked in distilled water at a sample-to-solution ratio of 1:10 (w/v) for two days to remove the non-collagenous materials. For solvent variation, the second and third sample was soaked in 0.1M NaOH and 0.2M Ca(OH)2. The insoluble body tissue was collected and washed with 200 ml of distilled water for an hour with gentle stirring. The body was then demineralized and disaggregated with 250 ml of 0.05 M Tris-HCl (pH 8) containing 0.5 M ethylene diamine tetra-acetic acid disodium salt (EDTA- 2Na) for 3 days. The mixture was filtered and washed with 500 ml distilled water for 2 days by changing the distilled water every 24 h. The mixture was further centrifuged at 3200 g for 30 miner to collect the collagen fibrils (crude extract).

For the second extraction process, the collagen fibril was suspended in 0.5 M acetic acid at a sample-to- solution ratio of 1:10 (w/v) and digested with pepsin at pepsin-to-collagen fibril ratio of 1:15 (w/w) for 2 days. The resultant viscous solution was filtered and centrifuged at 10000x g for 30 minutes. The resultant supernatant was then dialyzed against 0.02 M Na2HPO4 (pH 7.2) for 3 days by changing the solution daily.

2.2.2 Collagen Purification. The dialysate obtained from the previous step was further centrifuged at 10000x g for 30 h. The precipitate obtained was dissolved in 0.5M acetic acid. Salting out was then performed by adding NaCl to the mixture at a sample-to-solution ratio of 5:100 (w/v). The resultant precipitate was collected by centrifuging of the mixture at 20.000 × g for 30 h. The resultant precipitate was re-dissolved in 0.5M acetic acid before it was dialyzed against 0.1M acetic acid and distilled water, respectively. Dialysis against 0.1M acetic acid was performed for 2 days by changing the solution daily while the dialysis against distilled water was performed for 1 day along with the changed of the solution at every 5-6 h. This collagen obtained was referred to as pepsin-solubilized collagen (PSC).

2.2.3 Collagen Characterization. Collagen identification and characterization were carried out by UV spectroscopy, SDS Page, and amino acid composition analysis. Furthermore, the structure of pepsin-solubilized collagen (PSC) was compared to industrially manufactured collagen by Scanning Electron Microscopy (SEM) and quantified by Lowry method.

2.2.3.1 UV Spectroscopy. The collagen samples were dissolved in 0.5 M acetic acid solution with a sample/solution ratio of 1:10 (w/v) and the UV absorption spectra were recorded by a spectrophotometer (*Spectroquant Pharo 3000, Merck, Germany*) at room temperature.

2.2.3.2 Lowry Method. Lowry method was performed according to the methods of [25] with slight modification. Lowry method was carried out in duplicate. First, mix reagent A (1% KNaC4H4O6·4H2O, 8% Na_2CO3 and 2% CuSO4.5H2O), reagent B (4% NaOH 1 M), and reagent C (1% Sodium Dodecyl Sulfate) with the ratio 3:1:1. Then prepare standards from 0.25 mg/ml bovine serum albumin (BSA) by adding 40-300 microliters of distilled water to bring volume to 350 microliters/tube. Add 400 microliters of the reagent, mix thoroughly, incubate at room temperature for 10 minutes. Add 200 microliters 0.2 N Folin reagent very quickly, and vortex

IOP Conf. Series: Earth and Environmental Science **89** (2017) 012032 doi:10.1088/1755-1315/89/1/012032

immediately. Complete mixing of the reagent must be accomplished quickly to avoid decomposition of the reagent before it reacts with protein. Incubate for 30 minutes more at room temperature. Use glass or polystyrene cuvettes to read the absorbance at 650 nm. BSA concentration and the absorbance can be plotted as a calibration curve to obtain a formula:

$$y = mx + c \tag{2}$$

Where y is the absorbance value, and x is the protein concentration in mg/ml. To find the protein concentration in the sample, we repeat the steps above by replacing BSA with the samples. But first, the PSC solution was centrifuged at 10,000 × g for 2 min to remove the undissolved collagen fibrils. The supernatants were used in this analysis.

2.2.3.3 Sodium Dodecyl Sulfate Polyacrylamide Gel Electrophoresis (SDS-PAGE). Electrophoretic patterns of collagens were analyzed according to the method of [26]. The SDS-PAGE was carried out using discontinuous Tris-glycine buffer (Tris-HCl 0.5 M pH 6.8, 2% SDS, 25% glycerol, 0.1% bromophenol blue, and 5% β- mercaptoethanol) with the ratio 1:1 (v/v). The mixtures were then heated at 95-100oC for 3 minutes. After that, the mixtures were centrifuged at 13,000 rpm for 3 minutes to remove the undissolved collagen fiber. A total of 30 µg of each collagen sample was loaded on the gel (4% stacking gel and 10% resolving gel) and subjected to electrophoresis at a constant current of 20 mA/gel (Atto Mini Gel Electrophoresis, AE-6531, Japan). After SDS-PAGE, the gel was stained with staining solution containing 0.1% (w/v) Coomassie ® brilliant blue R-250 in 40% (v/v) methanol and 10% (v/v) acetic acid for 12 h. The gel was then de-stained with a de-staining solution containing 40% (v/v) methanol and 10% (v/v) acetic acid.

2.2.3.4 Amino Acid Composition Analysis. Amino acid composition analysis was performed at PT. Saraswanti Indo Genetech (SIG Laboratory), Bogor, West Java and Province Health Laboratory of Jakarta.

2.2.3.5 Scanning Electron Microscopy (SEM). The morphological characteristics of the isolated collagen were observed by Scanning Electron Microscopy (SEM) at LIPI Biology Research Center, Bogor. First, the collagen sample was dehydrated and coated with gold ion by ion coater iB_2 then it was ready to be observed.

3. Result and Discussion

3.1 Saponin Isolation

3.1.1. Solvent Selection for Saponin Extraction. This result showed that butanol gave the highest amount of crude extract (total mass of saponin) which means butanol is the best solvent for saponin extraction. However, UV spectroscopy analysis showed ethanol 96% as the best solvent. Qualitative analyses such as Liebermann Burchard and Froth test gave a positive result and proved the high saponin content in the extract. Purple coloration at Liebermann Burchard test showed that saponin contains aglikon/sapogenin triterpenoid [27].

IOP Conf. Series: Earth and Environmental Science **89** (2017) 012032 doi:10.1088/1755-1315/89/1/012032

Table 1. Crude extract with polar solvent variation.

Solvent	Sample (gram) (wet weight basis)	Evaporated weight (gram)	Crude Extract (%)
Butanol	25.04	15.74	62.86
Ethanol 96%	10.01	4.720	47.15
Water	25.01	11.56	46.24
Iso-propanol	10.05	1.120	11.14

Table 2. Yield of saponin from extraction with polar solvent.

Solvent	Weight Basis Initial Sample Weight (gram)	Total Mass of Saponin (mg)	Yield of Saponin (%)
Butanol	250,400	71,610	28,60
Ethanol	100,100	29,100	29,07
Water	250,100	4713	1,880
Iso-propanol	100,500	2,832	2,820

This result also showed that butanol gave the highest amount of crude extract (total mass of saponin) which means butanol is the best solvent for saponin extraction. However, UV spectroscopy analysis showed ethanol 96% as the best solvent. Qualitative analyses such as Liebermann Burchard (LB) and Froth test also gave a positive result and prove the high saponin content in the extract. Purple coloration at Liebermann Burchard test showed that saponin contains aglicone/sapogenin triterpenoid [27].

3.1.2. Scale up for Saponin Extraction

Scale up of saponins extraction using ethanol 96%, and obtained results yield saponins as shown in Fig. 1. It is seen that the second trial of saponin yield has lower than the first trial result (22.99%), and this second trial test yield (12.98%) is smaller than last solvent selection yield result. This is caused used starfish used at scale up, is different slice size, and it is provided this sample is not as small as solvent selection one (> 1 cm), so that reaction surface with ethanol 96% eigs reduced. This figure is shown extraction yield per day of saponins. It showed that a drop in yield saponins from day to day. However, in the first trial, yield has increased on the third day, and it caused of reaction time at the second day of extraction is happen only 12 hours while on the third day it took place 30 hours. The qualitative test result of scale-up extraction for proven of saponin existence is shown in Table 3.

Table 3. Froth and Liebermann Burchard Result.

Solvent		Froth Test	LB Test	Saponin Existence	Aglikon Type
Butanol		+	purple	+	Triterpenoid
Ethanol		+	purple	+	Triterpenoid
Water		+	purple	+	Triterpenoid
Iso-propanol		+	purple	+	Triterpenoid
	day 1	+		+	Triterpenoid
Ethanol 1	day 2	+		+	Triterpenoid
	day 3	+		+	Triterpenoid
	day 1	+		+	Triterpenoid
Ethanol 2	day 2	+		+	Triterpenoid
	day 3	+		+	Triterpenoid

Note: + shows positive result of Froth Test (foaming); and Liebermann – Burchard test (coloration)

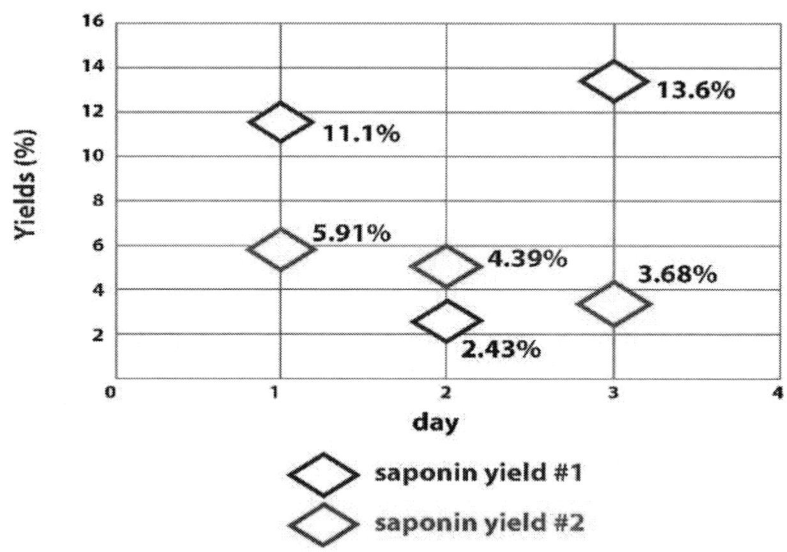

Figure. 1. Daily profile of yield from saponin Extraction.

3.1.3. Saponin purification by active carbon. The purpose of purification is to remove impurity materials which unintentionally carried from the extraction process, as well as removing pigment and protein content of *A. planci.* From 3 variations that were mentioned before, the first and second test (2.1.6(a) and 2.1.6(b)), protein, pigment, and other impurities were successfully removed from the extract. Unfortunately, the saponin content also removed. In third test (2.1.6(c)) with carbon and sample ratio 1:2 (w/v), batch process and retention time 20 minutes, was successful to purified saponin extract. As a result, protein and other impurities, e exceptf pigment, were successfully removed from the extract.

IOP Conf. Series: Earth and Environmental Science **89** (2017) 012032 doi:10.1088/1755-1315/89/1/012032

Table 4. Saponin mass, before and after purification (t = 20 minutes and activated carbon: sample = 1:2 (w/v)) from purification process of saponin extract using ethanol II.

Number of Tests	Sample (gram) (dry weight basis)	Tested Sample (mg)	Absorbance	Total Mass of Saponin (mg) (dry weight basis)
1	11.62	67,5	0.198	11,366
2	11.62	67,5	0.183	10,505
3	11.62	67,5	0.205	11,768

This resulted extract from the third test was further purified. 11,213 mg saponin in 16 ml of sample solution was achieved from the utilization of ethanol II on the extraction process. The purified saponin was freeze-dried and subjected to UV spectroscopy analysis (Table 4).

3.1.4. Sapogenin Isolation. Characterization sapogenin extract contained in *A. planci* can be done, using first sapogenin isolation from saponins. To get genuine sapogenin without the sugar chain isolation, the process can be carried out by hydrolyzing saponins and sapogenin precipitate separated from the sugar chain. The process of hydrolysis and precipitation with hydrochloric acid carried by the reflux temperature of the solvent is ethanol at 73 °C for seven hours [22]. The sample used is a purified sample of the entire sample extract as much ethanol II 8:32 grams. From reflux process, it obtained brown precipitate that found from precipitated sediment at the bottom of the container after separation using filter paper and followed by centrifugation. As a result of this difficult separation process, it is obtained precipitate is only ± 200 mg (217.50 mg) sapogenin mixture. A part of the sample which is suspected contain a mixture of sapogenin, can be analyzed using a spectrophotometer UV at 246.5 nm = λmaxEtOH = 16 [6]. One mg of the sample was tested quantitatively with spektrotrofotometer UV λmax = 246.5 nm in EtOH = 16 and resulted that sapogenin mass obtained 168.33 mg. Remaining samples were obtained can be tested further for proof sapogenin molecules using LC-MS testing tool to get a molecular weight of sapogenin.

3.1.5 Characterization of Saponin and Sapogenin. Fig. 2 showed chromatogram results before and after isolation sapogenin. In this chromatogram, a retention time of before isolation samples at 0 to 2 seconds, 6-10 seconds, and 14-17 contained spectral peaks, but it disappeared after isolation samples. This suggests that in isolation process, compounds or substances that un-isolated, have a compound chain breakdown and it is broken by hydrolysis. Allegedly, saponins, sapogenin, and a sugar chain are contained spectral peaks on the results before isolation samples and after isolation through hydrolysis, the remaining isolates sapogenin only without the sugar chain, so that saponin spectral peaks are disappeared after isolation.

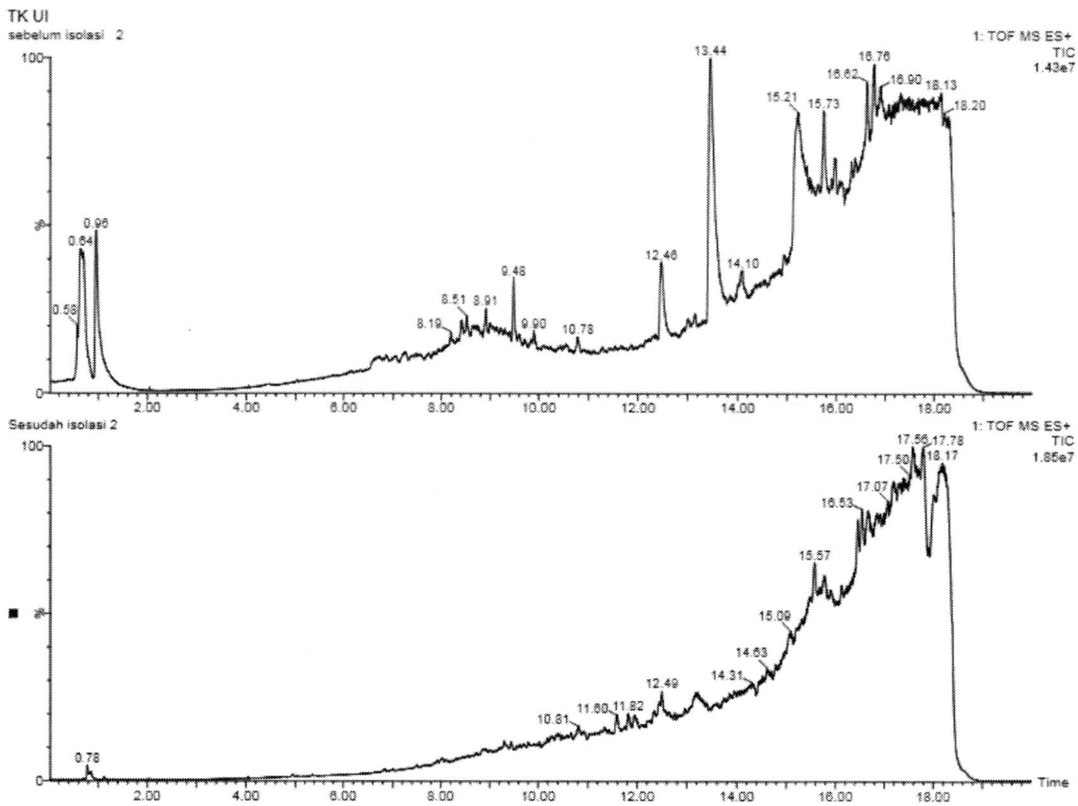

Figure. 2 Sample spectrum of saponin before and after isolation (documentation of PUSLABFOR).

In Fig. 3 are the results of *mass spectroscopy (MS)* of the sample either before or after isolation. In the results before isolation, LC-MS instrument at the Police H/Q Forensic Laboratory, searching peak - MS peak highest and search possibilities for the molecular formula based on molecular weight obtained. At spectrum with retention times above 11 seconds, which probably is the molecular formula molecule with a carbon chain under 10, there can be no saponins that have a carbon chain of at least 27 or 30 [3,28]. In the spectrum after isolation, the results of *MS* performed at a retention time over the time also shows the carbon chain molecules under 10, so either triterpenoid or steroidal sapogenin may not exist in the region. Although the results in the form of possibility MS develop a molecular formula that was alleged using computerization make uncertainty of a given molecular formula. Allegations that the region no saponin or sapogenin amplified by the shape of the spectrum is not clear where there are a lot of peaks and valleys also are attached to make the results contained in the area of retention time above 11 seconds is difficult to be analyzed and characterized by both computerized and manual with human vision. Therefore, saponin and sapogenin search centered on the spectrum with a retention time of under 11 seconds. Allegations that molecule can still be analyzed are those under 11 seconds, and precisely in the range 0 to 2 seconds because the peaks and valleys are also created on the curve very well. At the retention time range of 2 seconds both on the spectrum before and after isolation, there are many peaks of MS with a range of molecular weight of 200-800. Checking possibility of developing a molecular formula for each MS peaks, there is a molecule composed of a chain of the elements of carbon, oxygen, hydrogen, chloride, nitrogen and sulfur. According to [3,6-7] it does not only consist of saponin aglycone and sugar chains that are composed of chains of elemental carbon, hydrogen, and oxygen, but also bound with elements such as nitrogen, halogen, and sulfur. In the before isolation MS result, with suspected molecular formula of computer interpretation, there is a molecule that has a chain with a carbon number between 27-50. While the after isolation MS results sare molecules that have a carbon chain with a number between 26-40.

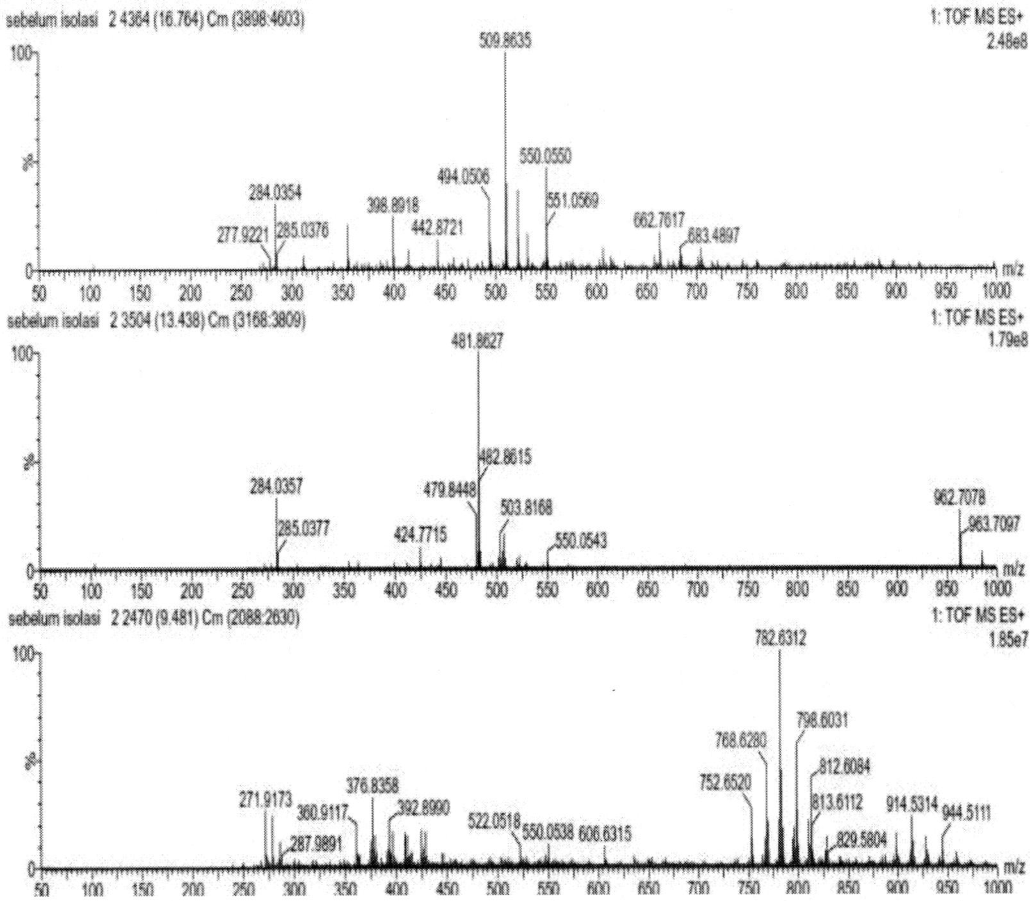

Figure. 3 *MS* Graph *before the isolation on the retention time of* 2.2470 - 2.436 seconds (result documentation from Forensic Laboratory Centre – Indonesian Police Headquarter).

Each MS peak with a chain that has a carbon number as enabled saponin or sapogenin, formulas tested one by one using the online search tool that is verified by the Royal Society of Chemistry, which was notified at www.chemspider.com. Unfortunately, the overall results of MS suspected saponin or sapogenin in the range of time under 11 seconds that shows no results on that website. Because of these conditions, to prove the presence or absence of saponin or sapogenin, it necessary to approach the LC-MS analysis of the other results. However, this analysis can only be done on the characterization of sapogenin while the characterization of saponins have been unable to do because of saponins can have a wide - range of species' sugar chain and it difficult to predict because of the diversity of possibilities molecular formula of sugar bound to the aglycone chain molecular weight also varies. Because of this, based on qualitative evidence sapogenin test using reagents Lieberman – Burchard, allegedly sapogenin present in *Acanthaster planci* is triterpenoids. This triterpenoids molecular weight based on open chemistry database *PubChem* is in the range of 424.7 ($C_{30}H_{48}O$ – *Triterpenoids* Taraxerone) to 752.9 ($C_{40}H_{64}O_{13}$ – *Triterpenoids* Cimiside B). In Fig. Four we can see that there is a range of molecular weights in the graph MS.

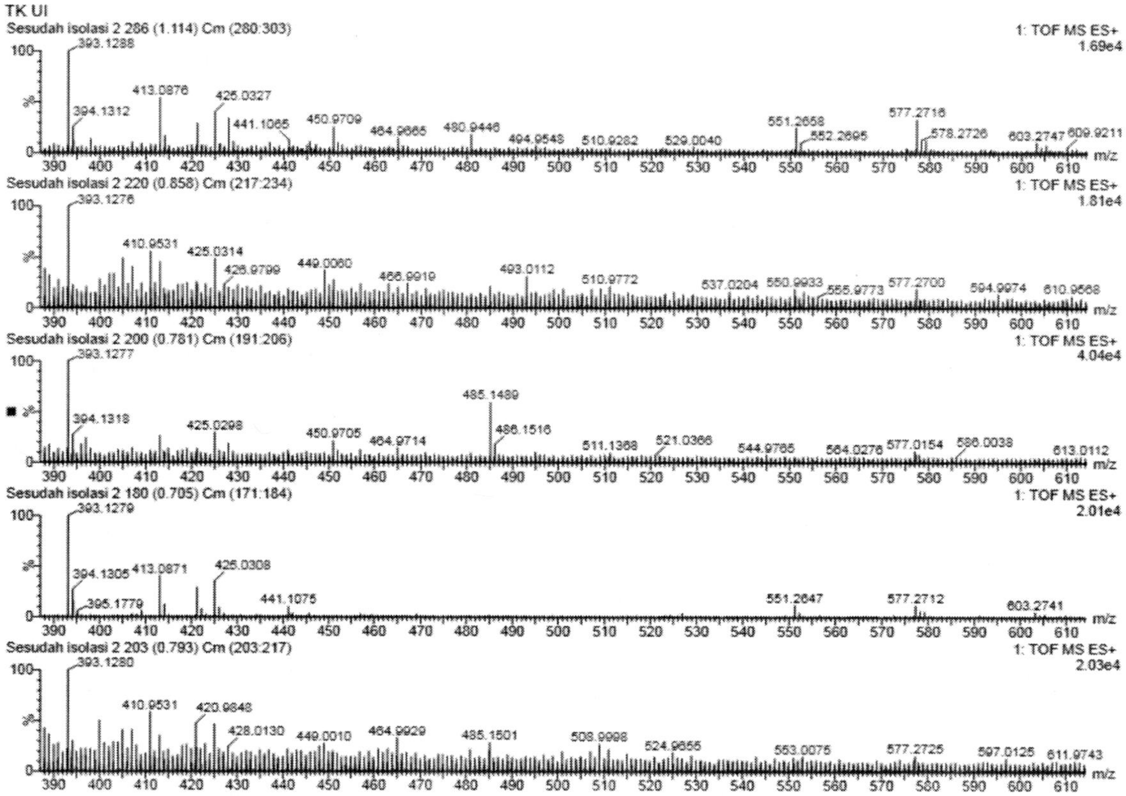

Figure. 4 *MS* Graph *after isolation of the retention time of* 0793-2286 seconds (result documentation from Forensic Laboratory Centre – Indonesian Police Headquarter).

Saponin extract with the concentration of 10, 20, 1, 2 mg/ml showed a good insecticide properties. However, sapogenin isolation by diluting the solid form with distilled water until the concentration reached 3 mg/ml was ineffective to exterminate termites. The average subsequent mortality using hypothesis testing, in which the pesticide is declared effective if the efficacy/ pest mortality reached 70%. In a hypothesis test using the t distribution, with a level of significance of 5%, and the mean termite mortality assumption is 70% of the sample is 3.5, it was concluded that the concentration of saponin insecticide 2mg/mL does not effectively kill termites.

3.2 Collagen Isolation

3.2.1 Collagen Extraction. Pepsin Solubilized Collagen (PSC) was successfully isolated from *A. planci* body with multistage extraction method. The highest yield of PSC was 2.26% (wet weight basis) while the crude extract was 86.75%.

Table 5. The Result of Collagen Extraction

Solvent	Crude Extract (%)	Yield (%)
Distilled Water	64.07	1.23
NaOH 0.1M	63.18	2.15
$Ca(OH)_2$ 0.2M	86.75	2.26

Higher quantities of PSC was recorded in the samples treated with Ca(OH)2 0.2M, followed by NaOH 0.1M, and distilled water. High values of protein from the samples treated with Ca(OH)2

0.2M signifies the mild hydrolytic effect of calcium hydroxide solution on the peptide bonds of PSC when compared with NaOH and distilled water. This mild hydrolytic effect will decrease the risk of protein fraction or denaturation during the extraction process. Therefore calcium hydroxide at high concentrations appears to be the best solvent for alkaline hydrolysis [29].

3.2.2 Collagen Characterization

3.2.2.1 UV Spectroscopy. The UV absorption spectrums of PSC at wavelength range 190–400 nm were showed in Fig. 5. PSC extracted from *A. planci* showed maximum absorption at 215-245 nm which is hb y the PSC from Edwards et al., which suggested that the polypeptides chains of collagen have the maximum absorbance at the wavelength ranges 190-230 nm. This is consistent with the previous researches which indicate that collagen commonly has a high amount of aliphatic amino acid, such as glycine, alanine, and valine, but the low amount of aromatic amino acids such as tryptophan, tyrosine, and phenylalanine. Due to the low content of tryptophan, tyrosine, and phenylalanine, there are no absorption bands between 250-290 nm [30]. This result is similar to those of collagens from sea cucumber *(Stichopus japonicus)* (220 nm), largefin longbarbel catfish *(Mystus macropterus)* (233 nm) and cobia skin *(Rachycentron canadum)* (221 nm).

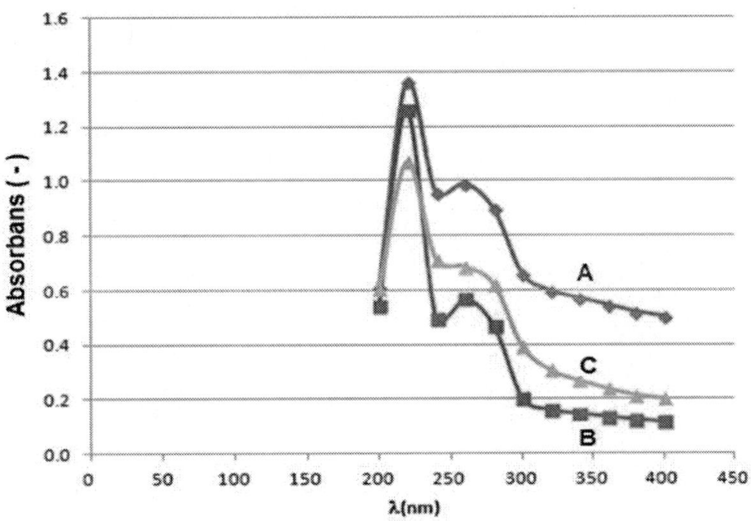

Figure 5. UV Absorption Spectrums of PSC with Solvent Variations: (a) Distilled Water, (b) NaOH 0.1 M, dan (c) $Ca(OH)_2$ 0.2 M.

3.2.2.2 Lowry Method. Protein concentration in the PSC was quantified by Lowry method. The highest amount of protein content was present in samples treated with $Ca(OH)_2$ 0.2M, followed by NaOH 0.1M, and distilled water.

Table 6. The Result of Lowry Method

Solvent	Absorbance	Concentration (mg/ml)	Protein Content (mg/ml)
Distilled Water	0.52	1.64	16.41
NaOH 0.1M	0.82	2.59	25.93
$Ca(OH)_2$ 0.2M	1.14	3.53	35.36

3.2.2.3 Sodium Dodecyl Sulfate Polyacrylamide Gel Electrophoresis (SDS-PAGE). Electrophoretic pattern of the PSC from *A. planci* body is shown in Fig. 6. Previous studies indicate that collagen type I commonly consist of two α chains (Shoulders and Raines, 2009). The SDS-PAGE of PSC extracted from *A. planci* shows that PSC consisted of α and β (dimer) chains. However, the α chains were the major component of PSC, as evidenced by the increasing band intensity at 100-200 kDa. The β-component of PSC composed of β11 and β12 chains. The β11 is a homodimer of α1 chains whereas the β12 chain is the heterodimer of α1 and α2 chains. These results suggested that PSC extracted from *A. planci* was most likely to be classified as type I collagen. itiAlso, electrophoretic pattern of PSC was similar to those reported for collagen from crown of thorns starfish [23], sea cucumber [32], carp skin [32], blacktip and bamboo shark [17].

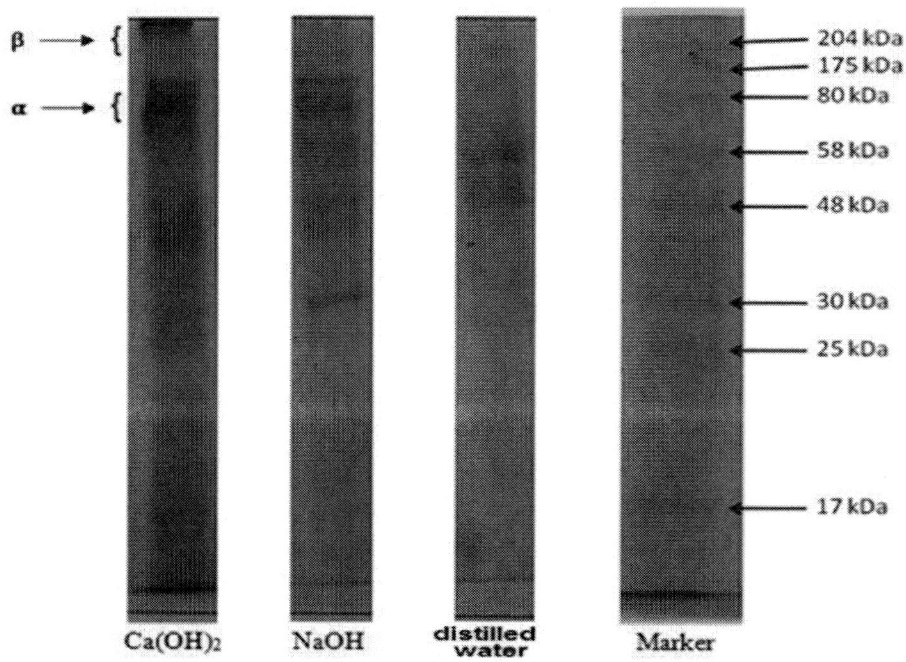

Figure. 6. SDS Page Pattern of PSC from the Body of Crown of Thorns Starfish with Variation of Solvent Extraction. Lane.

3.2.2.4 Amino Acid Composition Analysis. Glycine is the major amino acid in collagen because glycine could be found as being every third residue throughout the central region of the α chain. itiAlso, collagen also characterized by its high contents of alanine, proline, and hydroxyproline. However, in previous studies, glycine can be replaced by alanine because of their properties and structure similarity.

In general, the amino acid composition of PSC extracted from *A. planci* body had similar amino acid profiles with sea cucumber (*Stichopus japonicus*) [33], catfish (*Mystus macropterus*) [34], bamboo shark *(Chiloscyllium punctatum)* and blacktip shark *(Carcharhinus limbatus)* [17] as well as serta cobia skin (*Rachycentron canadum*) [35], freshwater fish scales [36].

Table 7. Amino Acid Composition of PSC from Crown of Thorns Starfish

Amino Acid	Composition (per 1000 residue)	
	NaOH 0.1 M	Ca(OH)2 0.2 M
Aspartic acid	n.d.	n.d.
Glutamic acid	n.d.	n.d.
Serine	166.29	79.63
Glycine	106.68	55.84
Histidine	n.d.	n.d.
Arginine	n.d.	n.d.
Threonine	n.d.	n.d.
Alanine	361.59	535.93
Proline	113.42	173.15
Valine	14.16	6.78
Methionine	n.d.	n.d.
Isoleucine	206.39	98.83
Leucine	8.29	3.97
Phenylalanine	18.79	9.00
Lysine HCl	77.01	36.88

n.d : not detected

3.2.2.5 Scanning Electron Microscopy (SEM). Morphological structures of the isolated collagen samples were observed under SEM micro- photography with lower and higher magnification. PSC had nodular like structures and tubular in nature, with irregular structures. Compared to PSC from other sources, such as the skin of sailfish *(Istiophorus platypterus),* PSC extracted from *A. planci* body are a thicker and largerze because it has not been lyophilized (Fig. 7). However, the PSC from *A. planci* showed a fibrillar structure with thick bundles and porous nature, which demonstrated that PSC from *A. planci* wlould have high wet ability and good fibril strength.

Pore size, porosity, and surface areas are widely recognized as important parameters for collagen to understand their biomedical importance. Other architectural features such as pore shape, pore wall morphology, and interconnectivity between pores of material have found to significantly influence cell behaviors such as cell adhesion, cell seeding, migration, growth, differentiation, mass transport, gene expression, and new tissue formation [37]. SEM microphotography with a magnification of ×3500 (Fig. 8) showed massive and porous nature as the non-collagenous substances in the skin were partially dissolved by NaCl solution. itiAlso, an almost uniform porous matrix of good interconnectivity was noticed. Based on the opreoceding account, PSC isolated from *A. planci body could be* used for biomaterial, cosmetic, biomedical application.

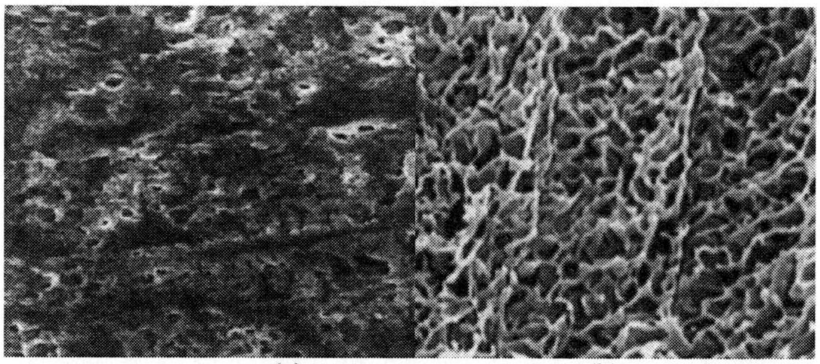

Figure. 7 Morphological Structures of PSC (Magnification ×1000) from (a) Crown of Thorns Starfish and (b) Sail Fish [38].

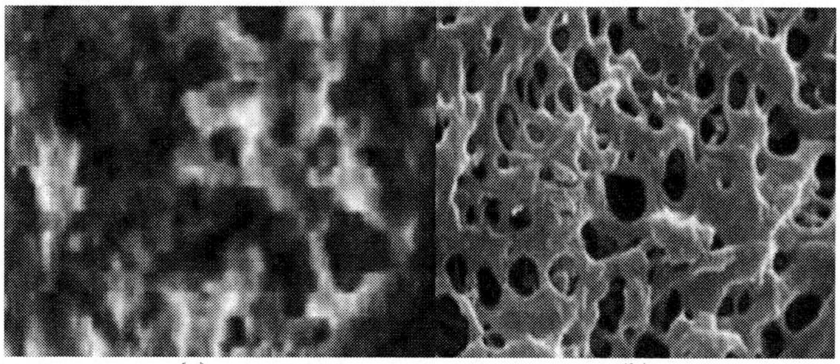

Figure. 8 Morphological Structures of PSC from: (a) Crown of Thorns Starfish (Magnification ×3500) and (b) Sail Fish (Magnification ×2000) [38].

3.2.3 Application and Further Development of Pepsin. Solubilized Collagen from Crown of Thorns Starfish (Collagen) has become a valuable and well-used component in biomedical, pharmacy, or cosmetic industry. Recent studies examined type I and III collagen content and distribution in the skin within the contexts of patient age [39]. As shown in Table 8, type I collagen content varied significantly among the four age groups. The result showed that total collagen declined with age, with the highest type I collagen content was noted in the youngest age group (adolescent).

Table 8. Type I Collagen Content in Normal Skin [39].

Age Group	Type I Collagen (μg/g)
Fetus	264.71 ± 5.88
Adolescent	279.12 ± 7.64
Adult	241.79 ± 8.23
Elderly	209.50 ± 14.31

Type I collagen is very fundamental in skin formation and repair, playing a crucial role in the maintenance of skin tensility and elasticity. Compared to collagen content in healthy human skin, collagen content in the PSC is very high (Table 10). Table 9 shown PSC isolated from the crown of thorns starfish.

Table 9. PSC isolated from Crown of Thorns Starfish [38].

Solvent	Collagen Content		
	µg/g	g/ml	%w/v
Distilled Water	7.847,50	0.016	1.64 %
NaOH 0.1M	13.561,03	0.026	2.59 %
Ca(OH)2 0.2M	19.623,48	0.035	3.53 %

Based on commercial collagen evaluation by Peng [40], collagen content in PSC from *A. planci* is similar to commercial collagen such as Collasol® (Croda Chemicals Ltd, Humberside UK), Collagen CLR® (Chemisches Laboratorium Dr. Kurt Richter, Germany), and AteloHelogen® (Medical, North Ryde, Australia).

Table 10. Analysis Results on Three Samples of Collagen Suitable for Cosmetic Applications [40].

Test	AteloHelogen®	CLRCollagen®	Collasol®
Protein content (%w/v)	1.05%	0.28%	4.00%
pH	4.7	3.8	4.2
pI	>8.9	>8.9	<8.9
Arsenic	<1 ppm	<1 ppm	<1 ppm
Heavy metals	<5 ppm	<5 ppm	<5 ppm
Dry weight	1.15%	5.07%	5.25%
Conductivity	0.45 ms	19.0 ms	42.0 ms
Ash content	<0.1%	1.2%	1.1%
Hydration regain	21%	3%	7%

These results indicate that extraction and purification of collagen from *A. planci* should be further optimized because of its high potential in the biomedical and cosmetic industry.

4. Conclusion

The best result of saponin extraction was achieved by using ethanol 96%, compared to the other polar solvents, such as butanol, isopropyl alcohol, and distilled water. The highest yield obtained was 29.07%. Triterpenoid is the specific type of sapogenin found in the saponin extraction from *A. planci,* Scale-up with ethanol 96% as the extraction solvent gave the resulting yield obtained were 22.99% and 12.98%, respectively. Purification method for saponin extraction was achieved by using activated carbon with a mass of carbon: sample volume = 1:2 (w/v) for 20 minutes. Sapogenin isolation was carried out by acid hydrolysis using hydrochloric acid. As a result, 168.33 mg sapogenin were obtained from the total of 217.50 mg precipitate. Saponin extract with the concentration of 10, 20, 1, 2 mg/ml showed a good insecticide properties. However, sapogenin isolation by diluting the solid form with distilled water until the concentration reached 3 mg/ml was ineffective to exterminate termites. Saponin extract with the concentration of 2 mg/ml was able to exterminate termites within 6 minutes. Based on the t-distribution analysis with 5% level of significance, only 3 of 10 samples have insecticide efficacy (EI) $\leq$ 70%. Therefore it is not a good/effective insecticide.

Type I collagen was successfully isolated from the body of the crown of thorns starfish with multistage extraction method. Three types of alkaline solvent were used in the extraction process. The best result was achieved by using Ca(OH)2 0.2 M as the solvent, followed by NaOH 0.1 M and distilled water. The result shows Ca(OH)2 0.2 M as the best extraction solvent because of its mild hydrolytic effect. Based on this method, 86.75% of crude extract and 2.26% of PSC yield was obtained. Characterization of PSC isolated from the body of the crown of thorns starfish was carried out by qualitative and quantitative analysis. Quantitative analyses consist of Lowry method and amino acid composition analysis. The result showed the highest protein content (37.19 mg/ml), was achieved by the utilization of Ca(OH)2 0.2 M as the extraction solvent. Qualitative analyses consist of gel electrophoresis, UV spectroscopy, and Scanning Electron Microscopy (SEM). These analysis methods showed the structure and properties of type I collagen in the PSC, as well as it's high potency for biomaterial, cosmetic, or biomedical application.

Acknowledgement
Deeply appreciation and gratefully thanks could be expressed to Directorate of Research and Community Engagement for PITTA research fund support.

References

[1] Reichelt R.E., Bradbury R. H., Moran P. J. (1990). The crown-of-thorns starfish, *Acanthaster planci*, on the great barrier reef, Mathematical and Computer Modelling, 13 (6), pp. 45 – 60. Fraser, B. R. N 2000 *Best Practice Guide for Crown-of-Thorns Clean-ups.* Jakarta: Coastal Project.
[2] Fraser, B. R..N 2000 *Best Practice Guide for Crown-of-Thorns Clean-ups.* Jakarta: Coastal Project.
[3] Chaleb, I 2010 Saponins as Insecticides: a Review. *Tunisian Journal of Plant Protection*, 39-50.
[4] Savitri I.K.E., M. Sahlan, F. Ibrahim, and A. Wijanarko 2012) Isolation and Characterization of Phospholipase A2 from Spines Venom of The Crown - of - Thorns Starfish Isolated from Papua Island. *International Journal of Pharma and Bio Sciences*, 3 (4), pp B603-B608.
[5] Hostettmann K., Mastons, A 1995 *Saponins.*
[6] Kitagawa I 1978 Saponin and Sapogenol. XXV. Steroidal Saponins from The Starfish *Acanthaster planci* L. (Crown of Thorn). Structure of Two Genuine Sapogenols, Thornasterol A and Thornasterol B, and Their Sulfates. *Chemical Parmacy Bulletin, 26*(6), 1852-1863.
[7] Komori, T 1997 Toxins from The Starfish *Acanthaster planci* and *Asterina pectinifera. Toxicon, 35*, 1537-1548.
[8] Tamura Y 2012 Application of Saponin-Containing Plants in Foods and Cosmetics. *Intech*, 85-101.
[9] *Budidaya Perikanan.* (2001, Maret). *download* November 12, 2013, dari Ristek Departemen Perikanan: www.ristek.go.id.
[10] Suara Pembaruan. (2012, April 18). *Serangan Rayap Rugikan Indonesia Rp 250 miliar.* Retrieved Maret 13, 2014, Suara Pembaruan: http://www.suarapembaruan.com/home/serangan- rayap-rugikan-indonesia-rp-250-miliar/19234.
[11] Brown, T.A., Nelson, D.E., Southon, J.R 1988 *Improved collagen extraction by modified Longin method.* Radiocarbon 30, 171-177.
[12] Zhang, Y., Liu, W., Li, G., Shi, B., Miao, Y. and Wu, X 2007 *Isolation and partial characterization of pepsin-soluble collagen from the skin of grass carp (Ctenopharyngodon idella).* Food Chemistry 103: 906-912.
[13] Sadowska M, Kołodziejska I, and Niecikowska C 2003 *Isolation of Collagen from the*

Skins of Baltic Cod (Gadus morhua). Food Chemistry, 81, 257-262.

[14] Kimura S, Omura Y, Ishida M, and Shirai H 1993 *Molecular Characterization of Fibrillar Collagen from the Body Wall of Starfish Asterias amurensis*. Comp Biochem Physiol B 104, 663-668.

[15] Mizuta S, Hwang J, and Yoshinaka R 2002 *Molecular Species of Collagen from Wing Muscle of Skate (Raja kenojei)*. Food Chemistry, 76, 53-58

[16] Nagai T and Suzuki N 2000 *Partial Characterization of Collagen from Purple Sea Urchin (Anthocidaris crassispina) Test*, International Journal Food Science and Technology. 35, 497-501.

[17] Kittiphattanabawon, P. Benjakul S, Visessanguan W, Shahidi F 2010 *Isolation and Characterization of Collagen from the Cartilages of Brownbanded Bamboo Shark (Chiloscyllium punctatum) and Blacktip Shark (Carcharhinus limbatus)*. LWT - Food Science and Technology, 43(5), pp.792–800

[18] Rigby, B.J 1968 *Amino-acid Composition and Thermal Stability of The Skin Collagen of The Antarctic Ice-Fish*. Nature, 219, 166-167

[19] Li, C.M., Zhong, Z.H., Wan, Q.H., Zhao, H., Gu, H.F.and Xiong, S.B 2006 *Preparation and Thermal Stability of Collagen from Scales of Grass Carp (Ctenopharyngodon idellus)*. European Food Research and Technology 223 (3-4), 236-241

[20] Astuti, S. M 2011 *Skrining Fitokimia dan Uji Aktivitas Antibiotika Ekstrak Etanol, Daun, Batang, Bunga, dan Umbi Tanaman Binahong (Anredera cordifolia X Steenis*). Bulletin Balai Besar Pengujian Mutu dan Sertifikasi Obat Hewan (BBPMSOH). 19 : 1-13(http://www.scribd.com/doc/217213144/Skrining-Fitokimia-Dan-Uji-Aktifitas-Antibiotika-Ekstrak-Etanol-Daun-Batang-Bunga-Dan-Umbi-Tanaman-Binahong-Sri-Murni)

[21] Sari, A. D. (2012). *Aktivitas Antioksidan dan Komponen Bioaktif Bintang Laut Culcita sp.* Bogor: Repository IPB

[22] A. D., Paton D., Krista Ballantyne K., Aubin, A.A 2002 Process for recovery and purification of saponins and sapogenins from quinoa (Chenopodium quinoa), US6355249 B2

[23] Tan, C.C., Karim, A. A., Latiff, A. A., Gan, C.Y. and Ghazali, F. C. 2013. *Extraction and Characterization of Pepsin Solubilized Collagen from the Body Wall of Crown of Thorns Starfish (Acanthaster planci)*. International Food Research Journal, 20(6), 3013–3020.

[24] Lee, K 2009 *Biochemical Characterization of Collagen from the Starfish Asterias amurensis*. Journal of the Korean Society for Applied Biological Chemistry, 52(3), 221–226

[25] Komsa-Penkova, R.; Spirova, R. and Bechev, B 1996 Modification of Lowry's method for collagen concentration measurement. *J. Biochem. Biophys. Meth.*, 32,33-43.

[26] Laemmli, U. K 1970 *Cleavage of structural proteins during the assembly of the head of bacteriophage T4*. Nature 227: 680-685.

[27] Webb, L. J 1995 A Preeliminary Phytochemical Survey of Papua - New Guinea. In L. J. Webb, *Pacific Science* (pp. 431 - 441).

[28] Suparjo. 2010 *Saponin: Peran dan Pengaruhnya Bagi Ternak dan Manusia.* Jambi: Fakultas Peternakan Universitas Jambi, Jambi (http://jajo66.files.wordpress.com/2008/06/ saponin.pdf.

[29] Usman, L.A. Ibiyemi, S.A., Oluwaniyi O.O., Ameen, O. M 2003 *Effect of acid and alkalinehydrolysis on the concentrations of albumin and globulin in Thevetia peruviana seed cake protein extract, Afrika Journal of Biotechnology Vol. 8 (22), pp. 6374-6378.*

[30] Lin, S.-H., Chen, C.-N. and Juang, R.-S 2006 *Extraction equilibria and separation of phenylalanine and aspartic acid from water with di(2-ethylhexyl)phosphoric acid.* J. Chemical. Technology and Biotechnology, 81: 406–412.

[31] Adibzadeh, N., Aminzadeh, S., and Jamili, S 2014 *Purification and Characterization of Pepsin- Solubilized Collagen from Skin of Sea Cucumber Holothuria parva.*, Applied

Biochemistry and Biotechnology, 173, pp.143–154.
[32] Jia, Y.J., Wang, H.B, Wang, H.Y., Li Y.Y., Wang M., Zhou J 2012 *Biochemical Properties of Skin Collagens Isolated from Black Carp (Mylopharyngodon piceus)*. Food Science and Biotechnology, 21(6), pp.1585–1592
[33] Cui, F., Xue, C., Li, Z., Zhang, Y., Dong, P., Fu, X., Gao, X 2007 *Characterization and Subunit Composition of Collagen from the body Wall of Sea Cucumber Stichopus japonicas*. Food Chemistry, 100, pp.1120–1125.
[34] Zhang, M., Liu, W., and Li, G 2009 *Isolation and Characterization of Collagens from the Skin of Largefin Longbarbel Catfish (Mystus macropterus)*. Food Chemistry, 115, 826-831
[35] Zeng, S. Yin J, Yang S, Zhang C, Yang P, Wu W 2012 *Structure and Characteristics of Acid and Pepsin-Solubilized Collagens From the Skin of Cobia (Rachycentron canadum)*. Food Chemistry, 135(3), pp.1975–1984.
[36] Zhang, F.X., Wang A.N, Li Z.H., He S.W., Shao L.J 2011 *Preparation and Characterisation of Collagen from Freshwater Fish Scales*. Food and Nutrition Sciences, 02(08), 818–823.
[37] Yang, S. F., Leong, K. F., Du, Z. H., & Chua, C. K 2001 *The design of scaffolds for use in tissue engineering*. Part 1. Traditional factors. Tissue Engineering, 7(6), 679–689
[38] Tamilmozhi, S., Veeruraj, A. and Arumugam, M. 2013. *Isolation and Characterization of Acid and Pepsin- Solubilized Collagen from the Skin of Sail Fish (Istiophorus platypterus)*. Food Research International, 54(2), pp.1499– 1505.
[39] Wang C., Rong Y.H., Ning F.G Zhang G.A 2011 *The Content and Ratio of Type I and III Collagen in Skin Differ with Age and Injury*. African Journal of Biotecnology, 10(13), pp.2524– 2529.
[40] Peng, Y., Glattauer V, Werkmeister JA, Ramshaw JA 2004 *Evaluation for Collagen Products for Cosmetic Application*. Journal of Cosmetic Science, 55(4), pp.327–41

IOP Conf. Series: Earth and Environmental Science **89** (2017) 012033 doi:10.1088/1755-1315/89/1/012033

Diversity and Abundance of Chlorophyta in Krakal Beach, Gunung Kidul, Yogyakarta

A.C. Putri[1], I.C. Nugroho[1], N.U.N. Firdaus[1], N.O.J. Puspita[1], S.A.R. Fajrin[1], S.D.A. Hamzah[1]

[1]Faculty of Biology, Universitas Gadjah Mada, Jalan Teknika Selatan, Sekip Utara, Yogyakarta 55281, Indonesia

Email: anggun.cinditya.p@mail.ugm.ac.id

Abstract. Chlorophyta plays an important role in energy flow as the main producer in marine food chain, material circulation, bioaccumulation, and bio-indicator in the intertidal zone. Several genera of Chlorophyta have been used by local society around coastal area of Gunung Kidul, Yogyakarta as local product, but the research about diversity and abundance information of Chlorophyta in Krakal beach was not yet understood. The aim of this study is to gain the information about diversity and abundancy of Chlorophyta in Krakal Beach, Gunung Kidul,Yogyakarta. This research was conducted in March 11th 2017. This research utilizes Line Transect and Stratified Random Sampling method which is used 1x1 m plot. There are six genera of Chlorophyta have been identified in this research, such as Chaetomorpha sp.; Boergesenia sp; Ulva sp.; Cladophora sp.; Enteromorpha sp.; and Halicystis sp. From 6 genera of Chlorophyta, the highest genera coverage is Enteromorpha sp. (9.88%). This research is expected to record data of macroalgae abundance especially Chlorophyta, in Krakal Beach. By this research, monitoring of macroalgae could be done and supported by government and local people to maintain the sustainability of Chlorophyta.

1. Introduction

Indonesia is an Archipelago country, with the sum of the island about 13.466, according to toponymy survey "*Tim Nasional Pembakuan Nama Rupabumi*" at 2007-2010. Based on *Badan Informasi Geospasial* (B.I.G.) data, Indonesia had a total beach line 99.933 km. Krakal Beach is one of beach in Indonesia which located in Gunung Kidul, Yogyakarta, exactly in Ngestirejo, Tanjungsari, Gunungkidul,

Yogyakarta, Indonesia (S8°8'44" E110°36'0"). Krakal beach had a big diversity potential of the marine organism. The ecosystem of Krakal beach is rocky, with this kind of ecosystem can be found a lot of coral flora and fauna communities, some of them are Algae, Bivalves, Coral Reefs, Fishes, and other marine organisms [1]. Macroalgae (seaweed) is eukaryotic organism capable of performing photosynthesis. This organism mostly found in the sea as primary producers in the ecosystem [2]. Macroalgae also provide shelter (habitat) for various types of marine life such as Mollusca, Crustacea, Echinodermata and coral fish [3]. Whereas in the economy, macroalgae are often made various products that have high selling value and wide market [4]. Some examples of macroalgae products which useful, look like food, medicine and important matter in industries [5]. Macroalgae can be divided into three divisions based on its color pigment: Chlorophyta (green algae), Rhodophyta (red algae), and Phaeophyta (brown algae) [6]. Green algae or Chlorophyta is abundant in warm (trophic) waters and recorded at least 12 genera of green algae [7].

According to many utilization of macroalgae, the conservation and development of it were very needed for the sake of macroalgae utilization sustainable and for to keep the marine ecosystem conservation. Therefore, the information of macroalgae diversity on the beach in particular on the Krakal beach, Gunung Kidul, Yogyakarta is needed.

2. Methodology

2.1. Materials

Chlorophyta that found in the intertidal zone of Krakal Beach was used as samples by purposive random sampling. Each sample was taken completely from the substrate for identification.

2.2. Instrumentations

Instrumentations, which were used in this research, were quadrat plot 1 m x 1 m with sub-plot 10 cm x 10 cm for measured the abundance and diversity of Chlorophyta, zip-lock plastic for storage, digital camera for documentation, FAO Species Identification Guide For Fishery Purpose for identification [9]. Thermometer and pH meter for measuring the environmental parameters such as temperature and acidity of seawater.

2.3. Sampling and Preparation

Sampling was done on March 11th, 2017 in Krakal Beach, Gunung Kidul, D.I. Yogyakarta (Figure 1) at 2 pm when it was minimum tides.

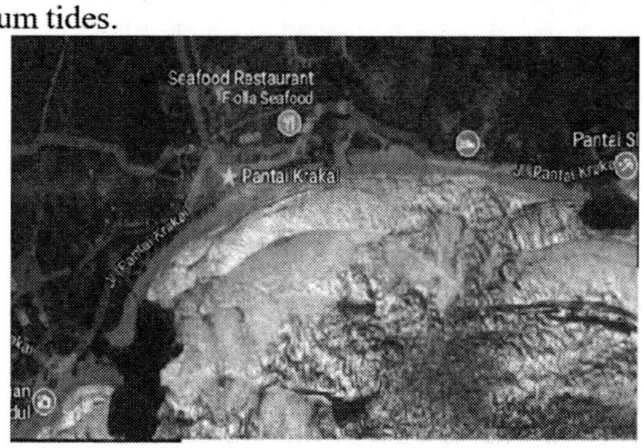

Figure 1. The study site, Intertidal zone of Krakal Beach (Source: Google Maps)

2.4. Measured of Seaweed Abundance

Chlorophyta diversity and abundance was determined in the intertidal zone of Krakal Beach by 1 m x 1 m quadrat plot with sub-plot 10 cm x 10 cm (figure 2). Four transects line were designed by differences in depth, transect A at the seaside, transect B and C in the middle, and transect D at the end of the zone. Then each transect was divided into five sampling areas. The distance between transects is 15 m and the distance of each sampling point is 20 m (figure 3). In each plot, environmental parameters were measured, temperature by thermometer and acidity of seawater by pH meter.

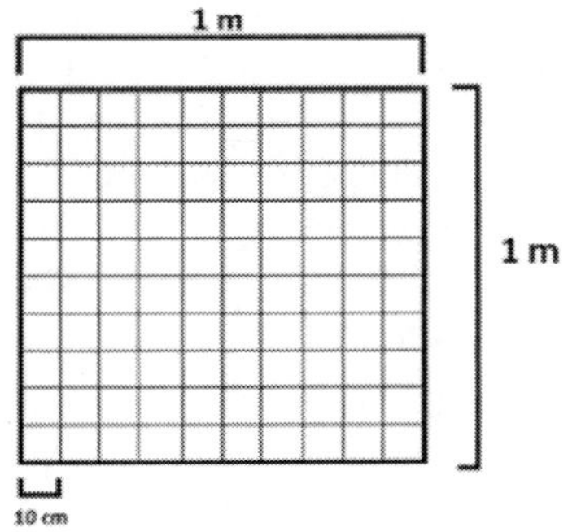

Figure 2. Quadrat plot with 100 sub-plotS

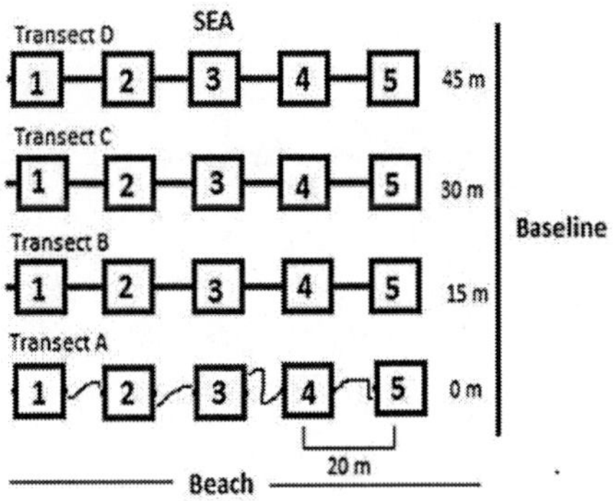

Figure 3. Sampling design

Percent cover of seaweed calculated using the following formula [9].

$$Pi = \frac{ai}{A}$$

Description:
Pi = covered species i (%)
ai = the total cover species i (%)
A = the total area covered macroalga (%)

2.5. Identification
Samples (Chlorophyta), which take by purposive random sampling in each quadrat plot stored in zip-lock plastics. Then, samples were cleaned and photographed in millimeters block paper for identification. The identification results show the seaweed diversity.

3. Result and Discussion
Based on a field study in Krakal Beach Gunung Kidul, obtain the percent cover (figure 4) and the abundance of Chlorophyta genera in Krakal Beach, Gunung Kidul, Yogyakarta. In figure 4, there are 6 genera of Chlorophyta found in Krakal Beach, they are Ulva sp., Halicystis sp., Enteromorpha sp., Cladophora sp., Chaetomorpha sp., and Boergensenia sp. the highest percent cover is owned by Enteromorpha sp. (9,88%) and the lowest coverage is owned by Halicystis sp. (0,14%).

Among the six Genera of Chlorophyta which is found in Krakal Beach, the highest abundance is owned by Enteromorpha sp. (34,85%), and the lowest coverage is owned by Halicystis sp. (0,49%). The diversity of Chlorophyta in Krakal beach consists of six genera they are *Boergesenia sp.; Chaetomorpha sp.; Cladophora sp.; Enteromorpha sp.; Halicystis sp.; Ulva sp.*

In this study, the most found Chlorophyta genera in each transect are different. Enteromorpha sp. is mostly found in transect A and *Chaetomorpha sp.* is mostly found in transect B and transect D. In transect 3, Cladophora sp. is the most found. The number of genera found in each transect is also different. In transect A and B there are six genera of Chlorophyta found (*Boergesenia sp.; Chaetomorpha sp.; Cladophora sp.; Enteromorpha sp.; Halicystis sp.; Ulva sp.*). In transect three there are four genera of Chlorophyta found (*Boergesenia sp.; Chaetomorpha sp.; Cladophora sp.; Ulva sp.*) and three genera in transect 4 (*Chaetomorpha sp.; Cladophora sp.; Ulva sp.*)

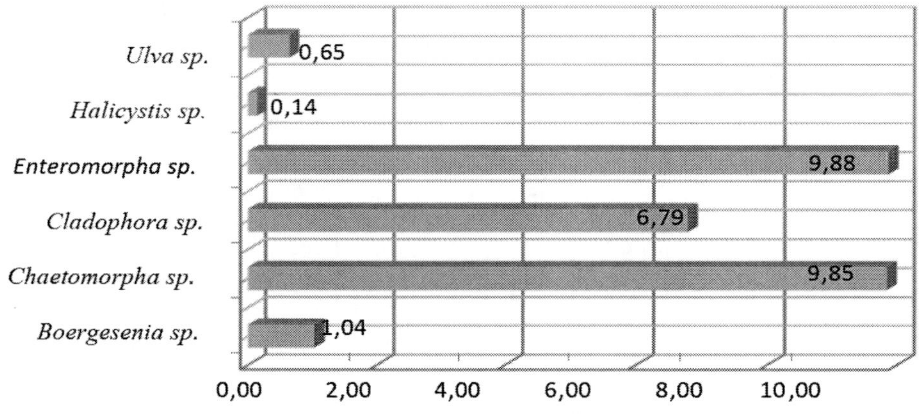

Figure 4. Percent Cover of Chlorophyta Genera in Krakal Beach

Air temperature, water temperature and pH of the water are measured. The average of air temperature is 35°C and air temperature is 31,6°C. The pH of the water is seven (neutral).

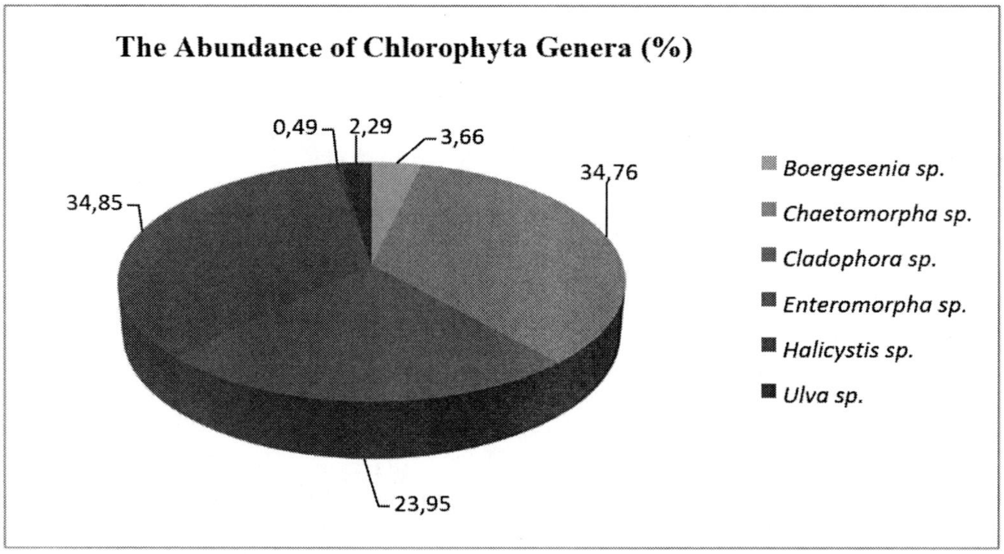

Figure 5. The Abundance of Chlorophyta Genera in Krakal Beach

There are six genera of Chlorophyta found in Krakal Beach that has been identified. They are *Ulva sp., Halicystis sp., Enteromorpha sp., Cladophora sp., Chaetomorpha sp., and Boergensenia sp.* The highest percent cover Chlorophyta in Krakal Beach is *Enteromorpha sp.* as much as 9,88% among the others six genera. *Enteromorpha* grew on rocky substrate in the lower intertidal to shallow subtidal zones or in shallow tide pools and exposed to air during low tide [9]. This genus is mostly found in transect A which has rocky and sandy substrate. Although *Enteromorpha sp.* has the highest coverage among six genera, this genus is only found in transect A. Suitable environmental conditions are the main reason *Enteromorpha sp.* grows well a long transect A.

Followed by *Chaetomorpha sp.* 9.85% coverage, *Chaetomorpha sp.* has wider tolerance limit. *Chaetomorpha sp.* is found in every transect from the shoreline (transect A) up to 45 meters to the sea (transect D). *Chaetomorpha sp.* has thallus in the form of a lumpy thread. The percent cover of *Cladophora sp.* is as much as 6,79%. Thallus of *Cladophora sp.* is green-dark green, forming compact cushions. It has 3-14 cm in diameter and up to 2 cm high, composed of entangled, stiff, often incurved axes [10]. *Cladophora sp.* is also found in each transect from transect A to D.

Halycistis sp. has the lowest percent coverage as much as 0,49%. *Halycistis sp.* is only found in transect A and transect B in low density. In those transect, *Enteromorpha sp.* and *Chaetomorpha sp.* are dominating.

Ulva sp. has the percent cover about 0,65%. *Ulva sp.* is in the bottom two of the lowest percent cover. Among the six genera, *Ulva sp.* is mostly used by local people. The local people in Krakal beach often take it to make a seaweed chips ("keripik Ulva") as a commodity trading. The condition of Krakal beach is damaged by human activity and the exploitation of natural resources now [1]. Macroalgae are vulnerable to changes and ecological pressures that occur in their environment. People activities can affect the existence of macroalgae [11].

Also, this research was conducted in March that is the rainy season. Rainwater can disrupt the salinity of seawater that causes macroalgae cannot grow well [12]. According to the previous research the highest coverage and density of seaweed in Krakal beach around July-August (Summer)[13].

4. Conclusion

Based this field study, six genera of Chlorophyta were inventoried in Krakal Beach; they are *Ulva, Enteromorpha, Boergesenia, Halicystis, Cladophora, and Chaetomorpha.* Among the six genera, *Enteromorpha* has the highest percent coverage as much as 9.88%. While *Halicystis* has the lowest coverage as much as 0.14% from 28.35% coverage of Chlorophyta genera in total.

Acknowledgement

We are grateful to Department of Algae, Crustacea, Pisces, Echinodermata and all of the members of Marine Study Club Gadjah Mada University for the support.

References

[1] Pujiastuti, P., 2009, *Pemanfaatan Secara Lestari Kawasan Perairan Pantai Krakal Sebagai Sarana Pembelajaran*, Universitas Negeri Yogyakarta: Yogyakarta.
[2] Petsut, N., A. Chirapart and M. Keawnern, 2012, A stability assessment on seasonal variation of seaweed beds in the Tart peninsula of Thailand. *Biodiversity Journal*, vol. 3 (3): 229- 230.
[3] MarianingsihP., Evi, A., Teguh, S., 2013, Inventarisasi dan Identivikasi Makroalga di Perairan Pulau Untung Jawa, *Prosiding Semirata FMIPA Universitas Negeri Lampung*.
[4] Praptinah, Muzayyinah dan Harlita, 2004, Keanekaragaman Rhodophyceae di Pantai Sundak sebagai Sumber Belajar Biologi Algae. *Bioedukasi*. Vol.1, No. 1, Hal. 13 – 19.
[5] Angka, S.L., dan M.G. Suhartono, 2000, *Bioteknologi Hasil Laut*, PKSPL-IPB: Bogor
[6] Pratama, W., S.C. Dewi, I.Z.R. Sari, A. Hardiyati and A.E. Wajong, 2015, Distribution and Abundance of Macroalgae in Intertidal Zone of Drini Beach, Gunung Kidul, DIY. *KnE Life Sciences*, vol. 2 (1) pp. 514- 515.
[7] Romimohtarto, K. dan S. Juwana, 2009, *Biologi Laut: Ilmu Pengetahuan Tentang Biota Laut*, Djambatan, Jakarta.
[8] Tuwo A, 2011,*Pengelolaan ekowisata pesisir dan laut: pendekatan ekologi, sosial-ekonomi, kelembagaan, dan sarana wilayah*, Brilian Internasional, Surabaya.

[9] Trono Jr, G. C. Seaweed. In Kent E. Carpenter and Volker H. Niem, 1998, *The Living Marine Resources Of The Western Central Pacific.Volume 1. Seaweed, Corals, Bivalves, and Gastropods*, Rome : Food And Agriculture Organization Of The United Nations, pp. 19-96,.

[10] Leliaert, F. & Coppejans, E., 2003, The marine species of Cladophora (Chlorophyta) from the.South African East Coast. *Nova Hedwigia*, vol. 76 pp. 45–82.

[11] Atmadja, W.S., 1999, Sebaran dan beberapa Aspek Vegetasi Rumput Laut (Makroalga) di Perairan Terumbu Karang Indonesia, Lembaga Penelitian Oseanografi LIPI, Jakarta.

[12] Nakajimaa, M., Hokoia, S., Oguraa D., Chiemi IBA, 2015, Relationship between Environmental Conditions and Algal Growth on the Exterior Walls of the Ninna-ji Temple, Kyoto, *Energy Procedia* Vol. 78 , 1329 – 1334,

[13] Pramesti, R., Susanto, A.S. Wilis., Ridlo A., Subagiyo, Oktaviaris Y., 2016, Struktur Komunitas dan Anatomi Rumput

[14] Laut di Perairan Teluk Awur, Jepara dan Pantai Krakal, Yogyakarta, *Jurnal Kelautan Tropis*, Vol. 19(2):81–94

IOP Conf. Series: Earth and Environmental Science **89** (2017) 012034 doi:10.1088/1755-1315/89/1/012034

Phytoplankton distribution and composition in the waters of Burung and Buntal islands, Kotania bay, western Ceram, Maluku, Indonesia

N V Huliselan[1], M A Tuapattinaja[1] , U N Pattimura[1]

[1]Fisheries 1 Department of Aquatic Resources Management, Faculty of Fisheries and Marine Science of Pattimura University

Email: nietteh@yahoo.com

Abstract. Start your abstract here Abstract. Phytoplankton distribution and composition was evaluated on February, April and June 2015, from Burung and Buntal islands in Kotania Bay. Altogether 42 species phytoplankton representing at least 27 genera from 15 families were collected. The high similarity of species was found in the waters of both islands, but the numbers of species were found in the waters of Buntal Island was higher than those in Burung island. Based on spatial distribution, the highest abundance of phytoplankton was encountered in the waters of Buntal island, while in June, the average abundance of phytoplankton reached the highest with the value of 1010225 cell/m3, and it was also in the waters of Buntal island. We speculate about the differences of a habitat of the two islands, and argue that the composition and abundant of species are related to the season and condition of the surrounding waters.

1. Introduction

The hydrography of Kotania bay is influenced by the monsoons, with southeastern winds prevailing between April and November and northwestern wind between December and March [1], and Kotania bay surrounding water is influenced by the mass waters of Banda Sea that rich of nutrients. Therefore, Kotania bay the waters have potentially a source of economic fishes, and the bay provides a significant income to fishermen and other coastal peoples. Currently, it is estimated that the catch activities has led to overfishing. Moreover, the artisanal fishers used poison and explosive fishing method that cause ecological disturbance to the waters. In some circumstance, the rapid changes of the waters qualities disrupt phytoplankton production [2].

Phytoplankton plays an important role in marine ecosystems, and they support and improve fisheries as a primary producer. Their role in the marine food chain is of paramount importance, and their abundance determine the amount and type of marine animals including fish which are caught commercially [3], Distribution and composition of phytoplankton depend upon nature of the water. Therefore, the abundance and number of species may vary according to waters conditions.

Burung and Buntal islands of Kotania Bay surrounding waters provide good habitat for those minute marine floating algae as both islands are dominated by different productive ecosystems. Seagrass ecosystem dominated the shallow regions of Burung, island, meanwhile, mangrove ecosystem covers around the island of Buntal. As the neritic areas, the waters are highly productive, since the bottom traps sinking organic particles and some of the regenerated nutrients and return to the water column [4].

Beside the distribution and composition of phytoplankton, their abundance is also important to the marine life, and they can bloom under favorable conditions. Since phytoplankton is present in the lighted region of the sea, therefore they are abundant in the marine regions especially in the neritic ecosystems, but the similarity of species phytoplankton may vary in both waters (Burung and Buntal) according to the different of water conditions. Temporarily, temperature and salinity are the important factors to phytoplankton distribution and composition, and the abundant are also influenced by monsoons [5]. In a natural situation, other factors may influence the composition of phytoplankton, e.g. nutrients and light [6]. The phytoplankton is also distributed vary in space and time [7], and nutrients especially nitrate and phosphates play a major part in controlling the abundant of phytoplankton [4]. Therefore this study presents an analysis of the spatial-temporal distribution and composition of phytoplankton at the waters of Burung and Buntal islands, Kotania bay.

2. Material and Method

This study was conducted between February to June 2015 in the surrounding waters of Burung and Buntal islands in Kotania bay (Figure 1). This sampling period covered west and east monsoons and transition between those, and the sampling was done in every two months (February, April, and June) at six fix stations. This study is primarily concerned with the factors controlling the temporal distribution of phytoplankton in Kotania bay by subject to marked monthly changes in hydrography associated with changes in monsoon conditions and to analysis the distribution, composition and the abundance of phytoplankton and also the community structure between two habitats.

Phytoplankton was collected by towing plankton net sled fitted with a 64 μm meshed with a mouth area of 0.3 m2 horizontally. The catch was preserved immediately in 4% formaldehyde in seawater. In the laboratory, phytoplankton samples were identified to species level whenever possible by following (8,9).

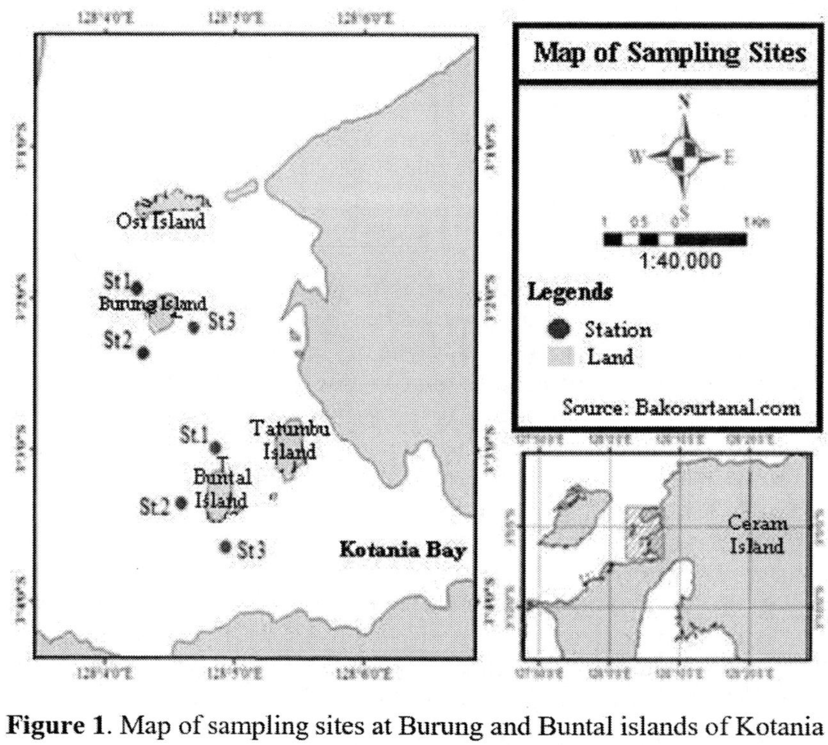

Figure 1. Map of sampling sites at Burung and Buntal islands of Kotania bay

The numbers of phytoplankton are calculated by computed the number cubic meter of seawater entered the net (V) and is analyzed based on the formula developed by [10] as follows:

$$A = \pi . r^2 . L \quad (1)$$

Furthermore, the abundance of phytoplankton was counted followed the formula as follow:

$$D = \frac{Nf.Vp}{V} \quad (2)$$

Where D is the abundance of phytoplankton, Nf is the numbers of phytoplankton in a drop of sample, Vp is the numbers of a total drop of the sample and V is filtered sea water. The Shannon-Wiener diversity indices calculated using natural logarithms (H') were used to describe species diversity and species richness within phytoplankton during the study period [11] as follow:

$$H' = -\sum_{i=1}^{s} Pi \ln Pi \quad (3)$$

Where pi is the proportion (n/N) of individual of one particular species found (n) divided by the total number of individuals found (N), and s is numbers of species. H' < 2 described as an unsteady community, while, H' 2 <H'< 3 is a fair community and H' > 3 is described as steady community Evenness Indices in the phytoplankton community is the degree to which individuals are split among species. The values can be determined [11] by using formula as follow:

$$J' = \frac{H'}{H'\text{Maks}} \quad (4)$$

Where J' (Evenness index) is constrained between 0-1, *H'* is the number derived from Shannon diversity index, *H'*max is the maximum possible value of H' (each species was equally likely), and S is a total number of species. Simpson's index (D) characterized species diversity in the phytoplankton community could be calculated by calculating the proportion of species *i* to the total number of species (*pi*) and squared.

$$D = \frac{n(n-1)}{N(N-1)} \quad (5)$$

N is the total number of species of phytoplankton, and n is the total number of organisms of a particular phytoplankton species. Simpson's index of diversity is 1-D.

Similarity indices (S) could be determinate by using Sorenson's index as follows:

$$S = \frac{2C}{A+B} \quad (6)$$

Where C is the number of species the two communities have in common, A is the total number of species found in community A, and B is the total number of species found in community B.

3. Results and Discussions

The month of February to Juni 2015 was represented west monsoon and east monsoon and the transition between those, therefore the condition of the water especially temperatures of surface water changed. In April to May 2014, the sea surface temperature of Banda Sea was 28.93-30.45°C, while in June, temperatures of the surface waters decreased to 27.39-28.27 °C [12].This study also found that the temperatures in all stations were 30-31°C in February, 28-32°C and 26-30°C in June (Table 1). Data of this study showed that the highest of temperature was in April (32°C) at Station 1 of the island of Burung and the lowest was also at those waters but at station 3 and it was also in June. The high temperature happened in the Banda Sea was in March to May during the dried monsoon [1].

Table 1. Temperatures of surface waters (°C) of Burung and Buntal islands (February to June 2015)

	Burung island			Buntal island		
Months	St. 1	St. 2	St. 3	St. 1	St. 2	St. 3
February	31	30	30	30	30	30
April	32	30	28	31	30	31
June	26	30	26	26	27	27

In the neritic waters, the change of temperature and salinity nearly occur together [6], even though, the fluctuation of the temperature was low. The ranges of salinities at all stations in the waters of both islands were reverse to the temperature (Table 2).

Table 2. The salinities of surface waters (‰) of Burung and Buntal islands in February to June 2015

	Burung island			Buntal island		
Month	St. 1	St. 2	St. 3	St. 1	St. 2	St. 3
February	32	31	33	31	32	32
April	30	29	31	30	29	29
June	34	36	33	36	33	34

In June, temperatures in the waters of both islands tended to decrease, while the salinity was on the contrary to temperature. The temperature in the waters varied considerably more than salinity; it found that the highest salinity and the lowest temperature were in June. The range of optimal temperature needed by marine organisms is from 20 – 30 °C [13]. The temperature and salinity at all stations of the study sites supported the marine organisms living on, as well as phytoplankton. Salinities range from stations of the island of Burung varied between 29 and 36 $^0/_{00,}$ while at stations in the island of Buntal were 29 – 36 $^0/_{00.}$ Those variation salinities value indicated that the water was categorised as shallow coastal waters [3] but have influenced by the open ocean as the Banda Sea. Partially, in the enclosed sea as Kotania bay salinities varied extremely, the condition indicated that the dynamic of hydrography parameters was influenced by several factors such as geography structure, water circulation, rainfall and precipitation [14, 4].

The highest salinity was at station 1 in the water of island of Buntal and station 2 of island of Burung. Both stations is closed to each others, therefore the stations had nearly the same salinity values. Furthermore, pH was the other hydrography parameters studied. It found that the range of pH at all stations was between 6 and 8 and most of the marine organisms including phytoplankton able to tolerate this condition, therefore nearly almost plankton organisms live in the waters with pH around 7-8.5 [13].

3.1. *Phytoplankton Composition*

There were 42 species of phytoplankton found at all stations, which consisted of 27 genera, five orders. Altogether they belong to four classes, e.g., Bacillariophyceae (26 species), Dinophyceae (12 species), Cyanophyceae and Cyanophyceae had only two species respectively. The result of this study found that phytoplankton species at all stations in the waters of Burung and Buntal islands were different in time and space (Table 3).The high number of species was in June at the station 1 of the island of Buntal. Accounting for about 42 species of phytoplankton encountered from both islands, as much as 41 species found in the waters of Buntal island which covered by mangrove ecosystem, *Ceratium candelabrum* did not occur at all stations in Buntal island and also species of *Ornithocercus serratus* and *Ceratium confortum* were absent at all stations of Burung island.

During the study, encountered 35 species of phytoplankton were present at all stations and it was in June. There were numbers of 28 species of phytoplankton encountered in February, while *Bacteriastrum hyalinium, Hemiaulus haucki, Eucampia zoodiacus, Streptoteca indica, Rhizosolenia imbricate, R. stolterforthii, Leptocylindrus danicus, Stephanophyxis turris, Thalasiosira diporocylus* occurred only in June. The differences species of phytoplankton occurred in the waters in the study sites, and they varied temporarily and spatially since the distribution of phytoplankton depends upon the conditions of the water.

Table 3. Numbers of phytoplankton species at all stations

Months	Burung island Number of species			Buntal island Number of species			Total
	St. 1	St. 2	St. 3	St. 1	St. 2	St. 3	
February	18	22	18	3	15	14	28
April	17	16	15	17	25	19	30
June	19	26	31	34	31	30	35
		40			41		

Phytoplankton belongs to class Bacillariophyceae, dominated the waters of both islands from February to June, it was about 70%, and 25 species encountered belonged to them (Figure 2), while the other classes were in less than 10%. This phenomenon showed that phytoplankton belongs to class Bacillariophyceae are capable of adapting the condition of the waters, where temperatures and salinities varied in time and space [3, 15].

3.2. The Abundance of phytoplankton

Quantitatively, the abundance of phytoplankton at all stations of the island of Burung and Buntal varied in numbers (Table 4). They occurred abundantly in June, especially at the stations in the island of Buntal.

Table 4. The abundance of phytoplankton in the waters of island of Burung and Buntal

Month	Phytoplankton of island of Burung (cell/m^3)			Phytoplankton of island of Buntal (cell/m^3)		
	St. 1	St. 2	St. 3	St. 1	St. 2	St. 3
February	123749	96907	292311	31392	209190	23749
April	188694	80050	121019	55359	164813	112202
June	84850	405596	1376706	1725478	1343039	1125682

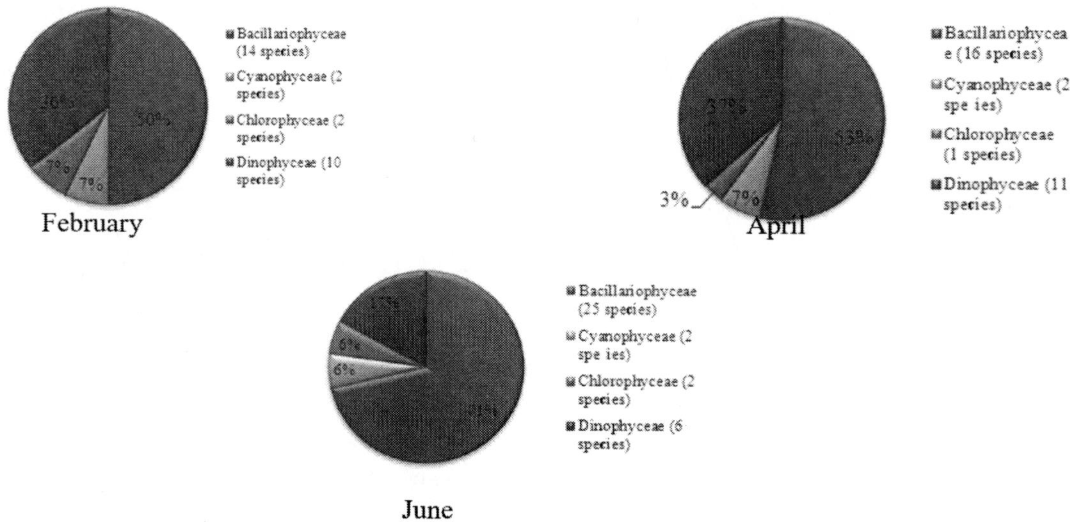

Figure 2. Composition of phytoplankton in the waters of the island of Burung and Buntal (February, April and June 2015)

The abundant of phytoplankton at all stations of both islands ranged between 23749 and 1725478 cell/m3. The high abundance was found in June and at station 1 of island of Buntal and the lowest was also at station 3 of island of Buntal and it was found on February. These phytoplankton numbers were higher than the result found from Ambon bay [5]. The abundance of phytoplankton in stations of the island of Burung was highest at station 3 and the lowest was at 2. Even though both islands were dominated by productive ecosystem, e.g., mangrove and seagrasses, it is argued that the mangrove which covered island of Buntal more productive than seagrasses in the shallow part of the island of Burung. This result reflected the important of the coastal ecosystem in supporting the abundant of phytoplankton as the foundation of the marine food chain [3].

Table 5. The correlation between hydrography parameters and the abundance of phytoplankton at all stations of island of Burung and Buntal (February to June 2015)

Variables	Regression (alpha = 0.05)		
	y = a + bx	r	R^2
Temp. (x) abundance (y)	y = 61278.72 - 1968,18x	-0.712	0.507
Salinity (x) abundance (y)	y = -45402 + 1552.773x	0.613	0.375
pH (x) abundance (y)	y = 30474.29 - 3694.76x	-0.389	0.151

IOP Conf. Series: Earth and Environmental Science **89** (2017) 012034 doi:10.1088/1755-1315/89/1/012034

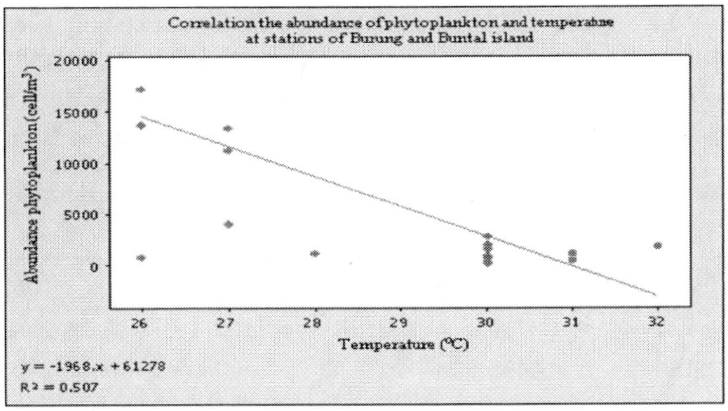

Figure 3. The correlation of temperatures and the abundance of phytoplankton at all stations

The result in Table 5 explained that the abundance of phytoplankton in the waters of both islands was influenced by hydrographic parameters dynamic, such as temperatures and salinities (Figure 3 and 4). Therefore, it can be stated that the variations of the abundance of phytoplankton in the waters of both islands influenced by monsoons and the temperature and salinity value depend on the weather or monsoons. This study detected that the temperature and salinity varied in time and space with was followed by the abundance of phytoplankton. Overall, the temperature was the most influenced the abundance of phytoplankton and also happened to the phytoplankton diversity.

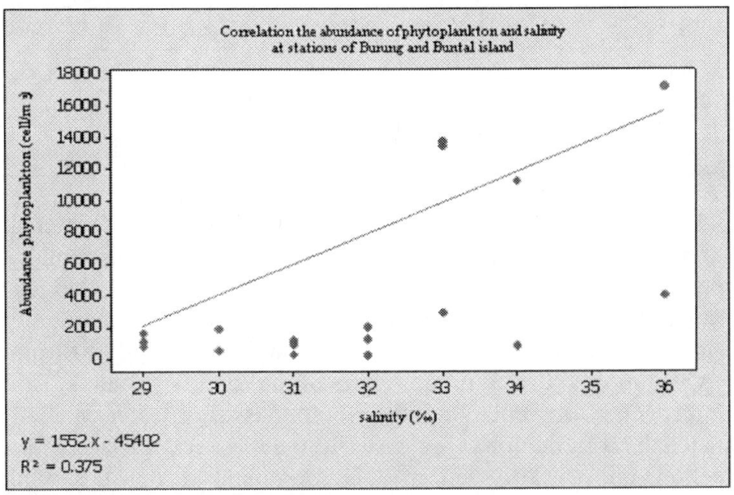

Figure 4. The correlation of temperatures and the abundance of phytoplankton at study sites.

3.3. *Diversity, Evenness, dominance, and Similarity of phytoplankton at all stations*

The steady-state of phytoplankton communities at all stations in the water of island of Burung and Buntal estimated by computing the diversity of species using Diversity index which based on the number of species of phytoplankton present (richness) in the stations and the number of individuals per species, therefore the more species found, the more diverse the phytoplankton found. The dominance and similarity of phytoplankton defined which species dominated the others, and the similarity of different communities, showed by numbers of species present in one community and not in others [11]. Phytoplankton has different types and quantities, therefore the diversity of species of phytoplankton should be analyzed to get more detail information of the community structure.

Table 5, shows that the diversity of phytoplankton community at station 1 in the island of Buntal higher than other stations. In common, it revealed that phytoplankton community of island of Buntal waters (2.181 – 2.872) more diverse than at stations in island of Burung (1.524 – 2.738).

Table 6. The diversity (H') of phytoplankton communities of island of Burung and Buntal

Month	Burung island			Buntal island		
	St. 1	St. 2	St. 3	St. 1	St. 2	St. 3
February	2.168	2.409	2.268	2.008	1.949	2.192
April	1.524	1.726	1.817	2.181	2.612	2.577
June	2.356	2.495	2.738	2.872	2.682	2.736

The diversity value at all stations (Table 6) were higher in June than other months especially at station 1 in the water of island of Buntal (2.872), it can be said that June is more productive than others months, even though the values of diversity index were not reached the value of >3 (steady-steady) because the presence of some species in high numbers. It seemed that the waters were in rich condition during the wet season since the runoff flow from the coastal area and the mangrove ecosystem reached the waters [2]. In the other side, the evenness of the phytoplankton communities at all stations of both glasses of water reached a high value (Table 6).

Table 7. The values of diversity index (H') of the communities at all stations

Month	Burung island			Buntal island		
	St. 1	St. 2	St. 3	St. 1	St. 2	St. 3
February	2.168	2.409	2.268	2.008	1.949	2.192
April	1.524	1.726	1.817	2.181	2.612	2.577
June	2.356	2.495	2.738	2.872	2.682	2.736

Evenness index (J') of the phytoplankton community in both glasses of water showed the equality of the phytoplankton community. The Phytoplankton from all stations was in equally as an individual and a total number of species. The highest was at the degree of 0.875 from 0-1, and it was at stations 3 in the water of island of Buntal (Table 7). While, the dominant species in the community of phytoplankton at both glasses of water were high at station 2 (0.462) in the water of island of Burung and it was in April, while the lowest was at station 3 (0.125) in the waters of the island of Buntal (Table 8).

The Simpson's index (D) of phytoplankton from both glasses of water characterized that there were some species dominated the communities. In February, at the waters of island of Buntal, *Trichodesmium erithraeum* (7097 cell/m^3 out of 23749 cell/m^3) were dominant at station 3, while in April, *Ceratium tripos* was dominant (20 937 cell/m^3 and *Trichodesmium erithraeum* (16642 cell/m^3) from the total numbers of 112202 cell/m^3). While in June, *Biddulphia sinensis* (197225 cell/m^3), *Chaetoceros decipiens* (154231 cell/m^3), *Rhizosolenia alata* (149454cell/m^3), and *Climacodium fraunfeldianum* (83258 cell/m^3) from the total of 1125682 cell/m^3. The phytoplankton community in the water of both islands had their characteristic species structure depended on their habitat [16].

Table 8. The degree of Evenness (J') index of phytoplankton at all stations

Month	Burung island			Buntal island		
	St. 1	St. 2	St. 3	St. 1	St. 2	St. 3
February	0.750	0.779	0.785	0.783	0.720	0.831
April	0.538	0.623	0.671	0.770	0.811	0.875
June	0.786	0.766	0.797	0.814	0.781	0.804

Table 9. Dominancy index (D) of phytoplankton communities at all stations

Month	Burung island			Buntal island		
	St. 1	St. 2	St. 3	St. 1	St. 2	St. 3
February	0.250	0.221	0.215	0.217	0.280	0.169
April	0.462	0.377	0.329	0.230	0.189	0.125
June	0.214	0.234	0.203	0.186	0.219	0.196

The similarities index of phytoplankton at all stations showed that the percentage of the similarity ranged from 73-96%. This data revealed that phytoplankton communities between stations in the water of island of Burung and Buntal in time and space were similar. This condition was supported by the value of Dominancy index which was low.

Table 10. The percentage of similarities index (S) between space (habitat) and times at all study sites

Dimension	Object	S	Percentages
Spatial	Burung and Buntal islands	0.963	96.3
Temporal	February and April	0.897	89.7
	February and June	0.730	73.0
	April and June	0.738	73.8

Phytoplankton communities in the waters of the island of Burung and Buntal were similar significantly, as well as the species between station and time. This condition reflected that almost phytoplankton species occurred in the waters of the island of Burung and Buntal supported the productivity of the surrounding waters.

4. Conclusion

The temporal and spatial distribution of phytoplankton and the abundance were influenced by the changes of hydrography parameters (temperature and salinity) which reflected by the value of coefficient correlation r = -0.712 for temperature and r = 0.613 for salinity. At the same time, the monsoon influenced the value of temperature and salinity. At least 42 species of phytoplankton encountered at all station, and the high abundance was in June. The mangrove ecosystem which covered the island of Buntal supported the abundance of phytoplankton compare to seagrass ecosystem in the shallow water of island of Burung. The species of phytoplankton community occurred in equally at all the study sites. Further, the phytoplankton community structure reflecting by diversity index in showed that the community was in a fair state.

References

[1] Zijlstra J.J, Baars M A, Tijssen S B, Wetsteyn J I J, Witte A G, Ilahude, Hadikusuma 1999 *J. Sea Res. 25 (4) 431-447*

[2] Suthers I M and Rissik D 2009 Plankton A guide to their ecology and monitoring for water quality. CSIRO Publishing. 150 Oxford, Colling wood VIC 3066 Australia p 256

[3] Lalli C M and Parsons T R 1999 Biological Oceanography An Introduction 2nd Edition Butterworth Heinemann p 314
[4] Castro P and Huber M E 2007 Marine Biology Sixth Edition Mc Graw Hill Newyork p 460
[5] Huliselan N V, Haumahu S and Tuapattinaja M A 2014 Peranan Plankton dalam Menyerap dan Menyimpan Karbon di Perairan Teluk Ambon Laporan Penelitian Unggulan Perguruan Tinggi Tahun Kedua Lembaga Penelitian Universitas Pattimura
[6] Barnesh R S K and Hughes R N 1999 Marine Ecology third Edition Blackwell Science Ltd Oxford p 286
[7] Angel M V 1994 Long-term. Large-scale patterns in marine pelagic systems In Giller P S, Hildrew A G and Rafaelli D G (eds) Aquatic Ecolog Scale Pattern and Process Blackwell Science *Oxford* pp 403-439
[8] Yamaji I 1979 Illustration of the Marine Plankton of Japan Hoikusha Publishing Co Ltd Japan p 537
[9] Tomas C R. 1997 Identifying Marine Phytoplankton Acad Press Harcourt Brace and Company San Diego California USA p 858
[10] Perry R 2003 Marine Plankton of Southern California 3nd Edition UCLA Ocean Globe and Malibu High School pp 21-22
[11] Magurran A E Measuring Biological Diversity 2nd Ed Oxford UK Blackwell Science p 264
[12] Tubalawony S, Purnama R and Ferdinandus J 2016 Proc. Pembangunan Kelautan dan Perikanan Berbasis Laut Banda IPB Press Bogor
[13] Effendi H 2003 Telaah Kualitas Air Bagi Pengelolaan Sumberdaya dan Lingkungan Perairan Penerbit Kanisius Jogyakarta
[14] Arinardi O H, Sutomo A B, Yusuf S A, Trimangningsih, Asnaryanti, Riyono S H 1997 Kisaran Kelimpahan dan Komposisi Plankton Predominan di Perairan Kawasan Timur Indonesia LON LIPI Jakkarta
[15] Levinton J S 1995 Marine Biology Function Biodiversity Ecology Oxford University Press Newyork p 420
[16] Nybakken J W 2001 Marine Biology An Ecological Approach Benjamin Cummings An imprint of Addison Weley Longman Inc San Francisco p 516

Potential Demand of Relatively New and Emerging Marine Ecotourism in Kei Islands

Ida I Dewa Ayu Raka Susanti, Wellem A. Teniwut, Syahibul Kahfi, Roberto M.K Teniwut, Simon Picaulima, Marselus Hungan, Meyske Rahantoknam, Cawalinya Hasyim, Siska D. Rahakbauw, M.R. Renjaan, Anna M. Ngabalin, Tati A. Ngangun1, Frischila Pentury, Kamilius D. Betaubun, A.K Ngamel, Elisabeth C. Ohoiwutun

[1]Fisheries Agribusiness Study Program, Tual State Fisheries Polytechnic, Langgur-Sathean Ringroad Km.6, Langgur, Southeast Maluku, Indonesia 97611

Email: ayususanty76@gmail.com

Abstract. One of the main resources for the economic development of coastal are marine ecotourism. Although for developing country such as Indonesia and especially for an area that located far from a big city like Kei Islands in Maluku, development of marine ecotourism might face a greater challenge than any other places. Considering the potential for multiplier effect on economic of a coastal community, the purpose of this research was to analyze and measure the economic value of new, raw and emerging marine ecotourism spot and determine the demand based on domestic tourist perspective. We pick five new and emerging marine ecotourism spots in this area as case study and used purposive sampling technique to target potential domestic tourist who have visited those five marine ecotourism spots and measure their preference and measure their demand potential by used willingness to revisit (WTR) and logistic regression for analyze the data, this research conducted in Kei Islands, Indonesia. This result can provide broad and comprehensive perspective on developing marine ecotourism in a mildly isolated coastal area like Kei Islands and place similar.

1. Introduction

Ecotourism is a tourism concept that involves the implementation of cultural and environmental factors are utilized with the aim of boosting the economy of surrounding communities [1]. A developing country that has the potential of natural sightseeing is more likely to look at the eco-tourism sector as an alternative source of income and at the same time enhancing the development of the economic potential of rural communities [2]. Thus for the archipelago region with less consistency on revenue can take this opportunity to utilize natural potential they possessed to increase region revenue.

The Kei islands astronomically extend from the ordinates 5^0 - 6^0 south latitude and 131^0 - 133.5^0 East Longitude and geographically, the Kei Islands have borders of the territory: the north is bordered by the Banda Sea and Papua, the south by the Arafura Sea, East with Aru Regency and the west by the Banda Sea. As archipelagic region, Kei Islands, seems to isolated from big cities one of the clear indictor related this argument is the cost of flight from either Jakarta or Bali to Kei Islands is more expensive than cost of flight from Jakarta or Bali to other countries in Southeast Asia Region, makes its extremely hard for local government in this matter to attract foreign and domestic tourist to visit Kei Islands especially domestic visitors. Based on data from tourism board of Southeast Maluku regency in 2015 there are 17 tourism spots in Southeast Maluku with 2 tourism spots in Tual City makes there are 19 identify marine ecotourism spots in Kei Islands.

IOP Conf. Series: Earth and Environmental Science **89** (2017) 012035 doi:10.1088/1755-1315/89/1/012035

Marine ecotourism spot in Kei Islands has been covered few times by national TV and the story been picked up international tourism website but it has relatively low impact on the number of visitors from outside the region and given the location of Kei Islands are far from big cities in the country makes high travel expenses is also become an inhibiting factor in the development of nature-based tourism sector especially to attract foreign visitors. Social media has drastically changed the conventional promotion system where social media provides access to customer or potential customer to interact with potential customers [3] though the real impact of social media on a product will depend on the characteristics products and markets [4]. Its widely known that promoting through social media can influence customer behavior [5] which in the tourism sector where getting many positive reviews on social media, magazine, also promotions on TV it will have a positive influence on perceived value and the level of consumer attachment to the tourist sites.

Enjoying the natural view may provide intangible benefits for the visitors, though basically, it has no real value in the sense that there is no basic standard for assessing the price of the sights of the natural-based tourism in the world [6]. Some studies using the willingness to pay (WTP) for measuring the economic value of tourist sites, such as the research conducted by [7] used to measure the WTP visitor perceptions on the value of money they wish to spend in maintaining the beauty and preservation of S. Georges Bay beach in Malta; [8] use the WTP to measure the economic value of the National Marine Park of Zakynthos in Greece coastal region then [9] use the WTP to measure the monetary value of the desire of visitors to preserve the beach in Croatia. Customer who have positive value in product and service that their experienced will increase the chance of them to convert their behaviour financially [10] because perceived value is related to a feeling that can be urge into visible behaviour, regarding to their intention to but a product and service, the more positive feeling they have then will increase their purchase intention where income as control variable.

In order to increase the willingness of the local community to pay to enjoy ecotourism spots in this region, then one factor to consider is the support infrastructure of the tourist spots in the region, which is related to an effort to increase of perceived value of the tourist because according to research conducted by [11-13] that better improvement on spot infrastructure includes transportation and foods will increase the perceived value perceived value of local community.

With many of raw and undeveloped marine ecotourism in Kei Islands, also, considering geographical constraint of Kei Islands, makes the role of local visitor to visit the tourism spots become very important. In order to measure the potential demand of new marine ecotourism spot in Kei Islands, we use willingness to revisit as the demand indicator, since most of the researchers tend to use willingness to pay (WTP) as indicator of economic valuation of tourism. WTP has its advantage but to obtain the purposed this research which was to analyze and measure the demand power of newly and undeveloped marine ecotourism spot in Kei Islands then willingness to revisit (WTRv) has better view. WTP and WTRv served same purpose since both measures the value of something but since we try to view the attraction of marine ecotourism spots in Kei Islands, then WTRv has a better advantage.

2. Methodology

This study located in Kei Islands, we used questionnaire to collect the research data. On 300 questionnaire only 270 questionnaires that valid to use for data analysis. Respondents of this study were local tourist who has been visited five raw and newly emerged marine ecotourism spots in Kei Islands namely ngur munwatwahan in Ohoidertutu Village, Traditional Village in Tanimbar Kei Village, Letman Reefs on Letman Village, Ancient trace on Ohoidertavun Villages and Waterfalls in Kei Besar Island (Figure 1). We measure their willingness to revisit with answer "yes" and "no" and analyze determinant factors that influence their decision. This study conducted from May to July 2017, we used purposive sampling to choose the exact respondents that meet criteria, where those who have their own income and able to bring themselves to the marine ecotourism locations on this study. By choosing these respondents, we hope can provide us the best sources for analyzing our data to obtain the objective of the study. For data analysis, we use logistic regression analysis with SPSS 24.

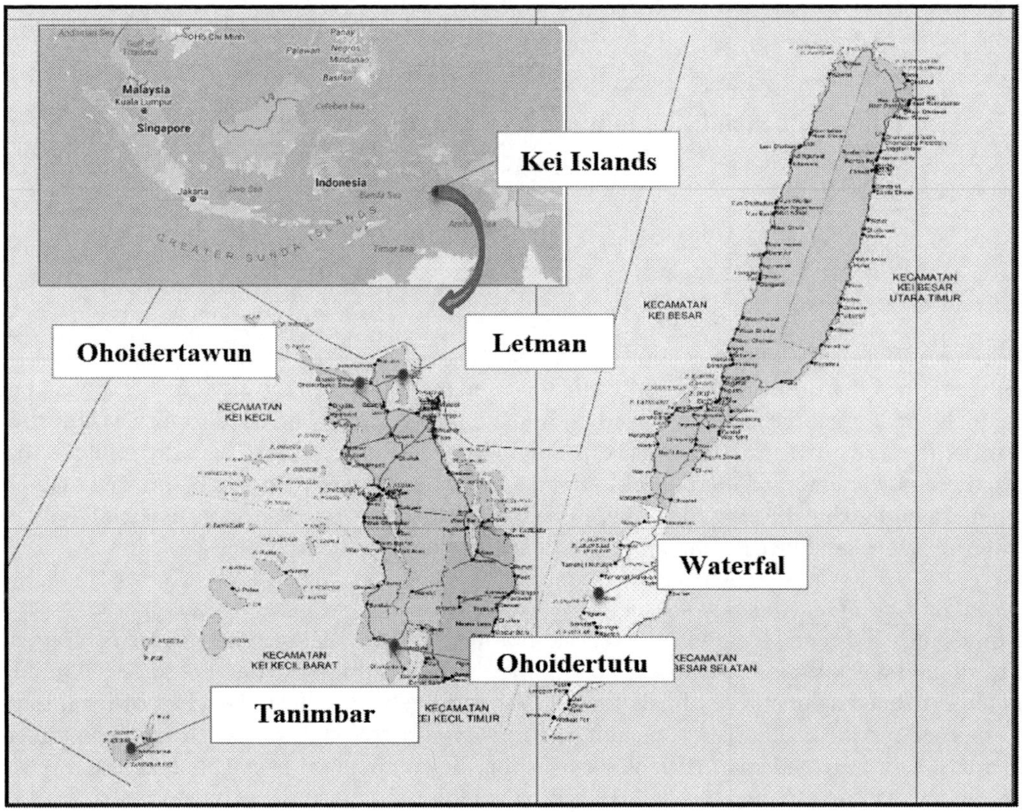

Figure 1. Study area

3. Result and Discussion

Based on logistic regression analysis, from 9 variables with total 24 indicators only 5 variables with 10 indicators that have significant effect on willingness to revisit of local visitors to newly emerged marine ecotourism sports in Kei Islands. On table 1 can be seen the result of significant effect of determinant factors to local tourist willingness to revisit. Based on the purpose of visiting newly emerged marine ecotourism spots, local tourism which has a lifestyle to travel tend to not revisit the spots. Based on our interview with the respondent, it showed that top marine ecotourism spot in Kei Islands likes Baer Islands, Ngubloat Beach or Hawang Cave, still attract them more that the newly emerged one. They mentioned the infrastructure and transportation that to freely available and costly as one of the factors they tend not to revisit although they admit it has a better natural view.

Table 1. Logistic Results on Willingness to revisit

	B	Sig.	Exp(B)
Purpose			
Lifestyle	-1.512	.057*	.220
Travel Companion			
Travel with Partner	1.912	.050**	6.767
Travel with Friend	1.841	.012**	6.301
Type of Marine Ecotourism			
Islands and Reefs	-2.204	.015**	.110
Traditional Coastal Village	-2.027	.025**	.132
Marine Tourism spot for Sports	-3.645	.013**	.026

Income			
IDR 1 million to IDR 2 million	1.157	.071*	3.182
IDR 2 million to IDR 3 million	1.056	.096*	2.876
Intangible Aspect			
Self-Challenge	.163	.043**	1.177
Preference	-.180	.083*	.835

*Significant on 10%; **significant on 5%

3.1. *Based on Travel companion*
As newly emerged marine ecotourism spots that located quite far from the city, makes travel companion become a choice to visit these spots. Based on logistic result, those who used to travel with friends and their partner (spouse, girlfriend and boyfriend) tent to revisit the location. This may relate to the cost of travel to these newly emerged marine ecotourism spots, when travel with others the cost may be lighter than when traveling alone. Respondent mentioned about the view of these spots especially the waterfall and beautiful white sand beach suited their purposes to entertain their partner.

3.2. *Based on type of marine ecotourism*
Based on islands and reefs, local visitors tend to not to revisit newly emerged marine ecotourism spots. Aside from the transportation and infrastructure, the reefs in Letman and ancient trace in Ohoidertawun still underperformed against Kei islands top reefs spot namely Baer Islands. Most of local tourist visit newly emerged marine ecotourism spots out of curiosity at first and made the comparison which makes it hard for newly emerged marine ecotourism spots to compete with established one especially on transportation and infrastructure related. Based on traditional village, local visitors also tend to not to visit this kind spots, because same as the previous reason, its about cost of travelling to this village, not to mention the location that far from city area, despite have been appeared several times on national TV but traditional coastal village in Tanimbar Kei village still find difficult to attract more local visitors. Same goes to newly emerged marine ecotourism spots that were used to sports activity, local tourist tends to revisit the locations.

3.3. *Based on income*
Based on income, can be seen on table 1 that local visitors with income from IDR 1.000.000 to IDR 4.000.000 tend to revisit the newly emerged marine ecotourism spots in Kei Islands. As it mentioned in the discussion, the travel cost plays a significant role on local visitors decision to revisit newly emerged marine ecotourism spots, therefore, those who have more to spend to able to make extra spending to revisit these newly emerged marine ecotourism spots.

3.4. *Intangible aspects*
Based on intangible aspect, local visitors who have the adventure spirit in them tend to revisit these newly emerged marine ecotourism spots. Since these locations meet the criteria of adventure spots makes it's understandable to find those who love a challenge to revisit the spots. In the other hand, those who have certain preference on marine tourism spots tend to not to revisit these newly emerged marine ecotourism spots. Its because they already have plan includes plan on scheduling and financial to spend on traveling makes it hard to revisit since high cost on traveling, in some time, we cannot visit some of the newly emerged marine ecotourism spots because of weather and season related to wind and sea wave.

4. Conclusion
In general, it's still hard for a local tourist to revisit to these newly emerged marine ecotourism spots, since it's all the infrastructure and transportation is not sufficiently available as [14] pointed out that destination with better quality on access has high chance to grow. In general these newly emerged marine ecotourism spots have potential as sources of local economic revenue especially on villages that located

far from city area, because as most of the established marine ecotourism spots on Kei Islands faced decade ago related to transportation and infrastructure, these newly emerged marine ecotourism spots has legitimate potential to be better perform in the future because most of the local visitors that been visit these newly emerged marine ecotourism spots give very positive review on the view, the problem is how to get there same as local government problem to attract more foreign and domestic visitors to Kei Island, they have to how to attract more local visitors to newly emerged marine ecotourism spots in Kei Islands.

References

[1] Scheyvens, R 1999 Ecotourism and the empowerment of local communities. *Tourism management*, 20(2), 245-249.

[2] Clifton, J., and Benson, A 2006 Planning for sustainable ecotourism: The case for research ecotourism in developing country destinations. *Journal of sustainable tourism*, 14(3), 238-254.

[3] Gensler, S., Völckner, F., Liu-Thompkins, Y., and Wiertz, C 2013 Managing brands in the social media environment. *Journal of Interactive Marketing*, 27(4), 242-256.

[4] Fischer, Marc, Franziska Völckner, and Henrik Sattler 2010 How Important Are Brands? A Cross-Category, Cross-Country Study. *Journal of Marketing Research*, 47, 5, 823–39.

[5] Hays, S., Page, S.J., and Buhalis, D. (2013). Social media as a destination marketing tool: Its use by national tourism organisations. *Current Issues in Tourism*, 16(3), 211239.

[6] Loomis, J., Kent, P., Strange, L., Fausch, K., and Covich, A 2000 Measuring the total economic value of restoring ecosystem services in an impaired river basin: results from a contingent valuation survey. *Ecological economics*, 33(1), 103-117.

[7] Blakemore, F. B., Williams, A. T., and Ozhan, E. (2000). Tourist valuation of Olu Deniz beach, Turkey, using contingent valuation and travel cost approaches. *Journal of World Leisure*, 4, 48e55.

[8] Togridou, A., Hovardas, T., and Pantis, J. D 2006 Determinants of visitors' willingness to pay for the National Marine Park of Zakynthos, *Greece. Ecological Economics*, 60, 308e319.

[9] Logar, I., and van den Bergh, J 2012 Respondent uncertainty in contingent valuation of preventing beach erosion: an analysis with a polychotomous choice question. *Journal of Environmental Management*, 113, 184e193.

[10] Li, G., Li, G., and Kambele, Z 2012 Luxury fashion brand consumers in China: Perceived value, fashion lifestyle, and willingness to pay. *Journal of Business Research*, 65(10), 1516-1522.

[11] Murphy, P., Pritchard, M. P., and Smith, B 2000 The destination product and its impact on traveller perceptions. *Tourism management*, 21(1), 43-52.

[12] Sirakaya, Ercan, Victor Teye, and Sevil Sonmez 2002 Understanding Residents' Support for Tourism Development in the Central Region of Ghana. *Journal of Travel Research*, 41 (1): 57–67.

[13] Chen, C. F., and Tsai, D 2007 How destination image and evaluative factors affect behavioral intentions?. *Tourism management*, 28(4), 1115-1122.

[14] Mill, R. C., and Morrison, A. M 1992 The tourist system: An introductory text (2nd ed.). Englewood Cliffs, New Jersey: Prentice-Hall.

Analysis of Successful Strategy to Develop Sustainable Marine Ecotourism in Gili Bawean Island, Gresik, East Java

M P Wardani[1], A Fahrudin[2], and F Yulianda[2]

[1]Department of Resource and Environmental Economics, Bogor Agricultural University, Bogor 16680, West Java, Indonesia
[2]Department of Aquatic Resources Management, Bogor Agricultural University, Bogor 16680, West Java, Indonesia

Email: mentariwardani@gmail.com

Abstract. The sustainability of resources and marine ecotourism in Gili Bawean Island is still developing to the current day. The management is conducted individualistically and is currently far away from being integrated and sustainable. It is important that stakeholders understand the island's condition and the urgency of coastal resources, to determine collective action, which leads to sustainable ecotourism on the island. This research aimed to discover stakeholders' involvement in determining key variables and formulate a strategy of marine ecotourism development based on possible future scenarios in Gili Bawean Island, Gresik Regency, East Java. The field study was done through an expert meeting of stakeholder representatives on March–April 2017. The data was analyzed using Participatory Prospective Analysis (PPA), a comprehensive and quick framework, which was designed to demand requests in structural anticipation and exploration and also to focus on interaction and consensus among stakeholders. The results of this research show that five main variables should be emphasized in developing marine ecotourism on the island, including tourist activities, institutions, and economic activities, as well as the quality of human and natural resources. Counting heavily on those variables, it is hoped to create an integrated marine ecotourism development. Coordination among stakeholders can be declared successful when the tourist objects are managed better, and the quality of tourist destinations and the number of tourist visits increase noticeably. Good governance of marine ecotourism contributes to increments in tourist amenities, boosts the welfare of local communities, and secures sustainability of local natural resources.

1. Introduction

As a product, ecotourism is one of the important attractions of Indonesia. The potentials and objects of ecotourism are unique, diverse and scattered in various regions. According to [11], the marine tourism sector is the most efficient sector in the marine field. Therefore the development of marine tourism should be prioritized; it can be implemented through the optimal utilization of tourist objects and attractions. Various objects and attractions which may be utilized are nature tourism (beach), and diversity of flora and fauna (biodiversity), such as ecotourism, business tourism, cultural tourism as well as sport tourism. The potential of marine tourism, which spreads across most of the coastal districts or cities will have a substantial, direct impact on the income of local communities and local governments. Marine tourism, one of the sectors in the marine field, which is becoming the foundation of marine development, is currently the most favored sector by the government.

Since the beginning of the millennium, people have been discovering the term "ecotourism," and the ministry in charge of the tourism sector began to explore and discuss it [17]. The selection of

IOP Conf. Series: Earth and Environmental Science **89** (2017) 012036 doi:10.1088/1755-1315/89/1/012036

ecotourism as a concept of regional development is based on several main elements: First, dependence on the quality of natural resources as well as on historical and cultural heritage. Second, the involvement of the community. Third, an increase in the awareness and appreciation of nature, and the values of historical and cultural heritage. Fourth, the growth of the ecotourism market at both national and international levels. Fifth, a means of realizing sustainable economy [15]. In other words, ecotourism offers a low investment and high-value concept for the resources and the environment while making it a powerful tool for the participation of the community, because the entire production assets use and belong to the local communities.

Gili Bawean Island is located at the coordinates of 05°47’59.63’’ LS : 112°46’14.086’’ BT. Based on the calculations of satellite imagery, the area of Gili Bawean Island is approximately 375,169 m^2 (37,517 ha) [10]. It has productive natural resources such as coral reef ecosystems, along with the biota living within the ecosystem, and environmental services such as tourism and recreation areas. The development of marine ecotourism may add to the regional economic opportunities and revenues as well as open new business opportunities for local communities. The object of marine ecotourism in the water of Gili Bawean Island consists of the objects of commodity and the objects of the ecosystem. The objects of the commodity include the potential of marine biota species and non-biological materials that attract tourists, such as thick sandy beach conditions with a soothing beach vegetation. In comparison, the objects of the ecosystem are the tourist attractions in the area, specifically the expanse of sandy beaches as well as coral reefs for diving and snorkeling.

The strategies needed to develop ecotourism should fulfill the basic principles that [19] proposed: ecotourism should include the knowledge of local communities, governments, non-governmental organizations, industry, tourists before and after traveling. It includes the carrying capacity of the resources and provides long-term benefits for them, as well as local communities and industries, for whom the benefits can be in the form of conservation, science and culture or economics. The sustainability of resources and marine ecotourism on Gili Bawean island is still developing. The management is conducted individualistically and is far away from being integrated and sustainable. It is important that stakeholders understand the island’s condition and the urgency of coastal resources to determine collective action, which leads to sustainable ecotourism on the island. This research aimed to discover stakeholders’ involvement in determining key variables and to formulate the strategy of marine ecotourism development, based on possible future scenarios on the island.

2. Methodology

The survey was conducted during March and April of 2017 (Fig.1). Respondents interviewed were stakeholders associated with marine ecotourism on the island using a purposive sampling method [16]. The sample includes the Technical Implementation Unit (TIU) of the Bawean Tourism Department (three persons), the TIU of the Marine, Fishery and Animal Husbandry Unit of Bawean (one person), two tour managers, the TIU of the Port and Conservation of Fisheries and Marine Resources of Bawean (one person), the Regional Development Planning Board of Gresik District (three persons), the Department of Tourism and Culture of Gresik District (three persons), the Fisheries Department of Gresik District (three persons), and the Department of Marine and Fisheries of East Java Province (two persons). The interview material includes the main variables which affect the development of marine ecotourism on the island.

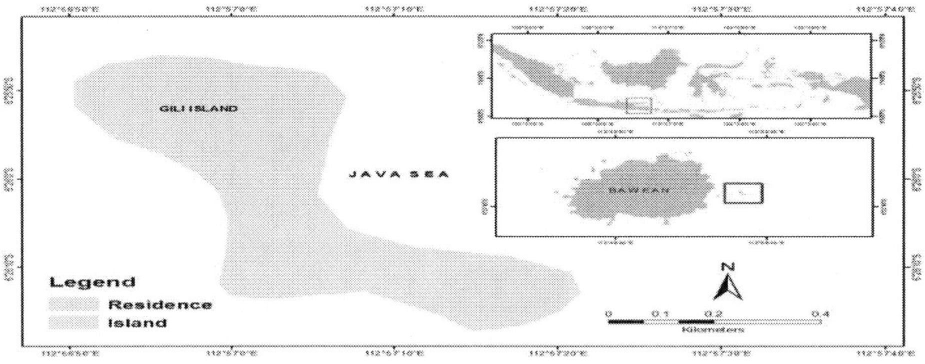

Figure 1. Map of research location

The type of data used in this study is the entire data, information, and opinions expressed by the experts (stakeholders' representatives) in the expert meeting. To achieve a focus of discussion, writers had set the definition of the system discussed, which was the development of marine ecotourism in Gili Bawean Island. This research used a participatory prospective analysis method which aimed to generate a consensus in the interest of planning [5]. Following [4], the participatory prospective study was conducted through expert meetings. In the meeting, experts or participants were required to identify the main variables which were considered most influential on the development of marine ecotourism in Gili Bawean Island. At this stage, the difference in the influence level of the variables on the system studied would be set, to determine the variables to be inserted as entry points for effective planning [2,5-6,7,9]

Based on the participatory prospective analysis, scores were obtained on the cross-linked effects between variables which are analyzed by a matrix [2] with the help of Microsoft Excel software. This process observed the influence/direct dependence, (I/D) of each variable with another variable, by using a consensual valuation approach. The valuation of direct effects on a variable towards other variables was obtained by using the scale from "0 = no effect" to "3 = strongly effects". The values that have been discussed and agreed upon by the participants are directly inserted into the I/D matrix. The score of the cross-linked effect on the results of the agreement is presented fully in the table of the effects. The results of the effect analysis among variables are presented in the form of graphics and tables [13].

3. Results and Discussion

3.1. Tourism and Marine Ecotourism on Gili Bawean Island

According to [19] on Tourism, "tour" refers to the activity of a person or group of persons visiting a certain place for recreational purposes, or for self-development, or to study the uniqueness of tourist attractions visited in a short duration of time. Tourism is a wide range of tour activities, and this is supported by various facilities as well as services provided by the community, entrepreneurs, government and local government. There are several types of tourism according to [12], including cultural tourism, health tourism, sport tourism, commercial tourism, industrial tourism, marine tourism and natural reserve tourism. Marine tourism exploits the potential of coastal and ocean environments both directly and indirectly, as expressed by [12]. Direct activities include boating, swimming, snorkeling, diving, and fishing. Marine tourism cannot be separated from the activities of nature tourism, often referred as coastal tourism activities, utilizing the potential of coastal environments as the main attraction.

Ecotourism activity in Gili Bawean Island is very diverse, consisting of natural tourism, marine tourism, religious tourism and cultural tourism. Marine ecotourism on the island is classified into attractions, objects and culinary. Attraction-based tourism consists of art and cultural attractions derived from the community. Meanwhile, natural attractions include coral reefs, seagrass, mangroves and beach sand. The management of ecotourism described above is conducted incidentally as it is not planned in an integrated manner (either spatially or temporally). Therefore, the study of strategies to optimize the sustainability of marine ecotourism on the island is urgently needed to establish the sustainability of resources and the well-being of local communities.

3.2. Participatory Prospective Analysis of Stakeholders in the Development of Marine Ecotourism on Gili Bawean Island

The stages of identifying the influencing variables in the development of marine ecotourism in Gili Bawean Island refer to the research conducted by [5]. The development consists of 40 variables, identified by the stakeholders' representatives of 18 persons. Following the identification of the variables, the stakeholders' representatives were required to define the variables by consensus. Evidently, in this process, a considerable number of variables are repetitive or similar to each other. Eventually, an agreement was accomplished to merge and dispose of several variables. Therefore 11 variables were defined by the consensus (Table 1). In this case, the most decisive variables in the development of marine ecotourism in Gili Bawean Island are not specified. The effects between variables cannot be described. Therefore all variables own the same interest and power within the system. Based on the 11 variables in Table 1, the stakeholders' representatives have discussed and consensually scored the cross-influences between the variables, which were analyzed in a matrix with the help of Microsoft Excel software. This process was carried out through structural analysis and group work, where the influence/direct dependence (I/D) analysis of each variable with other variables was conducted using a consensual valuation approach.

Table 1. Variables which affect the development of marine ecotourism in Gili Bawean Island

No	Variable	No	Variable
1.	Natural resources	7.	Environmental destruction
2.	Quality of human resources	8.	Population growth
3.	Tourism activities	9.	Infrastructures
4.	Institutions	10.	Carrying capacity of the region
5.	Economic activities	11.	Tourism facilities
6.	Zoning area		

The graphs of direct and indirect inter-variable effects (Fig.2) indicate the variable spreads within the four quadrants bounded by two axes. The depiction is based on the weighted I/D values of each variable calculated from the effect and dependency table. Interpretations of the results include the position of variables, the form of variable distribution, and the interpretation of direct and indirect results. Each quadrant corresponds to the particular characteristics of the variable. Quadrant I is the driving variable area. Quadrant II is the control variable area (leverage), characterized by effects as well as dependence. Several variables in this quadrant can also be classified as influential variables. Quadrant III is the output variable area, which is highly dependent and has little influence. Quadrant IV is the marginal variable area; this group will be immediately excluded from the analysis [5].

In Figure 2, from the presentation of the results after analyzing the direct and indirect, writers select the variables located in Quadrant I and II. The variables in the quadrant have a large influence on the system so that they can serve as the entry points for effective planning and management [2,8]. Thus, five variables which are the most influential on the system, namely tourism activities, institutions, human resource quality, economic activities and natural resources have been selected. This is supported by the weighted global strength value of each variable. Each of the five variables has higher values than the other six variables (as presented in Table 2). From the results, it can be concluded that variables 1-5 (Table 2) are selected as the most influential variables.

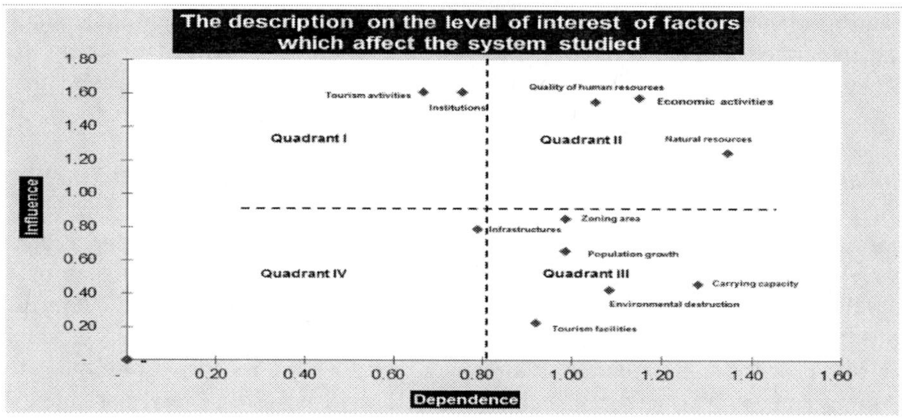

Figure 2. The analysis results of direct and indirect effects between variables

Table 2. Weighted global variable weight scores

No.	Variable	Weighted Global Strength	No.	Variable	Weighted Global Strength
1.	Tourism activities	2.08	7.	Infrastructures	0.72
2.	Institutions	2.00	8.	Population growth	0.48
3.	Economic activities	1.70	9.	Carrying capacity of the region	0.22
4.	Quality of human resources	1.67	10.	Environmental destruction	0.22
5.	Natural resources	1.10	11.	Tourism facilities	0.08
6.	Zoning area	0.73			

3.3. The Construction of Scenarios

The scenario development in Table 3 is conducted through structured brainstorming and group discussions. In the forum, participants are required to provide estimations based on the condition of each determinant variable in the future. This estimation is the opinion and reflection of the needs of stakeholders in the future [2-3,6,8-9]. From the estimation regarding the condition of these variables in the future, scenarios that may occur on the island of Gili Bawean have been illustrated. The scenarios are successfully constructed and presented in Table 4. Based on the scenarios that have been successfully established, it can be seen that the differences between scenarios Affect the required efforts in the development of marine ecotourism on Gili Bawean Island.

Table 3. The condition of the variables set by the participants in consensus

Factor	State	
	1A	1B
Tourism activities	Increasing, with the addition of tourism activities such as jet ski and banana boat. The addition of tourism activities provides optimal benefits for local communities, stakeholders and optimum satisfaction of the customers.	Current condition, no improvement in tourism activity. It only sees tourism activities such as diving, snorkeling, and beach tourism.
	2A	2B
Institutions	Increasing, with the presence of evident rules or policies on marine ecotourism management and development as well as coordination of the related inter-agencies.	Current condition, no evident rules from local government related to the management and development of integrated tourism potential at the district or provincial level; no coordination between the related agencies.
	3A	3B
Economic activities	Increasing, with coastal, fisheries and marine resources as well as environmental services which are not obtained by excessive exploitation, but should be consumed selectively.	It remains as it is today, the exploitation of environmental services, and this is related to coastal, fishery and marine resources by communities to gain economic benefits.
	4A	4B
Quality of human resources	Increasing, with the active participation of the community which consciously assists the government's program with initiative and reaction of commitment. Therefore the benefits of marine ecotourism are fairly distributed within the whole community.	It remains as it is today, passive participation raises the awareness of not conducting any activities which may harm the environment.
	5A	5B
Natural resources	Increasing. The management and development of marine ecotourism should include the carrying capacity of the resources.	Current condition, conservation becomes the foundation of the marine ecotourism development area.

Source: Outcomes from the meeting of experts
Information:
: Baseline (current condition) : Moderate scenario
: Optimistic scenario

Table 4. Scenario of marine ecotourism development on Gili Bawean Island

Baseline (Current Condition)		1B–2B–3B–4B–5B
No.	**Scenario**	**State**
1.	Optimistic	1A–2A–3A–4A–5A
2.	Moderate	1B–2B–3A–4A–5A

According to [1], the development approach in its implementation should be based on the blue economy as it will synergize with the triple-track strategy, namely pro-poor (poverty reduction), pro-growth, pro-job (employment) and pro-environment (preserving the environment). Therefore, the development of marine ecotourism on Gili Bawean Island is arranged according to the expected scenarios, which are optimistic after considering the key factors which are very influential in the management and development of marine ecotourism. These key factors are tourism activities, institutions, economic activities, quality of human resources and natural resources.

The optimistic scenario of marine ecotourism will occur when there is an addition of tourism activities in the form of Jet Ski and banana boat. The addition of tourism activities provides optimal benefits for local communities and stakeholders; the value of optimal satisfaction for tourists (1A); evident rules or policies on the management and development of marine ecotourism; as well as related inter-agency coordination (2A). These services are related to coastal resources, fisheries, marine and environmental services which are not obtained by excessive exploitation, but should be consumed selectively (3A). The active participation of society, consciously helping the government program with initiatives and commitment shows that the benefits of marine ecotourism are fairly widespread for all society (4A). The management and development of marine ecotourism should include the resource carrying capacity (5A). This optimistic scenario needs to be supported for it to occur.

3.4. Development of Marine Ecotourism in Gili Bawean Island

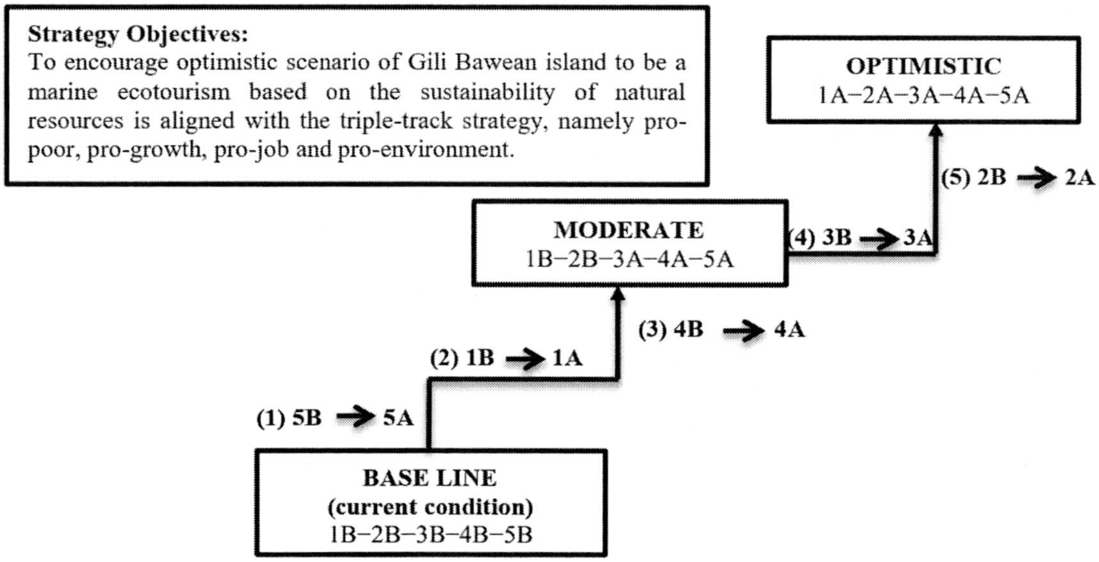

Figure 3. Formulation of marine ecotourism development strategy in Gili Bawean island

To manifest the optimistic scenario of marine ecotourism development in Gili Bawean Island, the required stages (Figure 3) are: the baseline (current condition) (1B-2B-3B-4B-5B), the first stage (1) is that the management and development of marine tourism need to be based on the sustainability of natural resources. A proper and successful ecotourism area should optimally be based on the aspects of maintaining the sustainability of the environment, including natural resources. The carrying capacity should be noted so that the spatial and temporal planning will be effective. The carrying capacity refers to both the ecological carrying capacity and physical carrying capacity. The ecological carrying capacity is the maximum level of an area's use. The physical carrying capacity is the maximum amount of usage or activity recommended in the area without causing damage or degradation of the quality. The second stage (2) is the addition of tourism activities such as jet ski and banana boat. The addition of tourism activities provides optimal benefits for local communities, stakeholders and optimum satisfaction of the customer. The third phase (3) is improving the quality of human resources, creating a robust quality of human resources in the field of marine ecotourism, either in skills, ability to innovate, adaptability when facing various external environment changes, work culture, education level, and the level of understanding the strategic problems and concepts which will be implemented. Local communities, especially indigenous people who live in tourist areas become one of the keys to ecotourism because they Zill provide the attractions as well as being the determinant of the tourism products' quality. The former local community members are involved in the management of tourism activities before the existence of planning, management and development activities. The increase in the three factors will evolve the ecotourism from the baseline (current condition) into a moderate scenario (1B-2B-3A-4A-5A).

Following the increment in the three factors will be the fourth stage (4) which is an economic activity that needs to be improved. About coastal resources, fisheries, as well as marine and environmental services, should not be obtained by excessive exploitation but by consuming it selectively. Therefore environmental sustainability is maintained, and it remains economically beneficial. In general, marine ecotourism activities in Gili Bawean Island have made economic

IOP Conf. Series: Earth and Environmental Science **89** (2017) 012036 doi:10.1088/1755-1315/89/1/012036

impacts on the community because of the rotation in money between tourists, business units, and manpower. The increasing number of tourists visiting the island will make similar impacts in the form of higher revenues to business units. The fifth stage (5) is the existence of evidence rules or policies on the management and development of marine tourism, as well as related inter-agency coordination. The management and development of marine ecotourism is the responsibility of the relevant stakeholders of the Indonesian Ecotourism Network, the Tourism Department, the Marine and Fisheries Department, the NGOs such as The Nature Conservancy (TNC) and World Wide Fund for Nature (WWF), tourism businesses, community leaders and so forth. The success of the management and development of marine ecotourism can be measured through two indicators, specifically the management of tourism destinations and their increasing quality, as well as and the numbers of tourists [14]. If these two indicators improve, the coordination between agencies can be said to have been successful, and the satisfaction of tourists will be expected to increase, the community is expected to encourage the improvement of welfare and the quality of the environment. With the increment of the five factors affecting the development of marine ecotourism into an optimistic scenario (1A-2A-3A-4A-5A) it marine ecotourism of Gili Bawean Island equal to pro-poor, pro-growth, pro-job and pro-environment.

4. Conclusions

There are five determinants of the success of ecotourism development in Gili Bawean Island: tourism activity, institutions, economic activity, human resource quality and natural resources. Based on the considerations of these five factors, a strategy for ecotourism development is needed, with the support of the government as the owner of authority in the arrangement, provision, and designation of various infrastructures related to the needs of tourism. The government's macro policy is a guide for other stakeholders in carrying out their respective roles and tasks.

References

[1] Adrianto L 2013 Presentation Materials of Blue Economy for Fisheries (Bogor: Bogor Agricultural University) (in Indonesian)

[2] Bourgeois R and Jesus F 2004 Participatory Prospective Analysis: Exploring and Anticipating Challenges with Stakeholders (Bogor: CAPSA Monograph) p 46

[3] Coates J, Durance P and Godet M 2010 Strategic Foresight Issue: Introduction Technol. Forecas. *Soc. Change* 77: 1423-1425

[4] Cornwall A and Jewkes R 1995 What is Participatory Research? *Soc. Sci. Med.* 41 (12): 1667-1676

[5] Damai A A, Boer M, Marimin, Damar A and Rustiadi E 2011 Participatory Prospective Analysis in Coastal Zone Management of Lampung Bay (Graduate Forum vol 34) (4): 281-296. (in Indonesian)

[6] Durance P and Godet M 2010 Scenario Building: Uses and Abuses. Technol. Forecas. *Soc. Change* 77: 1488-1492

[7] Godet M 2010 Future Memories Technological Forecasting & Social Change 77: 1457-1463

[8] Godet M and Roubelat F 1996 Creating the Future: The Use and Misuse of Scenarios. *Long Range Plann* 29 (2): 164-171

[9] Gray T and Hatchard J 2008 A Complicated Relationship: Stakeholder Participation and the Ecosystem-Based Approach to Fisheries Management *Marine policy* 32 (2): 158-168

[10] Ministry of Marine and Fisheries 2012 Survey on the Identification and Mapping of Resource Potential of Small Islands (Gili Bawean Island). Jakarta: PT Karya Nugraha (in Indonesian)

[11] Kusumastanto T. 2003. The Empowerment of Marine Resources, Fisheries and Sea Transportation in the 21st Century (Bogor: Coastal and Marine Resource Studies Center-Bogor Agricultural University) (in Indonesian)

[12] Pendit N S 2002 Tourism Science: The First Introduction (Jakarta: Pradnya Paramita) (in Indonesian)

[13] Salim H L and Purbani D 2015 Community Based Marine Tourism Development in Kaledupa Island, Wakatobi Regency, South East Sulawesi Province. *Human and Environmental Journal* vol 22 (3): 380-387 (in Indonesian)
[14] Santosa H 2013 The Harmonization of Sustainable Tourism in Wakatobi Through Destination Management Organization (DMO) Program. *Napoleonic Bulletin* 3: 1-4 (in Indonesian)
[15] Shelly R and G Wall 2001 Evaluating Ecotourism: The Case North Sulawesi. *Annual Tourism Research* vol 23) (1): 122-132
[16] Sugiyono 2013 Quantitative, Qualitative, and R & D Research Methods (19th Edition). Bandung: CV Alfabeta (in Indonesian)
[17] Teguh F and Avenzora R 2013 Ecotourism and Sustainable Tourism Development in Indonesia, Learning and Success Potential. Jakarta: Ministry of Tourism and Creative Economy-PT. Gramedia (in Indonesian)
[18] Law Number 10 of 2009 on Tourism (in Indonesian)
[19] Wright K 1993 Tapping into Market Potential for Ecotourism. *Annual Tourism Research* vol 45 (2): 56-67

Increase Economic Valuation of Marine Ecotourism Spots In Small Islands

Siska D. Rahakbauw[1], Wellem A. Teniwut[1], Meiskyana R. Renjaan[1], Marselus Hungan[1]
[1]Fisheries Agribusiness Study Program, Tual State Fisheries Polytechnic, Langgur, Indonesia, 97611

Email: wateniwut@polikant.ac.id

Abstract. Ecotourism is one of the fast-growing sectors especially in the developing country as a source of revenue. To get a sustainable development of ecotourism, it needs broad and comprehensive effort from central government and local government, perfect example in that regards in Indonesia is Bali and Lombok. For another area in Indonesia like Kei Islands which located in two administrative governments have a major problem to build a sustainable nature-based tourism because of the location of this area to the major cities in the country makes the travel cost is high. This situation makes the role of local community as the backbone of the growth and development of nature-based tourism is critical. By using structural equation modeling (SEM), we constructed a model to enhance local community perception on economic valuation of ecotourism spots in the area. Results showed that perceived quality as the mediation driven by the intensity of appearance on national television and the internet could increase community attachment to increase willingness to pay from the local community on ecotourism in Kei islands. Also, the result also indicated that WTP value for the local community on ecotourism in Kei Islands was $10.81 per trip, with average trip per month was 1 to 4 times.

1. Introduction

Ecotourism is one of the fast-growing sectors which has developed rapidly since the 2000s [20]. The sector is getting more support from governments, NGO and local community due to the ability to provide two benefits at once, first is economic development and second is sustainability [3]. The tourism sector can provide a major contribution to the total gross domestic product (GDP) of the country, and as an archipelago country, Indonesia has all the resources to be one of the leading countries on the developing of ecotourism. According to data from the Ministry of Tourism of Indonesia, since 2011-2015 the tourism sector in Indonesia ranks as the fourth largest foreign exchange earner in Indonesia. The most recently available data is of the annual number of foreign tourists visiting Indonesia in 2014-2015 showed an Increase of 7.78% on the previous annual total though the number is still concentrated in a few tourism locations [12]. Maluku is an archipelago region with massively natural resources for ecotourism including marine-ecotourism though currently, the contribution of the tourism sector remains to underachieve considering the potential of natural resources where agriculture and fisheries sectors remain the highest contributor to the regional GDP [18].

Kei Islands are one of the areas in Maluku province which also have the potential of natural beauty a very good example is Baer Island that shares similar characteristics with Raja Ampat in Papua, then a couple of beaches with white sand and sandbar lagoons are home to extinct pelicans species. The development of the tourism sector, including ecotourism, can provide a good multiplier effect for the economic development of rural coastal areas where increase on the number of visitors will have a positive effect on the growth of small businesses. The tourism sector has been widely known as a way to improve the welfare of community simultaneously [10].

Various efforts have been made by local governments in Kei Islands to promote tourist sites in the region, including by creating a national level festival, but has not made a significant impact on the number of foreign tourists and domestic. When referring to the past success story of tourist sites were managed in the Southeast Asia region which were in Phuket (Thailand) and Bali (Indonesia) wherein both these locations have the same characteristics on their success in the development of ecotourism was both locations originally built by expatriates who have been stayed at these two locations. Also, another similar trait of these two locations is both located close to the mass tourism which already has large visitors number and sufficient infrastructures [8]. Tourists who choose to visit nature-based tourism are those that inclined motivated on social factors in the sense that they want to meet other travelers also have high levels of the same desire, and also visitors on nature-based tourism are those who have a lot of spare time [4] which this is related to the readiness of infrastructure on the area. Financial factors also become one of the major candidates that affect the desire to visit travel [15].

Given these conditions, the role of local communities in Kei Islands become very significant to support the development nature-based tourism in the sense that local community is the first customer of the newly industrialized tourism sector in the area just to keep it running on their early stage. The local government has to find ways to encourage residents in the area to be willing to pay to enjoy the natural scenery. Characters of the local community of this are more likely to be referred as rural peoples where the level of attachment to the region is still high will provide positive opportunities for development of tourism sector [7]. Factors community attachment can be an advantage factor for local governments in order to change the behavior of local residents to pay in order to enjoy the natural scenery.

The aimed of the study to formulate a model to improve the development of tourism sector based ecotourism in the Kei Islands. Consider the great potential of the tourism sector in the region and the real indication of the desire of local governments to develop the tourism sector based ecotourism this region, so the research is expected to make a real contribution to the development process of ecotourism in the Kei Islands. This study uses a structural equation modeling approach to generate models to improve the local community's willingness to pay to enjoy ecotourism spots in the region. Furthermore, this paper is divided into three parts where the first part is the methodology used in the study, then the

2. Methodology

This research was conducted in the Kei Islands, Indonesia from August to November 2016 with the number of valid respondents used in this study was 244 respondents, we used structural equation modeling analysis tools which, according to [5] our respondent has met the minimum number of respondents. In this study, we used AMOS to test research model with CFA, and SEM approaches for data analysis [2]. The selection of respondents conducted based on criteria where local residents who already have their income and able to influence others to give them money also be able to visit tourist sites in the region by themselves. All variables in this research are measured by a questionnaire with 1 to 5 Likert scale except for community attachment which measured by answering 'yes/no" on questioning family relation to the particular village where tourism spot is located and degree on a willingness to involved on developing tourism in this area. In Figure 1 showed a conceptual framework of this study, which can be seen that in order to increase the economic value of ecotourism in the islands, measured by willingness to pay (WTP) then we used community attachment and perceived value with driving factors were the intensity on appearance on media includes TV, magazine, and spots of social media and infrastructure.

2.1. Data Collection

Table 1 showed that majority of local tourist in Kei Islands were female by 52.9 percent compared to male visitors who were 47.1 percent, whereas the age of the visitors was around 30-40 years old 52.4 percent. Most of the visitors hold bachelor degree 41.8 percent accordingly with the average income a month was between IDR 1 million to IDR 4 million 55.7 percent, also, a number of local visitors to visit ecotourism spot in this area at least once a month 45.1 percent.

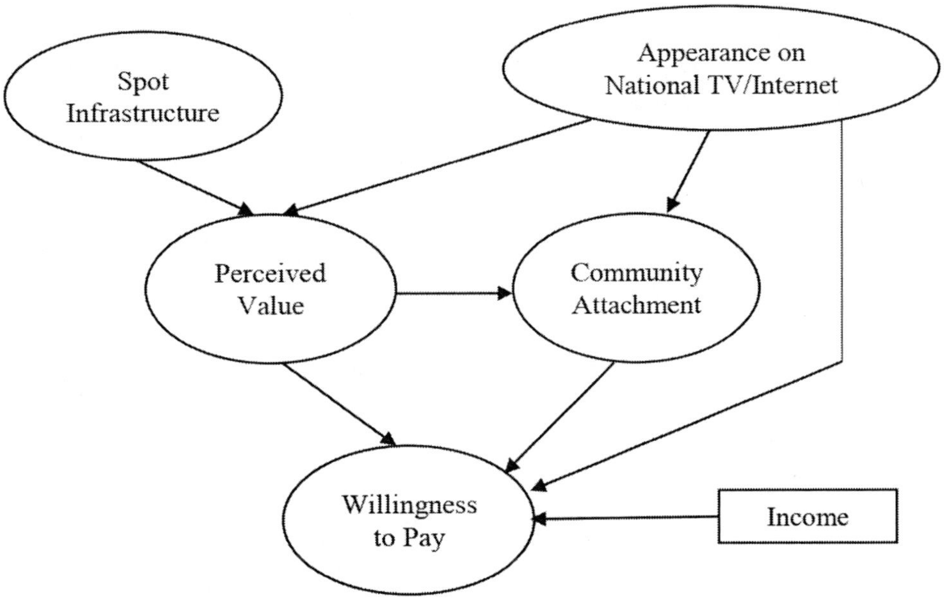

Figure 1. Conceptual Framework

2.2. Reliability and Validity Test

Reliability testing on the constructs in this study used Construct Reliability (CR) [6] and for the testing the validity this study used Average Variance Extracted (AVE) [1]. On Table 2 showed that all values of composite reliability above 0.7 and AVE value for all items above 0.5 which indicated all constructs on this research passed the reliability and validity of test with the results for the models produced a list of the goodness-of-fit indices, Including AGFI (0831), CFI (0906), and RMSEA (0.070).

Table 1. Respondents Characteristic

Characteristic	Frequency	Percentage
Gender		
Male	115	47.1
Female	129	52.9
Age		
20 - 30	73	30
31 – 30	128	52.4
41 – 50	34	13.9
Over 50	9	3.7
Education		
Elementary School	4	1.6
Junior High	4	1.6
High School	83	34
Diploma	38	15.6
Bachelor	102	41.8
Post Graduate	13	5.3
Income*		
< Rp. 1.000.000,-	71	29.1
Rp. 1.000.000,- Rp. 4.000.000,-	136	55.7
> Rp. 4.000.000	37	15.2

Frequency of visiting Ecotourism spot in the area (per month)		
Once a month	110	45.1
Two to Four times a month	92	37.7
Over Five times a month	42	17.2

Table 2. Reliability and Validity Test

Measurement Items	Factor Loadings	CR	AVE
Spot Infrastructure		0.844	0.576
SI1	0.788		
SI2	0.759		
SI3	0.773		
SI4	0.713		
Access Transportation		0.782	0.546
AI1	0.740		
AI2	0.761		
AI3	0.673		
Appearance on National TV/Social Media		0.856	0.666
Prom1	0.878		
Prom2	0.803		
Prom3	0.762		
Perceived Value		0.828	0.620
PV1	0.773		
PV2	0.922		
PV3	0.643		
Willingness to Pay		0.747	0.608
WTP1	0.669		
WTP2	0.822		

3. Results and Discussion

Prior to testing the hypothesis is necessary to test the model fit on the structural equation modelling (SEM) of this studies, where the result showed that GFI (0917), AGFI (0881), CFI (0917) and RMSEA (0.068), based on [13] that the SEM models in this study was fit. Furthermore, in table 3 and figure 2 showed the results of hypothesis testing done, the first hypothesis accepted where the spot infrastructure significantly affect the perceived value .697 (<0:01). Second hypotheses in this study also accepted where the appearance on National TV and internet significantly and positively influence the perceived value .163 (<0:01), community attachment .184 (<0:01) and the appearance on national TV and the internet significantly and positive effect on the willingness to pay .266 (<0:01). The third hypothesis point (a) in the study accepted as perceived value significantly and positively affect the community attachment .271 (<0:01), where third hypothesis point (b) is rejected as perceived value significantly and negatively affect the willingness to pay -.112 (<0.1). Based on the test results also showed income controls the willingness to pay .087 (<0.1). The fifth hypothesis on study accepted, the community attachment significantly and positively affects the willingness to pay .135 (<0.1).

On Table 4 described values of willingness to pay (WTP) for the local community to visit the ecotourism sites in the area where most of the visitors willing to pay IDR 50,000 ($ 3.75) per trip (45.5%), then followed by IDR 100,000 ($ 7.51) per trip (23.8%), IDR 200,000 ($ 15.02) per trip (11.9%), £ 300,000 ($ 22.53) per trip (7.8%), IDR 500,000 ($ 37.55) per trip (6.1%) and IDR 400,000 ($ 30.04) per trip (4.9 %) respectively. Furthermore, based on the average calculation the WTP value of the local community to spend on visiting ecotourism sites in the area was IDR 144 057 ($ 10.81) per trip. From research result, we also found that the most favorite tourist sites preferred local communities were Ngurbloat Beach, Ohoililir Beach and Ohoidertavun Beach. The main reason local community choose these three sites to compare with other ecotourism spots in Kei Islands were the access, the beauty and the already famous because it is often covered by national TV and have plenty reviews on the internet.

The results of this study also showed the behavior of the customer in the rural area to the tourism sector, where it showed that infrastructure has a major contribution to the perceived quality of the visitors. [13, 18] found that a positive image of a tourist a destination is a sign of a positive assessment of the visitors on the tourism infrastructure when consumers have a positive on a tourist location it will have a positive influence on the perceived value of the tourist site [16]. Despite the fact that the lack of sufficient infrastructure at tourist sites in the Kei Islands, results of this study provide a real contribution to the importance of attention to the development infrastructure on ecotourism spots in the Kei Islands for local government.

An era where distance and access to information are no longer a significant problem due to the use of the internet has created a fairly reliable promotion channels. This study confirms that promotion through national TV and later discussed by the audience on social media especially on a positive comment have a positive effect on perceived value. Appearances on national TV and the tourism review sites also have a positive influence on the local community attachment on tourist sites in the Kei Islands. Likewise, with the positive effect of perceived value to the local community attachment, it is because when their judgments on the quality of ecotourism in their area are good, then they tend to have more attachment out of pride as [19] suggested that the higher community attachment, the higher their support for tourism development. Thus, the important role of local community attachment is becoming increasingly significant in the development of ecotourism sector in this region, so that it can be the key factor to be considered by the local government in the future development of ecotourism in the region.

From the model formulated in this study, a direct effect of perceived value on a willingness to pay (WTP) is negative where it contradicts with past studies by [9] and [14] which suggests a positive influence of perceived value on a willingness to pay. This study shows otherwise that perceived value has a negative effect on the local community willingness to pay for visiting ecotourism spots in this area. This mainly because the perception of good quality will be associated with the higher price that's why the local community in the rural coastal area with an average income per month is between $100 to $300 a month tends to avoid to pay more. Whereas the intermediating effect from perceived value through community attachment will have a positive and significant effect on the local community willingness to pay and it cemented the important role of community attachment factor to developing of ecotourism sector in this region.

Value of willingness to pay for the local community to visit ecotourism spots in this area, based on this study was at $10.81 and most of the local community would prefer to visit nearest beaches in the city namely Ngubloat Beach, Ohoililir Beach and Ohoidetavun Beach because they said it was easy to access and have beautiful view also one important reason that drives the local community to pay was these beaches have made more appearance on national TV and pictures these beaches are the flagship Kei Islands on internet. This study also indicated that one of the barriers from the local community to visit spots like Bair Islands, Ngurtavur Sandbar and few fountains in the area is the infrastructure and access, as [11] pointed out that destination with better quality on access has high chance to grow. Based on model developed by this study the local governments can generate more local visitors by providing better infrastructure and access to Accelerate development of ecotourism in the Kei Islands.

In general Kei Islands have a huge potential of ecotourism but in reality, this area faces major constraints that way above their reaches, such as high travel cost, seasoning that can limited access and period to enjoy ecotourism spots and constraints that within their reaches such as infrastructure and access to ecotourism spots. This paper contributes several critical points. First, the important role of infrastructure as given factors on developing ecotourism and tourism in general. Second, as rural coastal area, local government can focused on increasing the local community attachment on ecotourism spots on this area by doing more promotion to increase the awareness of local community on ecotourism in the area, because by doing that can increase local community to pay to visit ecotourism which almost decade ago still free to have entry access to any ecotourism spot in the area.

Table 3. Summary of hypothesis testing

Hypothesis	Casual paths	Estimate	P value	Result
H1:	Spot Infrastructure --> Perceived Value	.697	0.000***	Supported
H2a:	Appearance on National TV and internet --> Perceived Value	.163	0.006***	Supported
H2b:	Appearance on National TV and internet --> Community Attachment	.184	0.000***	Supported
H2c:	Appearance on National TV and internet --> Willingness to pay	.266	0.000***	Supported
H3a:	Perceived Value --> Community Attachment	.271	0.000***	Supported
H3b:	Perceived Value --> Willingness to pay	-.112	0.092*	Not Supported
H4:	Income --> Willingness to pay	.087	0.083*	Supported
H5:	Community Attachment --> Willingness to pay	.135	0.082*	Supported

***sig. at 99%; **sig. at 95% and *sig. at 90%

Table 4. Economic valuation of Ecotourism from local community in Kei Islands

Willingness to pay	Number	Percent	Average
IDR 50.000 ($3.75)	111	45.5	
IDR 100.000 ($7.51)	58	23.8	
IDR 200.000 ($15.02)	29	11.9	IDR 144.057
IDR 300.000 ($22.53)	19	7.8	($10.81)
IDR 400.000 ($30.04)	12	4.9	
IDR 500.000 ($37.55)	15	6.1	

*Currency rate at 2017-02-15 16:04 local time

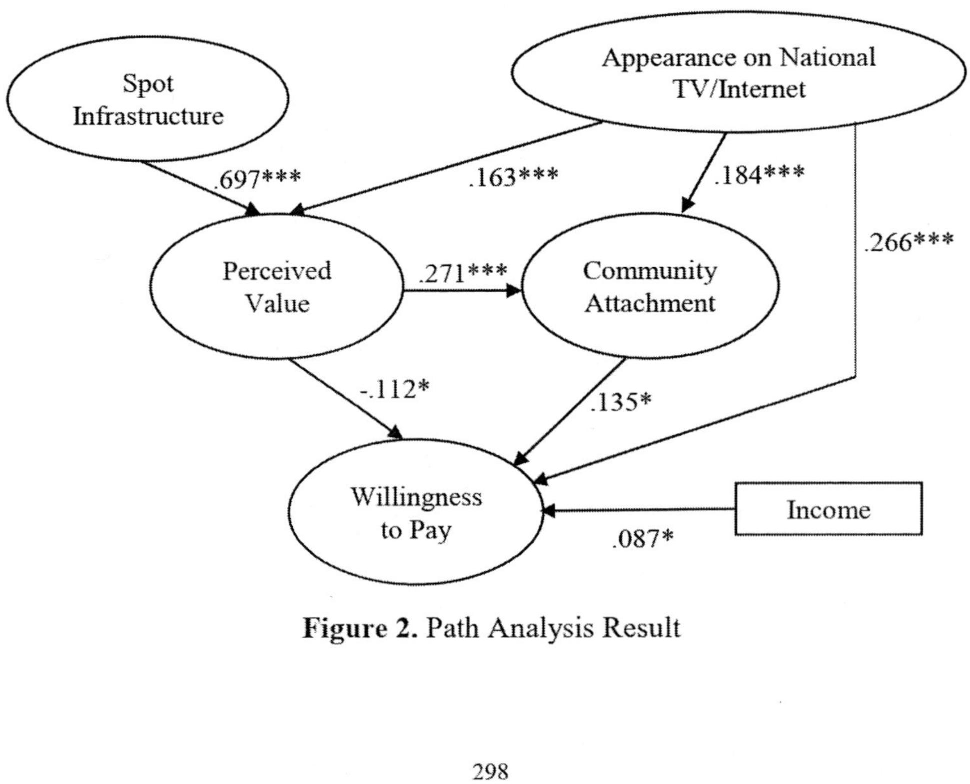

Figure 2. Path Analysis Result

4. Conclusion

Findings of the study show the important role of promotion on national TV and the Internet to give a positive effect on the perceived value in order to increase of community attachment, the results of the study also showed that the increase in perceived value on the ecotourism spot is not directly increased local community to pay more, although the mediating effect from perceived value through community attachment will increase the willingness to pay for the local community to visit ecotourism spots in this region. This study also has not used the comprehensive measurement of perceived quality that further research is expected to use a more comprehensive approach to the determinant factors of perceived quality by using the concept service quality to identify quality criteria desired by the local community. Further study can also make comparisons between local tourist who've been on similar ecotourism spots in other areas to measure the quality of services and perception gap in order to give a complete input to the local government about the position of ecotourism in this region than in other areas based on customer perception.

Acknowledgement

Authors would like to express their gratitude to head of fisheries agribusiness study program and students who help authors to collect the data.

References

[1] Anderson, J.C. and Gerbing, D.W 1988 Structural equation modeling in practice: a review and recommended two-step approach *Psychological Bulletin*, Vol. 103, No. 3, pp.411–423.

[2] Arbuckle, J 1997 Amos Users' Guide, Version 3.6. Chicago: Small Waters Corporation, SPSS Incorporated.

[3] Cater, E., 2002. Spread and backwash effects in ecotourism: implications for sustainable development. *International Journal of Sustainable Development* 5 (3), 265–281.

[4] Eagles, P. F 1992 The travel motivations of Canadian ecotourists. *Journal of Travel Research*, 31(2), 3-7.

[5] Hair, Jr., J. F., Black, W. C., Babin, B. J., Anderson, R. E., and Tatham, R. L 2010 Multivariate Data Analysis 7th edition, Pearson new international edition.

[6] Hair. J.F., Sarstedt, M., Ringle, C.M. and Mena, J.A 2012 An assessment of the use of the partial least squares structural equation modeling in marketing research, *Journal of the Academy of Marketing Science*, Vol. 40, No. 3, pp. 414-433.

[7] Harrill, R 2004 Residents' attitudes toward tourism development: A literature review with implications for tourism planning. *CPL bibliography*, 18(3), 251-266.

[8] Kontogeorgopoulos, N 2003 Towards a Southeast Asian model of resort-based" mass ecotourism: evidence from Phuket, Thailand and Bali, Indonesia. *ASEAN Journal on Hospitality and Tourism*, 2(1), 1-16.

[9] Li, G., Li, G., & Kambele, Z 2012 Luxury fashion brand consumers in China: Perceived value, fashion lifestyle, and willingness to pay. *Journal of Business Research*, 65(10), 1516-1522.

[10] Long, P. T., R.R. Perdue, and L. Allen 1990 Rural resident Tourism, Perceptions and Attitudes by Community Level of Tourism. *Journal of Travel Research* 28(3):3-9.

[11] Mill, R. C., and Morrison, A. M. (1992). The tourist system: An introductory text (2nd ed.). Englewood Cliffs, New Jersey: Prentice-Hall.

[12] Ministry of Tourism of The Republic Indonesia 2015 Ranking National Foreign exchange income of Tourism Sector of Export Goods 2011-2015, retrieved from: http://www.kemenpar.go.id/userfiles/devisa2011-2015.pdf

[13] Murphy, P., Pritchard, M. P., and Smith, B 2000 The destination product and its impact on traveller perceptions. *Tourism management*, 21(1), 43-52.

[14] Netemeyer, R. G., Krishnan, B., Pullig, C., Wang, G., Yagci, M., Dean, D., ... and Wirth, F 2004 Developing and validating measures of facets of customer-based brand equity. *Journal of Business Research*, 57(2), 209-224.

[15] Pennington-Gray, L. A., & Kerstetter, D. L 2002 Testing a constraints model within the context of nature-based tourism. *Journal of Travel Research*, 40(4), 416-423.

[16] Sanchez, J., Callarisa, L., Rodriguez, R. M., and Moliner, M. A 2006) Perceived value of the purchase of a tourism product. *Tourism management*, 27(3), 394-409.

[17] San Martín, H., & Del Bosque, I. A. R. 2008 Exploring the cognitive–affective nature of destination image and the role of psychological factors in its formation. Tourism management, 29(2), 263-277.

[18] Teniwut W. A., 2016 For sustainable revenue of fisheries sector in small islands: evidence of Maluku, Indonesia. AACL Bioflux 9(3):722-732.

[19] Vesey, Catherine, and Frédéric Dimanche 2001 Urban Residents' Perceptions of Tourism and Its Impacts: An Application of the TIAS Scale. Proceedings of the 2001 Travel and Tourism Research Association Annual Conference, pp. 151−58.

[20] Weaver, D 2002 Asian ecotourism: patterns and themes. *Tourism Geographies* 4 (2), 153–172.